*f*P

The

BIOLOGY

of

VIOLENCE

How Understanding the Brain, Behavior,
and Environment Can Break the
Vicious Circle of Aggression

DEBRA NIEHOFF, Ph.D.

THE FREE PRESS

THE FREE PRESS
A Division of Simon & Schuster Inc.
1230 Avenue of the Americas
New York, NY 10020

THE FREE PRESS and colophon are trademarks
of Simon & Schuster Inc.

Designed by Carla Bolte

Manufactured in the United States of America

10   9   8   7   6   5   4   3   2   1

Library of Congress Cataloging-In-Publication Data

Niehoff, Debra.
    The biology of violence: how understanding the brain, behavior,
and environment can break the vicious circle of
aggression / Debra Niehoff.
        p.   cm.
    Includes bibliographical references and index.
    ISBN 0–684–83132–5
    1. Violence—Physiological aspects.   2. Aggressiveness
(Psychology)—Physiological aspects.   3. Brain chemistry.
I. Title.
RC569.5.V55N53   1998   98-34673
616.85′82—dc21   CIP

Illustrated by Michael S. Linkinhoker, M.A., Department of Art as Applied to Medicine,
Johns Hopkins University School of Medicine, Baltimore, Maryland; illustrations copyright Johns Hopkins
University.

To my father, Henry Niehoff

Contents

Preface ix

1. Seeds of Controversy   1

2. The Vicious Circle   31

3. From Wilderness to Lab Bench:
   Animal Models of Aggression   54

4. All the Right Connections   79

5. Bad Chemistry   115

6. Raging Hormones   150

7. Out of Their Minds   188

8. Leading the Charge   225

9. Picking Up the Pieces   259

Notes   297
Index   333

Preface

Our society has looked in many places for an answer to the problem of violence. We have dissected the family and the community, scrutinized the media, and searched our schools. We have probed for deficiencies in our moral values and debated the hazards of guns. We have charted incomes, counted jobs, and weighed attitudes.

We have looked everywhere but inside ourselves.

Why do people hurt each other? This book looks for an answer to that question in a novel but thoroughly logical place: the brain. Drawing on twenty-five years of progress in neuroscience research, it explains how an adaptive behavior, aggression (the use of physical or verbal force to counter a perceived threat), escalates into a maladaptive behavior, violence (aggression directed toward the wrong target, in the wrong place, at the wrong time, with the wrong intensity). Rather than plumbing our distant past for the evolutionary antecedents of violent behavior, it focuses on physiological processes that operate in the here and now—processes open to intervention.

Aggression is a social behavior. For too long, the study of violence has separated those who commit violent acts and those on the receiving end. This book reunites the two. Recognizing the possibility that anger and fear may have something in common, it analyzes not only the biological origins of violent behavior in aggressors but the biological consequences of aggression for victims.

Using the words *biology* and *violence* in the same sentence is a risky proposition. To many, a biological perspective is equivalent to a genetic perspective, conjuring up memories of past atrocities and concerns about future discrimination. To others, locating violent behavior in the brain rather than the soul means placing it outside the reach of the law. I hope that this book will correct both misconceptions. It begins with an overview of the historical foundations of the antipathy toward biology, acknowledging that initial attempts to understand the neural underpinnings of violence and translate this knowledge into practical application were often awkward at best and calamitous at worst. As a consequence, biological explanations of aggression came to be viewed as deterministic, oppressive, and hurtful, not helpful.

Today we know that the neural basis of behavior is more than a genetic program or a killer instinct. Beginning in Chapter 2, I explain why the nature-nurture dichotomy that has long pitted biological and sociological approaches to understanding violence has not merely been resolved; it has been superseded. Advances in our understanding of brain development emphasize how the environment begins to shape the nervous system even before birth and, conversely, how the innate features of the brain begin the process of defining the way in which each of us perceives and reacts to the environment. Over the course of a grand tour that encompasses anatomy, the ever-changing neurochemistry that is the language of the brain, sex and stress hormones, the impact of disease and drugs, and the surprising plasticity of genes, I hope to convince you that behavior is the result of an ongoing developmental process.

Biology is *not* destiny. Because behavior is flexible, it can be changed or, better yet, steered in the right direction from the beginning. In the final chapter of the book, I offer a neurobiologist's opinion of what we can do to reduce and prevent violence, strategies designed to work with human nature and not against it. By breaking the destructive link between environmental cues and violent responses, we can break the vicious circle that lies at the root of human violence.

Acknowledgements

I could not have written this book without the support and participation of my scientific colleagues, who generously took the time to answer my questions, describe their research, offer advice, and read and critique the manuscript along the way. Foremost among these many advisors is Klaus Miczek, of the Department of Psychology at Tufts University, once my teacher and always my mentor. I am indebted to him as a source of inspiration and an example of scientific craftsmanship at its finest. In addition, I thank (in alphabetical order): Steven Arnold, Russell Barkley, Fred Berlin, Ira Black, Robert and Caroline Blanchard, Doug Bremner, Robert Cairns, Gregory Carey, Angela Creager, Ira Chasnoff, Emil Coccaro, Joseph Coyle, Ted Dawson, Joseph DeBold,

Mark D'Esposito, Frans de Waal, Burr Eichelman, Patricia Goldman-Rakic, Fred Goodwin, Richard Kavoussi, John Krystal, Michael Kuhar, Joseph LeDoux, Bruce McEwen, Michael Meaney, Karen Overall, Robert Post, Adrian Raine, Tony Rostain, Robert Sapolsky, Alan Siegel, Solomon Snyder, Matthew Stanford, Elizabeth Susman, George Uhl, Jan Volavka, David Wasserman, Daniel Webster, Rachel Yehuda, and Errol Yudko.

I'm also indebted to the many people outside the laboratory who contributed to this book, broadened my perspective, and added a human dimension to the science. Special mention goes to Connie Bastek-Karasow, director of Libertae; Susan Hauser, director of A Woman's Place; Eve Marschak, who taught me everything I know about sheep-herding and herding dogs; and all of those who were courageous enough to share their personal stories with me. Out of respect for the privacy of these individuals, I have changed their names and the identifying details of their personal lives.

If one picture is worth a thousand words, Michael Linkinhoker's superb illustrations are surely worth millions. Capturing the essence of difficult concepts, he has made them accessible; skillfully recreating on paper the details of an intricate anatomy, he has given readers an outstanding road map to help them chart a course through the brain.

Cheryl Skeates and Lucy Domingo provided expert secretarial assistance, and Trish Hand worked heroically to transcribe interviews. Chris Hand was the finest research assistant a writer could ask for; his efforts were supplemented by neuroscientists-in-training Julia Wenniger and James Kozloski.

My agent, Regula Noetzli—and her patience, wisdom, and intuition— helped me bring an idea to life. Stephen Morrow, my editor at The Free Press, deserves applause for his willingness to champion a controversial project, as well as for his expert editorial direction.

My family—husband David and daughters Jennifer and Haley—made coffee, endured hours of arcane scientific musing, did extra chores, kept quiet during phone interviews, read drafts, corrected typos, wrote witty messages on my screen saver, encouraged me to start riding again, and kept me going in a thousand other ways. Without them, I could not have finished—or survived—writing this book. Nor could I have made it without the support of friends who encouraged me: Kristin Carlson; Miriam Sexton; Marian Plunkett; John and Caitlin Matthews; Connie Bastek-Karasow; and Penny Silcox, Susan Clarke, and the rest of the "family" at Hickory Run Farm.

The

BIOLOGY

of

VIOLENCE

One

SEEDS OF CONTROVERSY

Janice is the first to speak. She is safe now, and that makes her bold enough to tell her story to the other women and their children gathered in the courtyard. But the breathy choke in her voice recalls a time when she was afraid, a time when her daughter, not yet a year old, had already learned how to babble and coo at Dad to distract him. If the baby could get him to laugh, he stopped hitting Janice.

She still has no idea what he was so angry about. "He lashed out at anything," she whispers. "But he could only reach as far as me."

Kathy follows Janice. Donna follows Kathy. Sharon follows Donna. The wind rises and the October night, crisp at seven o'clock, chills to a temperature undeterred by even the heaviest sweater. Those without coats shiver from the cold as well as emotion.

When the stories are over, the group lights candles, as they have done here for more than a dozen years, to shame the men who have beaten and terrorized them, to shame a society that closes its eyes. In 1996, approximately seven thousand women in this county had babies.[1] That same year, fifteen thousand—more than twice as many—were victims of domestic violence.[2] To look at the faces of these victims is to know that the mirror image of anger is not compassion but fear.

When people talk about violence, they're often thinking about crime—about murder and rape and armed robbery. But the men who hurt Janice and Kathy, Donna and Sharon are not convicts. The harshest sentence most will

1

receive is a restraining order. People tell themselves they're safe, because violence makes its home in crumbling neighborhoods fouled by poverty. But the women in the courtyard live nearly an hour from central Philadelphia, in quiet communities that enjoy some of the highest per capita incomes in Pennsylvania. People believe that violence is the work of strangers, and so they teach their children to run away from unfamiliar faces. But these assailants live in the same houses as their victims; this violence, says Susan Hauser, director of the county women's shelter, "comes from a person who once said, 'I love you.'" People recognize the violent; they make headlines. But every act of aggression involves two people. People insist that the violent are evil, as if that word alone could expatriate them to some dark and unknown place. But the sad truth is that the violent are only human, and they live right among us.

Behind the crowd, rows of T-shirts, hung from a clothesline that circles the courtyard, ripple in the candlelight. Some memorialize women murdered by their abusers. Others have been created by the women with candles, outward symbols of their determination to escape and survive. One shirt depicts the flashing red and blue light that tops police cars. Above the picture, the woman who marked time by the arrival of those cars has written, in bold, angry letters, "My children's night light."

THAT SAME FALL, another group of antiviolence protesters had gathered outside the Aspen Institute in Queenstown, a small town on Maryland's Eastern Shore, to light candles and remember the past.[3] But it wasn't violent behavior itself that inflamed these demonstrators; it was some people's answer to the problem. Inside the institute's Wye Conference Center, some seventy scientists, philosophers, bioethicists, lawyers, and sociologists had come together to debate the moral, medical, and social implications of research on genetics and criminal behavior. The conference, organized by University of Maryland legal scholar David Wasserman, was taking place nearly three years later than originally scheduled. During that time, it had come to stand for one of the most enduring conflicts between science and society of this century, an ethical debate as bitter as the tragedy it aimed to correct.

The trouble began with a speech made to the National Advisory Mental Health Council on February 11, 1992, by Dr. Frederick Goodwin, then director of the Alcohol, Drug Abuse, and Mental Health Administration (ADAMHA), to mark the announcement of a new federal strategy for coordinating research on the origins of violence. Commenting on the importance of this "violence initiative" in an America besieged by rising crime rates, Goodwin pointed to the deterioration of social structure in "high-impact inner-city areas," suggesting that "maybe it isn't just careless use of the word when people call certain areas of certain cities jungles."[4]

Someone in the audience called the *Washington Post*.

A media feeding frenzy ensued, in which Goodwin's observation was reported as a comparison of black Americans to monkeys in the jungle. Prompted by an aggressive "educational" campaign engineered by Maryland psychiatrist Dr. Peter Breggin, a long-time critic of the psychiatric community, African-American political leaders concluded that the "violence initiative" was actually a euphemism for a government-backed racist policy to control crime by scapegoating the black community.

Events came to a head when the National Institutes of Health (NIH) awarded a grant to Wasserman for his conference, then titled "Genetic Factors in Crime: Findings, Uses, and Implications." For critics of the violence initiative, this was the ideological last straw. In a July 4, 1992, interview on Black Entertainment Television, Breggin savaged the conference, comparing it to "the kind of racist behavior we saw in Nazi Germany."[5] Outraged members of the Congressional Black Caucus demanded cancellation of the conference, an end to the violence initiative, and a global moratorium on violence research. The NIH first refused, then conceded to the pressure, freezing Wasserman's $78,000 grant on July 20.[5] Eight months later, John Diggs, NIH deputy director for extramural research, notified the University of Maryland that the agency had formally canceled support for the beleaguered conference, citing statements in the brochure that had "significantly misrepresented the objectives of the conference" and that gave "the distinct impression that there is a genetic basis for criminal behavior."[6]

Academic leaders were appalled. "Actions of this kind put a chilling effect on the conduct of science," argued Robert Rosenzweig, president of the Association of American Universities, in an interview with the influential journal *Science*.[5] Over the next two years, Wasserman and the University of Maryland restructured the conference to focus specifically on the allegations of racism, in the hope that this move would persuade the NIH to reconsider its decision. Ultimately, Wasserman convinced the Program on the Ethical, Legal, and Social Implications of the Human Genome Project, a branch of the health agency's National Center for Human Genome Research, to reinstate funding for the politically corrected conference, even expanding the original award of $78,000 to $133,000.[3]

In an interview with the *New York Times* three days before the conference finally began on September 22, 1995, Wasserman emphasized that "the conference has changed from three years ago. We're placing a greater emphasis on the issue of race and racial tensions. . . . We're going to give it greater prominence."[7] He noted that every effort had been made to include not only genetic researchers but also vocal critics of such research, such as sociologist Dr. Dorothy Nelkin of New York University and Dr. Andrew Futterman, a psychology researcher from Holy Cross College in Worcester, Massachusetts. In fact, the *Times* suggested that some biologists declined Wasserman's invitation to participate on the ground that "the conference was so weighted toward the ethics and implications of the work that it slighted the science."[7]

And the critics—despite Wasserman's overtures—remained wary and pessimistic. "I haven't a clue what to expect, but I do have serious questions," Futterman admitted beforehand. "I think that people will probably try to set up the biology and crime relationship as a straw man. They'll attack it and come away feeling virtuous." Nelkin was blunter. "It's a silly conference. Nothing will come out of it," she insisted. Organizers of the vigil still condemned the idea of discussing genetics and crime in the same sentence as "racist pseudoscience" and berated Wasserman on the Internet as the chief advocate of a "genetic/biochemical approach to urban youth violence [that] is . . . sucking away resources from tackling the ROOTS of urban youth violence, including RACISM and CLASSISM."[8]

Afterward I asked Wasserman if he felt that the result had been worth the anguish, if the biologists had become sensitized and the sociologists educated. His answer was a qualified yes. "I think the conference was tremendously valuable," he replied. "I think some in the audience wondered what all the fuss was about. Some were relieved to hear that scientists weren't doing the kind of simple-minded research they thought. And there were some good dialogues between people who came away with a lot of mutual respect for each other." He paused for a minute, then added, "But I wouldn't do it again."

The Missing Link

Our frustration with the widespread violence of American society is well documented; by now, the reports are as mind numbing as the statistics. But news reports and homicide figures tell only part of the story, for much cruelty goes unrecognized and unpunished by the criminal justice system. Daily life has degenerated into a threatening and occasionally terrifying cacophony of belligerent coworkers, disgruntled patriots who issue death threats to judges and forest rangers, and school board members who threaten to tear a noisy constituent "limb from limb and enjoy the process." The Internet is clotted with hate mail; the roads swarm with agitators who view their cars as weapons.

Deluged with facts and figures, we still understand neither violence nor the violent. The belief that many, if not most, of the individuals responsible for the lurid headlines are somehow "sick," members of a shadow society that exists outside the mainstream, is widespread, yet we cannot explain exactly what it is that makes them different. Social critics, like Dr. Nelkin, point to poverty and racism but fail to explain how these factors twist behavior, or why the incidence of domestic violence, child abuse, hate crimes, and road rage has escalated in comfortable neighborhoods far from the inner city. We can predict prison costs and project handgun sales, but we become confused when we look at a group of troubled teenagers and try to forecast who will reform and who will degenerate. Crime victims initially inspire pity, then frustration. Why can't they get on with life? Why doesn't she just move out?

One of the women who cried at Janice's story came to the vigil to light a candle for her friend Sandra. Only a few weeks earlier, Sandra had told her belligerent husband she'd finally decided to move out. He tried to use force to stop her. Alarmed, the friend had asked Sandra's priest to intervene. The priest refused, maintaining that because no one had actually witnessed the alleged assault, the behavior and motives of both husband and wife remained open to question. "After all," she concluded, "we don't know what went on inside their heads."

This is the heart of the problem. Our violent behavior bewilders us because we lack crucial information. Countless newspaper articles, books, and television programs chart the social dimensions of violence: poverty, racism, the breakdown of the family, the pervasive influence of television, the ready availability of guns. But the outer world is meaningless until it enters the inner world, the dimension governed by brain and perception, thought and emotion, nerve and tissue. Until we know as much about this inner dimension as we do about the outer one—what goes on inside the heads of aggressors and their victims—we are not prepared to analyze the problem of violence effectively.

Columbia University neuroscientist Eric Kandel has written, "The central tenet of modern neural science is that all behavior is a reflection of brain function."[9] Beginning in the nineteenth century, our growing knowledge of brain structure and function has fueled recognition of the neural mechanisms underlying behavior. Indeed, at the turn of the century, science seemed poised on the edge of a biological revolution to equal the practical advances of the Industrial Revolution.

Why, then, has neuroscience been excluded from public debate about the problem of violence, banished as too dangerous for public consumption? One source of the controversy is the age-old conflict between behavioral science and ideology that continues to fuel widespread misunderstanding of both the character of biological science and its motives. A second answer lies in the popular misconception that scientific progress is a continuum, with discovery overlapping discovery in a steadily advancing wave that crests to one clear and perfect answer: the steam engine, the personal computer, antibiotic therapy, a cure for cancer. In reality, however, science proceeds by fits and starts, wandering into dead ends, overlooking the obvious, daydreaming, stumbling into a hidden pit. False starts and wrong turns are in fact a hallmark of newly emerging disciplines, as researchers struggle to mark a path through muddy, uncharted territory. Ground rules must be legislated, new tools and techniques invented. What initially seemed solid may suddenly shift; a single question may fragment into dozens of answers.

First efforts to widen the track by translating laboratory results into practical applications are equally likely to encounter obstacles. Delays and frustrations are inevitable. Patience and a critical eye are essential. For if impatience, poor judgment, or societal demands rush this learning process, scientists may reach conclusions before they even understand the question. The history of biology's fall from grace as an explanation for human violence is a cautionary tale of the ease with which inspired discovery can die at the crossroads of human ignorance and human need.

Form and Function

We can see and measure the physical, and so the search for the difference between those who kill and those who are content to seethe began with attempts to correlate physical features and personality traits by an ambitious physician and anatomist, Franz Joseph Gall (1758–1828).[10,11] Gall began speculating about the anatomical substrates of behavior even before he completed his formal training at the medical school of Vienna in 1795. Building on observations that dated from his childhood, when he noted a correspondence between the mental abilities of his classmates and the size and shape of their eyes, Gall proposed a working relationship between verbal memory and the frontal lobe, the brain area protruding into the bony cavity overlooking the eye sockets. His ideas ultimately evolved into *phrenology*, a theory that assigned precise neural addresses to a wide range of behaviors, personality traits, and mental characteristics, and also proposed that the relative importance of each faculty in a particular individual could be divined from the size and shape of the bumps in the skull overlying that functional region (Figure 1.1). A skilled practitioner familiar with a map such as that shown in Figure 1.1 could readily detect an antisocial mind-set merely by running his hand over the villain's head.

To the modern mind, Gall's system of correspondences appears farfetched. Even in his own time, critics were quick to point out that his conclusions were supported by only the barest shreds of actual clinical evidence. For example, "destructiveness" was associated with a bump located just above the ear because,[10] "First, this is the widest part of the skull in carnivores. Second, a prominence was found here in a student who was 'so fond of torturing animals that he became a surgeon.' And third, this region was well developed in an apothecary who later became an executioner."

Gall and his predictions, however ill founded they may have been, found a sympathetic audience in the Viennese public. But church and civil authorities proved less enthusiastic. Decrying phrenology as a materialistic assault on free will, they forbade Gall to teach and finally forced him to leave Vienna in 1805 for the less repressive environment of Paris. Here, he and a student, Johann Spurzheim (who would eventually conduct a worldwide public relations effort

Figure 1.1

The localization of higher function. Phrenological diagram showing the alleged location of various behaviors and personality traits.

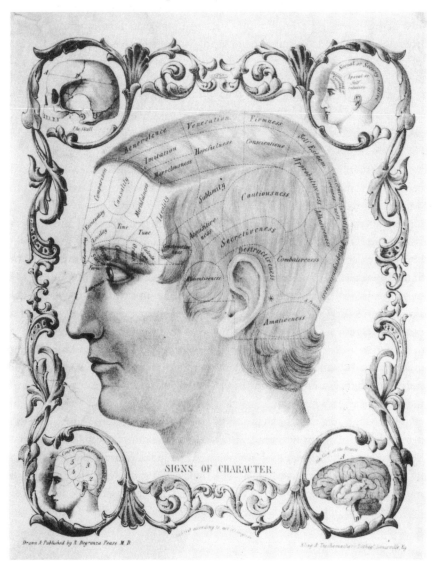

Courtesy of the National Library of Medicine.

on behalf of phrenology), extended their investigations, enjoyed the admiration of Parisian society, and attracted an international following.

One supporter, American physician John Bell, was so impressed by Gall that he founded the Central Phrenological Society upon his return to the United States in 1822.[10] In one of the earliest clinical reports linking brain function to violence, Bell delivered a lecture to the society describing Spurzheim's phrenological evaluation of a group of thirty women who had killed their own children. According to Spurzheim, twenty-six of these homicidal mothers had an underdeveloped brain center for "philoprogenitiveness," or love of children. "The implication," science historian Stanley Finger notes, "was that their crimes resulted from a physically defective brain."[10] Similarly, in an 1846 treatise, *Rationale of Crime,* Eliza Farnham provided this description of a phrenologist's opinion on the violent outbursts of one C.P., a woman jailed four times for crimes ranging from petty larceny to assaulting a prison warden with a carving knife: "In her head, destructiveness is enormously developed, with large secretiveness and caution, and very defective benevolence and moral organs generally."[12]

Interest in phrenology had waned by midcentury, weakened by the evidence of hands-on observations and increasingly scathing criticism on both sides of the Atlantic. But the idea that psychological makeup was reflected in physical characteristics resurfaced in the work of Italian criminal anthropologist Cesare Lombroso (1836–1909). Lombroso examined the physical features of prisoners, the insane, and cadavers. Based on these observations, he thought he could detect certain features—a sloping forehead, long arms, full lips, a twisted nose—that seemed to occur more often in criminals than in law-abiding citizens.[13,14] He concluded that such features, or *atavisms,* were evidence of regression to a more primitive stage of development. The violent, in other words, were not only different, but physically, mentally, and morally inferior, closet Neanderthals passing for civilized human beings. Moreover, they were born that way, cursed from the very beginning with a weak will and a native propensity to wreak havoc.

Today, the idea that the shape of our eyes or the width of our foreheads can predict behavior seems ludicrous. But we are visual creatures, and the temptation to rely on the strongest of our five senses is great. We still judge character by what we see, and if clothes, body language, or facial expression look too much like the lead story on the evening news, we rarely hesitate to draw what may well be an unwarranted, unreasonable, or even dangerous conclusion.

Purely Born

Phrenology was criticized in the scientific community for its failure to back up its claims with hard clinical evidence. Outside the laboratory and the clinic, Gall's theory also raised larger questions about the ethics of examining human behavior from a physical rather than a moral or spiritual perspective. Gall's most vitriolic critic, the renowned French neurophysiologist Marie-Jean-Pierre Flourens, derided Gall as a madman who "suppresses free will and wishes that there is a soul."[15] Such reproaches did more than exclude Gall, Spurzheim, and their supporters from the elitist society of nineteenth-century academic science. Moral objections to "functional localization"—the mapping of particular behaviors to specific brain regions—and Lombroso's atavisms sowed the seeds of an antipathy to behavioral biology that would soon be nourished by revolutionary ideas about a predictor for human behavior far more permanent than physical features: the genome.

CHARLES DARWIN'S MOST dangerous idea was not his well-known theory of evolution and natural selection, but the suggestion that human beings can tinker with the process. Although Darwin did not openly propose any direct meddling, concern about the evolutionary implications of the celebrated fertility of the Victorian lower classes led him to suggest cautiously that encouraging society's more exemplary members to rear larger families might have merit.[16] It was Darwin's younger cousin, Francis Galton, who breathed life into this idea that humans should control human reproduction, arguing that mental and moral characteristics—including a range of vices, character defects, socially undesirable habits, and "crimes of violence"—fell under the same hereditary authority as physical traits like height and eye color. He concluded that the solution to rampant crime rates and the exploding population of the urban poor was to control human breeding in much the same way that farmers bred "the best to the best" among their livestock.[17,18] Science historian Diane Paul sums up Galton's attitude: "Those highest in civic worth should be encouraged to have more children; the stupid and improvident, few or none."[17]

Galton called his proposition *eugenics,* after the Greek word *eugenes,* or "good in birth." Eugenic practice actually predated the Greeks; it has, in fact, been a sordid aspect of human reproduction since the dawn of time. Oxford scholar Allan Roper, in a 1913 essay on eugenics in the ancient world, observed, "The preface to a history of Eugenics may be compiled from barbarism, for the first Eugenicist was not the Spartan legislator, but the primitive savage who killed his sickly child."[19] Roper speculated that the precarious relationship between early human clans and an often unforgiving environment must have made care of the deformed and the infirm seem a dangerous luxury. But the comforts of civilization did not end dissatisfaction with less-than-perfect infants. The classical Greek texts that may have inspired Galton extolled the selection of a healthy, vigorous marriage partner as a chief civic responsibility. Their observation that "noble children are born from noble sires, the base are like in nature to their father," seemed logical to an audience familiar with agriculture and animal husbandry, who knew that breeding a weakling to a weakling typically produced inferior offspring. And population quality control did not end with the marriage contract. When nature failed to live up to expectations, civic-minded parents, like any sensible animal breeder, might well elect to dispose of the error. As Roper put it, "Infanticide saved the Greeks from the problems of heredity."

Infanticide eventually fell out of favor in most countries. But when the emergent science of genetics provided a mechanism for natural selection, eugenics sought to exploit this knowledge to control the threat to evolutionary progress posed by the unchecked propagation of "the reckless, the vicious, and the otherwise inferior."[16]

Galton urged such reproductive management by what Paul calls "positive eugenics": encouraging society's best and brightest to have more children.[17] But later champions of the eugenic cause advocated an aggressive "negative eugenic" approach that focused not on the most fit but on the least. Stressing the high cost to society of caring for the poor, the infirm, the insane, and the violent, leading turn-of-the-century eugenic advocates, such as Karl Pearson in Britain, Charles Davenport in the United States, and Fritz Lenz in Germany, gathered hereditary information on hundreds of thousands of people. Based on these heredity data, they initiated "genetic hygiene" measures ranging from segregation in "work colonies" to compulsory sterilization.[17]

American support for eugenics swelled during the early decades of this century in response to the rising tide of immigrants. Alarmists saw in the huddled masses yearning to be free a more ominous urge: to outbreed and overrun "native" Americans. Academicians and politicians weighed the reputed high fertility rates of so-called low-standard immigrants from southern and eastern Europe to the modest rates among *Mayflower* descendants and predicted social catastrophe, blaming this torrent of the socially inferior for "increasing levels of crime, insanity, and pauperism, and thus for the financial burden of custodial care, for urban political corruption, for strikes and other forms of labor militancy, and for unemployment, among other social ills."[17]

The self-appointed eugenic guardians whipped public fear of "race suicide" into a frenzy, adding a biological urgency to Progressive pressure to curb immigration. Worse, they introduced race as a critical factor in human genetics. According to these advocates, superior individuals who sought to control the immigration and reproduction of Jews, blacks, and eastern Europeans were not prejudiced, but expressing "a natural antipathy which serves to retain the purity of the type."[20] Such propaganda culminated in restrictive immigration laws, such as the Johnson-Reed Act of 1924, which severely curtailed the number of immigrants from southern and eastern Europe, and in miscegenation statutes that prohibited the interracial marriages so dreaded by the eugenicists. But it simultaneously promoted a truly dangerous and degenerate union: the link between behavioral genetics and racism.

Europeans also idealized the social value of racial purity and supported eugenic methods for achieving it. Under Hitler, however, eugenics degenerated into genocide. Public education, segregation, and voluntary measures gave way to compulsory sterilization, enforced abortion, and death camps. Violent, illegal, or dissident behavior—characterized as "social feeblemindedness"—emerged as a particular target of Nazi eugenicists.

The postwar disclosure of Nazi atrocities constituted a death sentence for eugenics, in America as well as in Europe. Genetic researchers renounced their high-profile public face and retreated to the lab. The eugenic societies that had once dispatched armies of field workers to collect volumes of trait data withered, despite contrite efforts to disentangle themselves from their racist past. Practices such as compulsory sterilization, once accepted as not only permissible but essential, fell to public outcry.

But the newfound revulsion toward eugenics did not stop geneticists from quietly taking over biology. In *Refiguring Life,* science historian and philosopher Evelyn Fox Keller proposes that the genetic annexation of biology, the conviction that "with the gene comes life," gradually confined biological research to a genetic prison.[21] The language of biology came to reverence a DNA that dictated not only physical structure, but every facet of existence from the moment of fertilization onward, and cursed biology with an enduring and unpalatable determinism.

Eugenics poisoned the study of the biological foundations of behavior with this same determinism. If gene action was the sole basis of function, as well as structure, then aberrant behavior reflected a genetic defect, a string of "bad code" that crashed the biological program. A biology that believed that life was written in the genes had to conclude that people were either "born good" or "born bad." Learning, experience, and perception were irrelevant. Character was decided in utero, and the birth certificate became a life sentence.

Thanks to eugenics, Galton's legacy was not a utopia of geniuses but an unbridgeable rift between nature and nurture. Nature, discredited by eugenics, lost validity as an explanation for aggressive behavior. On the other hand, nurture explanations for behavior became not only intellectually fashionable but morally obligatory.

ON JUNE 23, 1993, a small box arrived at the Tiburon, California, home of University of California geneticist Charles Epstein.[22] Epstein failed to note the box's unfamiliar northern California return address, perhaps imagining it contained lab samples, a gift, this month's unsolicited selection from yet another record club. The oversight proved deadly. Instead of test tubes or CDs, the package contained a pipe bomb that exploded in the researcher's hands, severely injuring him. It was Charles Epstein's brutal introduction to the Unabomber.

Beginning in May 1978, when a professor at Northwestern University turned over a suspicious package found in a university parking lot to local authorities, the elusive Unabomber mailed, delivered, or left behind sixteen of the deadly boxes in a seventeen-year killing spree that inspired the costliest manhunt in FBI history.[22,23] Only one eyewitness claimed to have seen the mysterious fugitive. No one understood why the bombs targeted university classrooms, computer stores, and airlines.

The arrest of Montana recluse Theodore Kaczynski in April 1996 finally put a human face on the enigmatic Unabomber. And Kaczynski himself had explained the motive for the attacks six months earlier, when the *New York Times* and the *Washington Post* printed his 35,000-word diatribe against the Industrial Revolution.[24] Science and technology, the manifesto claimed, have ruined our social and physical environment; they are the root cause of the stress and frustration that have "destabilized society, have made life unfulfilling, have subjected human beings to indignities, have led to widespread psychological suffering (in the Third World to physical suffering as well) and have inflicted severe damage on the natural world." Presumably Epstein's selection as a target identified genetics as one of the evil scientific advances contributing to this "disaster for the human race."

Those who believe that problem behavior is the result of a damaged society, rather than damaged genes, are as capable of political excess as the rabid proponents of racial hygiene. Charles Epstein's fate is a chilling reminder that science and ideology are always a dangerous combination, regardless of whether scientists initiate or receive the first overture.

The Meaning of Aggression

Before Darwin was a geneticist, he was a naturalist. His theory of natural selection was not the result of laboratory research, but grew out of detailed observations of the natural world. French biologist Isidore Geoffroy-Saint-Hillaire, a contemporary of Darwin, christened this "fly-on-the-wall" approach to the study of animal behavior *ethology*, from the Greek word for "character" or "characteristic."[25]

Ethology, like eugenics, owes more than its name to the Greeks. They were the first to describe the annual spring outbreak of avian civil war, then as now, a conflict centered around choice nesting sites.[26] Once found and claimed, such a territory will not be ceded willingly.

In my own backyard, a pair of lucky house sparrows hurries to cement their ownership of a birdhouse wedged into an ornamental pear tree. While the female snips choice bits of greenery for their new furnishings, the male stations himself at the top of the tree, where he swaggers and taunts potential competitors with an enthusiasm that puts the posturing of his human neighbors to shame. Most rivals get the hint. But one upstart decides this choice space is worth a serious confrontation. He lands on a branch just above the birdhouse and defies the first male to dislodge him.

The enraged resident charges the invader, shrieking and screaming. The second bird parries and dodges. Were he just slightly larger, he might have won. But his opponent catches him off balance and he ricochets down through the tree to the ground, bounces a few steps, then retreats, the victorious resident shadowing him all the way.

Bird watching continued to inspire the first systematic studies of aggression in the wild. From contests like the one described above, behavioral scientists following in Darwin's footsteps suggested that personal space is one critical reason why animals fight. Noting that many birds defend not only a hunting territory or nesting site, but a perching space, some researchers argued that aggression had evolved as a spacing mechanism, a way of limiting the population density.[26]

Other bird watchers contended that animals resort to aggression to defend resources and reputation rather than space. By comparing how many times a chicken pecked other members of the flock to the number of pecks it received in turn, Norwegian behaviorist Schjelderup-Ebbe discovered the infamous "pecking order."[26] Subsequent researchers noted that high-ranking individuals not only had a competitive edge in quarrels, but also enjoyed special privileges: the first helping at meals, the most comfortable sleeping spot, sexual carte blanche. The underlings, like anxious middle managers threatened with downsizing, typically recognized and accepted the dominant animal's superiority, minimizing the outbreak of potentially lethal confrontations. However, the lust to displace the leader and claim the top spot for oneself could also fuel rebellion.

During the 1930s and 1940s, many behavioral researchers viewed dominance hierarchies (as the pecking order came to be known) as the key factor in motivating, as well as containing, aggression.[26] As a result, the factors that governed the outcome of power struggles figured significantly in both ethological field studies and studies of aggressive encounters between laboratory animals. For example, scientists discovered that in rodent pecking orders, an animal that won one fight often went on to win more; these victors ultimately became dominant. They also noted that a rat or mouse fighting on familiar territory enjoyed a home field advantage, an observation that linked dominance and territoriality.[27]

Still other ethologists, influenced by psychological theories of motivation, saw aggression as a response to the call of powerful inner mandates, or instincts. They differentiated *appetites*—instinctively motivated behaviors aimed at getting something, such as food, water, or sex—and *aversions*—similar behaviors designed to avoid an unpleasant or painful stimulus.[26] In a 1928 paper provocatively titled, "Why Do Animals Fight?" Wallace Craig, an early advocate of this explanation, classified aggressive behavior as an aversion, arguing that "animals fight to rid themselves of the interference or thwarting of an instinct."[28]

Craig's link between hostility and disappointment was familiar to psychologists acquainted with Freudian psychoanalysis. They saw the frustrations that followed a failure to obtain anticipated rewards as a bridge between psychoanalytic theory and Pavlovian conditioning. We want something; someone stops us from getting it; we lash out. A concept that's well known to anyone who has ever refused a toddler's escalating demands in a toy store, tried to contest a telephone bill, or found themselves trapped in traffic, the idea that aggression is caused by frustration had a profound theoretical effect on aggression research in social and clinical psychology, as well as ethology.

Some behavioral researchers cited the frustration-aggression association as evidence that violent behavior can be learned. They exploited this possibility to develop animal "models" of aggressive behavior that transferred the study of animal behavior from the field to the psychology laboratory. Robert Hutchinson of Southern Illinois University and his colleagues, Roger Ulrich and N. Azrin, for example, trained rats to press a bar for a food reward and then frustrated the animals by turning off food delivery.[29] The outraged rat's attacks on the apparatus served as a measure of the intensity of the animal's aggression.

The Illinois researchers showed that other aversive stimuli, including physical pain, could also provoke angry reactions. Beginning in 1939, Hutchinson and his coworkers published dozens of studies that demonstrated fierce fighting between animals after electrical shocks were applied to their feet or tails (Figure 1.2).[30,31] Pain was so effective at provoking aggression in these studies that shocked rats and monkeys could be induced to attack a doll, a cloth-covered tennis ball, or even a rubber hose if a living victim wasn't close by.

A tormented monkey biting a rubber hose may remind us of our exasperation with cars that won't start, VCRs that eat tapes, and programs that aren't compatible with Windows 98. But the aggression that most concerns society involves other human beings. Devoid of social meaning, shock- and frustration-induced simulations of animal aggression seemed better suited to measuring pain thresholds than aggressive behavior. As a result, these initial attempts to construct a laboratory model of aggression did little to improve public confidence in either the relevance or the sensitivity of aggression research.

Killer Instincts and Selfish Genes

If gene action was a physiological engine driving behavior from the inside, ethology was the radar system tracking its course. Behavioral biology, like other areas of biological research, placed an increasing emphasis on the impact of behavior on survival. This emphasis on issues that contemporary behaviorist John Alcock calls "why" questions—"the 'evolutionary,' or *ultimate*, reasons for why an animal does something"—was reflected in the preoccupation of aggression researchers with such survival-oriented activities as resource distribution, territory, and mating privileges.[32] Questions about underlying mechanisms—the practical, or *proximate*, explanations of "how an individual manages to carry out

Figure 1.2

Shock-induced fighting.

Reprinted with permission from K. A. Miczek, The psychopharmacology of aggression, in *Handbook of Psychopharmacology*, V. 19, ed. L. L. Iversen, S. D. Iversen, and S. H. Snyder (New York: Plenum Press, 1987), 183–328.

an activity"—as well as questions about the short-term payoff of behaving aggressively took a back seat to the bigger question of evolutionary significance. Ethology and genetics worked hand in hand to foster the idea that behavior was as hard-wired and unchangeable as height or eye color.

Territorial birds and frustrated rats may have introduced the research community to the ethology of aggression, but Konrad Lorenz made it public. The 1963 publication of his best-selling book, *On Aggression,* astonished millions with its suggestion that the key to human violence lay in the behavior of animals.[33] This was an introduction with ominous overtones. Instead of ushering in a new era of self-awareness, Lorenz's book intensified opposition to biology as a legitimate means of explaining behavior and reinforced the idea that nature is incompatible with nurture.

Lorenz agreed with earlier ethologists that animals used aggression to optimize population density, accumulate and defend resources, and protect themselves and their young. But Lorenz emphasized that aggression was not simply a response to an instinct but was itself an innate, driving force:[33] "Knowledge of the fact that the aggression drive is a true, primarily species-preserving instinct enables us to recognize its full danger: it is the spontaneity of the instinct that makes it so dangerous. . . . The aggression drive, like many other instincts, springs 'spontaneously' from the inner human being." In Lorenz's eyes, aggression was not an aversion but an appetite—and a ravenous one at that. Humans were not only born bad, but born helpless, at the mercy of a "killer instinct" that bubbled up from a dark corner of the mind like oil and that needed only the match of some trivial insult to ignite. Contained for too long, it would combust spontaneously. The fragile defenses of ritual, culture, and morality were barely a match for this seething flood of instinctive rage.

Evolutionary biologists welcomed the challenge to the tyranny of nurture. As Harvard biologist E. O. Wilson (soon to instigate a biological revolution of his own) put it, "Lorenz has returned animal behavior to natural history."[34] But nurture advocates were hardly ready to retreat. They charged that Lorenz's idea of an aggressive drive not only discounted rational and moral control over behavior but actually justified violence—dangerous ideas for an Austrian living in the shadow of the Holocaust. In the eyes of these critics, Lorenz had merely revived, updated, and psychologized biology's fatalistic reputation. The controversy swept through universities and laboratories as well as newspapers, as biologists rushed to confirm Lorenz and social scientists rushed to prove him wrong.

Other biologists began to ask fewer questions about aggression and more about its absence, for right in the teeth of the Darwinian struggle for survival of the fittest were animals risking their own lives or renouncing their fair share to help others. Birds who teamed up to fight off predators. Wolves who ceded their right to reproduce to the leader of the pack. Lionesses who shared their catch with malingering sisters. How could such altruism be adaptive?

The answer, these researchers theorized, is that a generous spirit has rewards on earth as well as in heaven. Family members, for example, share a similar genetic makeup as well as a common history. By aiding and abetting kith and kin, a helpful relative ensures that at least some of the family genes make it to the next generation. Altruism can also benefit the good Samaritan directly. If I rescue you or share my dinner with you today, you are likely to do the same for me in the future, with a reproductive payoff for both of us.[25,35]

Good or bad, biologists reasoned, behavior had adaptive value and an evolutionary significance. *Sociobiology*—the term E. O. Wilson used to describe this union of social theory and evolutionary biology—thus brought even our cherished notion of moral superiority under genetic control.

Wilson's book *Sociobiology: The New Synthesis,* like Lorenz's *On Aggression,* provoked a firestorm of controversy. The heart of the problem lay not in the book's premise but in its scope. By extending sociobiological theory to encompass the evolutionary implications of human social behavior, Wilson outraged nurture-oriented scientists and activists. To these modern-day antieugenicists, sociobiology was Galton revisited, complete with the attendant risk that socially sanctioned forms of oppression, such as racism or sexism, would be tolerated as "adaptive," while undesirable behavior, such as a penchant for murder, would be the genetic burden of the socially inferior.

Wilson himself was horrified by these charges. "I had no interest in ideology," he declares in his autobiography, *Naturalist.* "My purpose was to celebrate diversity and to demonstrate the intellectual power of evolutionary biology." But he ruefully admits, "Perhaps I should have stopped at chimpanzees when I wrote the book."[34]

Sociobiologists after Wilson magnified the problem by characterizing individual effort as a slavish devotion of behavior to the interests of a "selfish" genome. The organism itself, human or animal, had value only as an expendable gene factory, and society was an illusion, a collective of opportunistic individualists at the mercy of genetic programs designed to maximize reproductive fitness. Empathy was not evidence of social evolution but a deceptively clever example of genetic manipulation. "Gene-centered sociobiology," as ethologist Frans de Waal calls it, seemed to revel in taunting critics, rather than striving to find a theoretical common ground between the warring camps of nature and nurture.[25]

Selfish genes and deadly instincts only reinforced the widening split between biological and social mechanisms of behavior, and, rather than broadening biology's perspective, cemented it more firmly to genetics. Aggression was inborn and immutable, driving the individual to act in socially unacceptable ways as a way of guaranteeing his or her reproductive success. In aggression, social behavior mutated into an asocial compulsion, as the "killer instinct" transformed ordinary people into genetic bounty hunters.

"Great and Desperate Cures"

Evolutionary biology sparked a debate over the genetic control of behavior. Franz Joseph Gall's theory of phrenology ignited an equally acrimonious debate over the anatomical control of behavior. The question Gall raised—did the brain act as a whole or the sum of its parts?—marked the first shot in a hundred years' war that began at the dawn of the nineteenth century and was not fully resolved until well into the twentieth. Ultimately a combination of persuasive debate, forceful personalities, and methodological refinements brought an end to overt hostilities, but it was an uneasy peace. On the one hand, the mapping of behavior to specific brain regions, or *functional localization,* formed the basis of modern neurological diagnosis. The downside was an oversimplified anatomy that would spawn new efforts to control violence— efforts that were as questionable as eugenics.

GALL'S CHIEF PROTAGONIST, Marie-Jean-Pierre Flourens (1794–1867), was an elder statesman in the French scientific community, a member of the prestigious Académie des Sciences. In contrast to the phrenologists, he was no theorist but a hands-on researcher who stressed the importance of carefully conducted laboratory studies:[15] "Everything in experimental researches depends upon the method; for it is the method that produces the results. A new method leads to new results; a rigorous method to precise results; an uncertain method can lead only to confused results."

Based on experiments in which he surgically removed a chunk of brain tissue from a pigeon, then noted the pigeon's postoperative behavior, Flourens could accept the idea that certain areas orchestrate critical body functions, such as breathing and heart rate. But he drew the line at so-called higher mental functions—reason, will, and emotion. He reported that when he removed the cerebral cortex, the folded helmet of gray tissue that forms the surface layer of the brain, his pigeons lost "all perception, judgment, memory, and will."[11] From this result, Flourens concluded that the entire cortex was necessary for these activities, an idea that directly contradicted Gall's belief in a neural division of labor.

The battle lines were drawn. On one side, neuroscientists had the "globalist" view espoused by Flourens, in which the brain, or at least the cortex, functioned as a single entity in the social, mental, and emotional realms. Or they could accept Gall's idea of a geographically organized brain, with a one-to-one correspondence between location and function.

Resolution lay in Flourens's maxim: the development of experimental methods that yielded precise results. His own studies, like many of the earliest anatomical experiments, were based on ablation, a technique in which part of the brain is destroyed and its functional role deduced from the resulting deficit. However, neurosurgery in the mid-nineteenth century was anything but the "rigorous method" so prized by Flourens. With little more than a superficial knowledge of the anatomy of the interior of the brain, no technique for positioning the head, no orderly way of describing location, and only the crudest of surgical tools, experimenters had little control over the size and extent of the lesion, nor could they ensure that exactly the same area was removed in every subject. The end result resembled an attempt to determine the impact of Harvard on scientific research in Boston by bombing the north shore of the Charles River and charting the subsequent decline in intellectual activity. Trauma and postsurgical infection further biased results, and a dearth of techniques for actually measuring behavior meant that experimental "results" often consisted of little more than a cursory examination of the unfortunate subject in the brief interval between surgery and death.

Fortunately, nature provided what the surgeons could not. The first convincing evidence in favor of functional localization came not from the laboratory but from the neurology clinic, in the form of autopsy data that demonstrated a clear correlation between damage to a circumscribed area of the brain and loss of a specific function. Speech was the first function mapped in this fashion. Beginning with the widely publicized case of an epileptic named "Tan," after the only intelligible word he spoke, the renowned French physician Paul Broca (1824–1880) observed a total of nine patients who had lost the ability to speak (a deficit known to neurologists as *aphasia*) but could still understand spoken language.[9,10] In each case, postmortem examination of the patient's brain revealed damage to the same region in the left frontal lobe of the cortex, a neighborhood known today as Broca's area.

The discovery of the brain's electrical properties by a self-effacing Italian physiology professor, Luigi Galvani (1737–1798), allowed neuroscientists to study the localization of function in an intact brain.[10,36,37] Rather than cutting out a piece of brain tissue, scientists could instead apply an electrical current to excite the region of interest. As a result, the investigator could analyze behavior directly instead of guessing about function from its absence.

Stimulation studies at first were nearly as crude as early lesion studies. For example, Galvani's nephew, Giovanni, having learned that he could produce movements of the face and eyes of an ox by applying an electrical current to the exposed brain, progressed to testing the freshly severed heads of criminals sentenced to the guillotine.[10] He proudly reported that brief shocks passed through the ear and mouth or applied directly to the human brain excited the same facial muscles he had activated in cattle.

In the 1850s, two German scientists, physician Eduard Hitzig (1838–1907) and anatomist Gustav Fritsch (1838–1927), constructed fine-gauge electrodes to restrict the applied current (provided by a crude battery) to a more precisely defined spot of brain tissue.[10,11,36] Funding and laboratory space being nearly as constrained for young investigators then as it is today, Hitzig and Fritsch were forced to construct a makeshift surgical suite on a dressing table in Hitzig's bedroom. Here, in a series of painstaking studies that provided the first truly convincing experimental evidence for functional localization, the two mapped the cortical sites responsible for controlling muscle movements of the paw, leg, face, and neck of the dog. They called these addresses "centers" and speculated that "certainly some psychological functions, and perhaps all of them, in order to enter matter or originate from it, need circumscribed areas of the cortex."[10] An American physician, Roberts Bartholow (1831–1904), extended Hitzig and Fritsch's work to the living human brain.[10,36] Bartholow inserted two wire electrodes into the brain of one of his patients, a young woman with a cancerous ulcer of the scalp that had eaten away a large hole in her skull. Before the woman died two weeks later (and the outraged citizens of Cincinnati drove Bartholow out of town), he demonstrated a correspondence between electrical stimulation of discrete brain regions and specific muscle contractions of the legs, hands, arms, or neck similar to those documented by the German researchers in the dog.

Experiments such as those of Fritsch, Hitzig, and Bartholow, coupled with observations of the changes in function that accompany brain injury or disease, ultimately turned the tide of neuroanatomical opinion in favor of functional localization. But the concept of "centers" implied a dangerously oversimplified one-to-one correspondence between location and function. This entry-level understanding of neuroanatomy, in which functions could be matched to patches of tissue like eggs arrayed in a carton, was not a recipe for progress, but for disaster.

NO OTHER GROUP of medical specialists was quicker to grasp the practical implications of functional localization than the neurosurgeons. The maps constructed by the localizationists allowed them to pinpoint the site of injury or disease based on the correspondence between symptoms (such as paralysis of a particular limb) and the brain region associated with the affected function. As a result, they could operate to remove tumors, abscesses, or scars with a degree of spatial confidence never before possible. Clinicians who saw that physical symptoms, such as pain, numbness, or seizures, could be eliminated or controlled by surgical removal of dysfunctional brain tissue began to wonder if surgery might also relieve intractable and debilitating psychological symptoms—depression, extreme anxiety, or hallucinations. Some even went so far as to speculate that surgery could control the agitation and belligerence that often characterize mental illness and reduced caring for them into a battle between patients and caretakers.

A fortuitous meeting at the 1935 International Neurological Congress in London provided the critical inspiration for testing the idea that socially unacceptable behavior could be controlled with brain surgery.[10,38] During one of the most popular seminars of the meeting, Yale University researchers Carlyle Jacobsen and John Fulton described how they had effected a radical personality change in an unruly chimpanzee named Becky. Becky had threatened to upstage Jacobsen and Fulton's research on the anatomical basis of learning and memory by throwing violent temper tantrums during experimental sessions. But after surgical removal of the frontal segment of her cerebral cortex, she became a model subject, as docile and cooperative as the researchers' other chimpanzees.

In Becky's surgical domestication, Portuguese neurosurgeon Egas Moniz, already renowned for his role in the development of radiologic techniques for imaging blood flow in the brain, saw both a novel solution to the problem of human neurosis and an opportunity to advance his own reputation. Following the presentation, he tried to persuade Jacobsen and Fulton to join him in determining whether surgery "should now be attempted to reduce severe anxiety and delusional states in humans."[10] The appalled Yale scientists refused. But

Moniz was undeterred. Three months later, assisted by fellow surgeon and long-time collaborator Pedro Almeida Lima, he drilled two small holes into the forehead of a severely depressed patient and injected enough alcohol to kill the exposed brain tissue. Impressed by her postoperative tranquility, they declared the operation a success.[10]

Over the next several months, Moniz and Lima refined their surgical procedure, fashioning a special knife, or "leucotome," by bending the tip of a steel needle to form a loop. With this new instrument, the surgeons removed up to four "cores" of tissue from both sides of the brain. Prefrontal leucotomy, as the procedure came to be known, ushered in a new era of medical treatment for the delusional, the suicidal, and the agitated. Only a year after the fateful meeting in London, Moniz published results from an initial group of twenty patients: fourteen had either recovered or improved substantially as a result of the surgery.[10] Agitated or depressed patients seemed especially responsive; the surgery reduced their hostility, brightened their mood, and enhanced their tractability.

Encouraged by this early success, Moniz performed about one hundred prefrontal leucotomies over the next eight years.[38] Neurosurgeons in Europe, South America, and the United States also adopted the procedure, and many confirmed the Portuguese surgeon's glowing reports. But outside the medical literature, disturbing inconsistencies quietly surfaced. Moniz followed his patients closely for only a few days after surgery and ignored them after their discharge from the hospital. A colleague who did conduct long-term investigations, however, documented not only numerous relapses but many deaths.[10] Other physicians began to question whether leucotomy was an "effort to maim and destroy the creative functions."[38] Moniz himself fell victim to a primitive malpractice suit, shot by a dissatisfied former patient in an assault that left him partially paralyzed.[10]

To a world that had little to offer the mentally ill, Moniz was still a hero and leucotomy a humanitarian advance. The medical community chose to overlook the negative reports, awarding Moniz the 1949 Nobel Prize in Medicine and Physiology for his pioneering efforts to develop the surgical solution to violence he christened "psychosurgery."

American neurologists Walter Freeman and James Watts of the George Washington University Hospital made Moniz's operation a household word. Freeman and Watts modified the procedure by opening the skull at the temple, inserting a scalpel, and swinging the knife in a wide arc to mow down both nerve cells and the fibers that connected them.[10,38] Their goal was to sever completely the frontal cortical feedback loops that Freeman believed to be the anatomical basis of a "circuit of emotion" responsible for the confused thinking and agitated behavior of the mentally ill. Cutting the circuit, he argued, freed the tortured patient "from the tyranny of his own past, from the anxious self-searching that has become too terrible to endure."[38]

Watts became increasingly uneasy about Freeman's growing zealousness and self-promotion and bowed out of the partnership. But Freeman continued to streamline the operation he called "lobotomy" on his own. Substituting electroconvulsive shock for standard surgical anesthesia, he rapidly slashed the offending cortical pathways by hammering an ice pick through the bone surrounding the eye socket. Freeman aimed to transform lobotomy into a simple office procedure, boasting that a competent surgeon should be able to perform ten to fifteen such operations in a single morning.

The return of thousands of psychologically damaged veterans after World War II stretched the limits of the already crowded mental hospitals and created a growth market for the burgeoning lobotomy industry. In fact, the Veterans Administration, foreseeing the problem, issued a memorandum in 1943 calling for "all consulting and staff neurosurgeons at its neuropsychiatric installations to receive training in prefrontal lobotomy operations."[38] Fueled by such directives and the lack of effective alternatives, physicians and families caring for the mentally ill were only too ready to embrace what chronicler Elliot Valenstein termed "great and desperate cures."[39] Between 1942 and the mid-1950s, lobotomy "liberated" perhaps as many as forty thousand American psychiatric patients from the tyranny of their past; over thirty-five hundred were "cured" by Freeman alone.[38]

But by the late 1950s, lobotomy had gone out of fashion. The 1957 introduction of chlorpromazine, first of the modern psychoactive medications, reduced both the risk and the cost of treatment and did much to temper enthusiasm for psychosurgery. In addition, psychosurgery had failed to live up to its initial promise. Long-term studies showed that although the surgery did relieve anxiety and agitation, it had little effect on the delusions and hallucinations that led to hospitalization in the first place.[38] Worse, reports of deficits similar to those observed after Moniz's operations began to surface, suggesting that the price of the surgery was a subtle but profound loss of motivation, insight, and organization, an inability to maintain interest in all but the most rudimentary of tasks. Clinicians called it the "prefrontal syndrome." The public thought of it as surgical voodoo, a mind-control method that reduced patients to zombies and punished disruptive behavior by neural mutilation.

Psychopharmacology euthanized lobotomy, but its spirit lived on. Locked into a constricted view that partitioned the brain into clearly demarcated centers, some neurologists maintained that the failures of psychosurgery were the result of poor resolution rather than a faulty premise. They insisted that more precise search-and-destroy procedures that targeted only the site of the putative damage and spared as much of the surrounding tissue as possible could control emotion safely and effectively, with benefits for both the patient and society. As a result, they worked to refine localization techniques. European neuroscientists, for example, pioneered surgical procedures for permanently implanting small-caliber electrodes, permitting repeated stimulation and observation of conscious animals. And at McGill University in Montreal, neurosurgeon Wilder Penfield produced exquisitely detailed maps of areas regulating speech, vision, hearing, memory, and even emotion in the human cortex by recording the responses of patients as he touched electrodes to dozens of sites across the surface of the brain. But his investigations were limited to the artificial environment of the operating room, his subjects to patients sedated and restrained in preparation for surgery.

The invention of the transistor liberated electrical stimulation studies. Surgical implantation of miniature transistor radio receivers replaced cables and insulated test chambers and permitted researchers to study a freely moving animal in its natural environment. Remote-control stimulation was a boon to aggression research, thanks largely to the promotional efforts of a flamboyant Spanish expatriate, José Delgado. Radio-stimulated cats and monkeys became militant terrorists or gentle pacifists at Delgado's electrical bidding. A pulse of electricity that activated no more than a spot of tissue evoked an entire behavioral sequence. For example, a brief jolt to a point deep in the brain of a female monkey named Ludi sparked a series of events that "began with a change in facial expression, followed by her turning her head to the right, standing up, rotating to the right, walking upright, touching the walls of the cage or grasping a swing, climbing a pole, descending, uttering a low tone, threatening subordinate monkeys, and finally, changing her aggressive attitude to a peaceful one and approaching members of the colony in a friendly manner."[36]

Delgado welcomed publicity and encouraged controversy. He returned to Spain, donned a toreador's costume, and subdued a charging bull not with a cape but with a push of a button. He equipped a chimpanzee named Paddy with a "stimoceiver" that relayed electrical signals from the animal's brain to a nearby computer that radioed responses back to the brain, triumphantly announcing to the newspapers, "We are now talking to the brain without the participation of the senses."[36] In a lecture at the American Museum of Natural History in New York, he declared, "It may certainly be predicted that the evolution of physical control of the brain . . . will continue at an accelerated pace, pointing hopefully toward the development of a more intelligent and peaceful mind of the species."[36]

By combining Delgado's state-of-the-art electrical mapping and high-resolution microsurgery, the new psychosurgeons of the 1960s sought to achieve this "intelligent and peaceful mind." Precision, or stereotaxic, psychosurgery began as an attempt to manage aggressive patients who also had intractable neurological or psychiatric conditions, particularly temporal lobe epilepsy. But it quickly came to be seen specifically as a "cure" for aggressive behavior:

> It was originally our intention to investigate the value of amygdalotomy upon patients with temporal lobe epilepsy characterized by psychomotor seizures . . .

as well as marked behavior disturbances such as hyperexcitability, assaultive behavior, or violent aggressiveness. The indications . . . were then extended to include patients without clinical manifestations of temporal lobe epilepsy, but with EEG [electroencephalogram] abnormalities and marked behavior disturbances. Finally, cases of behavior disorders without epileptic manifestations . . . were also included in the series.[40]

The cure gathered support with the publication in 1970 of *Violence and the Brain,* by Harvard neurosurgeons Vernon Mark and Frank Ervin. In the book, Mark and Ervin identified a new psychiatric disorder they termed the *dyscontrol syndrome,* which was characterized by four symptoms:

(1) a history of physical assault, especially wife and child beating; (2) the symptom of pathological intoxication—that is, drinking even a small amount of alcohol triggers acts of senseless brutality; (3) a history of impulsive sexual behavior, at times including sexual assaults; and (4) a history (in those who drove cars) of many traffic violations and serious automobile accidents.[41]

Using the remote stimulation and recording procedures developed by Delgado, Mark and Ervin reported that they could detect local electrical abnormalities in patients with dyscontrol symptoms, sites they christened "brain triggers of violence." They believed that by destroying these trigger sites, they could eliminate the unpredictable bursts of violent behavior. As evidence, they cited the case of a twenty-one-year-old woman with a history of seizures, panic attacks, and episodes of violent rage.[41] During one such episode, she had stabbed another woman in the lounge of a movie theater. A second culminated in an attack on a psychiatric nurse. Exhaustive trials of "all known antiseizure medications, as well as the entire range of drugs used to help emotionally disturbed patients," years of psychotherapy, and over sixty electroshock treatments had failed to curb her unpredictable rages.

Julia, surgically implanted with one of Delgado's "stimoceivers," was to become infamous in both the psychiatric literature and the popular press. Readers of *Life* magazine, for example, observed firsthand the graphic consequences of an electrical spark delivered by the stimoceiver to Julia's presumed "trigger site," and saw for themselves "the angry grimaces which included lip retraction and baring of the teeth, the ancient 'primate threat display,' "[41] followed by a sudden, savage attack on the wall of her room. Bursts of abnormal brain waves churned from the second channel of her radio receiver. Could anyone doubt the wisdom of cutting this behavioral cancer out of poor Julia's brain?

Mark and Ervin reported a dramatic decline in Julia's rage attacks after destruction of the trigger site on one side of her brain, a result they touted as convincing evidence of the antiaggressive effect of high-resolution psychosurgery.

Close examination of some of their other patients, however, told a different story. A summary of the outcome in ten patients similar to Julia showed that in at least five, initial improvement was followed by a gradual return of their aggressive behavior.[38] Only one—a woman whose symptoms could be traced to a severe head injury—was a clear success.

The rise of social unrest during the turbulent 1960s and early 1970s activated a familiar desperation to "do something" about the problem of violence. Psychosurgeons had a ready answer. Collaborator and advocate William Sweet, in his introduction to *Violence and the Brain,* recommended that "knowledge gained about emotional brain function in violent persons with brain disease can be applied to combat the violence-triggering mechanisms in the brains of the nondiseased."[41] Radio-controlled brain surgery, in other words, represented a reasonable and cost-effective alternative to long-term incarceration, argued advocates like California neurosurgeon M. Hunter Brown: "Each violent young criminal incarcerated from 20 years to life costs taxpayers perhaps $100,000. For roughly $6000, society can provide medical treatment which will transform him into a responsible well-adjusted citizen."[38]

The similarity of such statements to the economic arguments of the eugenicists did not trouble enthusiasts like Sweet and Brown. They found a receptive audience in get-tough legislators and law enforcement officials, who readily persuaded Congress to designate over $500,000 to support psychosurgery research.[38] Opponents, however, quickly made the connection, particularly after a 1967 letter by Mark, Sweet, and Ervin to the *Journal of the American Medical Association* suggested that psychosurgery could solve the problem of urban riots: "We need intensive research and clinical studies of individuals committing the violence. The goal of such studies would be to pinpoint, diagnose, and treat those people with low violence thresholds before they contribute to further tragedies."[42]

Mark and Ervin did not overtly target any racial or ethnic group in their writing, but their comments easily resurrected the malevolent link between biology and racism forged by the eugenicists, and the suggestion that violent behavior was the result of having a "low violence threshold" reinforced the behavioral determinism that critics had read into Lorenz and Wilson. And if the sorry consequences of lobotomy were any indication, the "cure" for this illness was almost as bad as the disease. Once again, biological explanations for violence had proven hurtful, not helpful.

Explanation or Excuse?

The crowds began to arrive by midmorning. But unlike the typical outdoor gathering in this community of Philadelphia's fittest families, they hadn't come to watch a prep school field hockey final or an equestrian trial. The spectator sport that occasioned this tailgate picnic was a high-tension standoff between Delaware County police and local multimillionaire John du Pont.

This live-action docudrama had actually begun two days earlier, on January 26, 1996, when du Pont shot and killed wrestler David Schultz, a former Olympic gold medalist who lived and trained at du Pont's Foxcatcher estate.[43] After the murder, du Pont barricaded himself in the steel-lined library of his mansion and waged a forty-eight-hour war of wills with authorities. The siege finally ended when the local SWAT team thought of turning off the heat, and du Pont came out shivering.[44]

According to postarrest reports, du Pont was an accident waiting to happen. In fact, if anyone ever fit a crime novelist's portrayal of a man gone over the edge, it was John du Pont.[44-46] He patrolled the 200-acre estate, which he believed to be the holy land of the Dalai Lama, in a tank. He mounted infrared cameras in the corners to watch for ghosts, threatened to shoot his wife in the belief she was a Russian spy, and feared that the clocks in the gym were taking him back in time. Other athletes admitted that du Pont's behavior was so bizarre that he would have been barred from the wrestling community had he not contributed so lavishly to the sport.

Nonetheless, rumors that du Pont's defense would center on his mental competency provoked responses ranging from disgust to outrage. Schultz was well-liked in the wrestling community, and he left behind not only a wife but two young children. Du Pont's hasty retreat after the crime suggested to many that far from being oblivious to the implications of his actions, he had actively attempted to evade capture. And local papers hinted that with a personal fortune estimated at $200 million, du Pont could well afford to "buy" an insanity plea.[47]

Was du Pont mentally unfit even to stand trial? The decision hinged on two criteria: whether he was lucid enough to understand the charges against him and whether he was rational enough to assist his lawyers in planning his defense. By the time the legal dust had settled and a competency hearing finally got under way eight months after his arrest, nearly everyone agreed that John du Pont failed on both accounts. Defense lawyers Richard Sprague and William Lamb (ultimately dismissed by du Pont because he believed they were part of a conspiracy headed by the CIA) argued that hundreds of hours of conversation with their client had failed to yield a single cogent discussion.[48] Three defense psychiatrists, as well as two appointed by the court, were in "universal agreement . . . that du Pont was actively psychotic."[46] A videotape from their interviews showed du Pont insisting that he was the Dalai Lama, describing a plane flight in which he'd commanded Bulgarian pilots, and reiterating his fears that he was the victim of a government conspiracy.

On September 24, 1996, Delaware County court judge Patricia Jenkins ruled that du Pont was incompetent and ordered him committed to Norristown State Hospital for treatment.[46] But she had no intention of leaving him there. Jenkins ordered an update on the millionaire murderer's condition within sixty days—and every ninety days after that until she could justify putting him on trial. On December 3, only two months after he was sent to Norristown, Jenkins was satisfied that du Pont had recovered sufficiently to meet the legal criteria for competency.[47]

John du Pont would have to answer for his actions after all. But what would really be on trial was not his behavior but his state of mind. Du Pont's guilt was a given. His new defense team knew they would never convince a jury that he was innocent; they could only hope that jurors would buy the argument that du Pont still belonged in a hospital rather than a prison.

THE MEDICALIZATION of violence by researchers such as Delgado, Mark, and Ervin offended some because it appeared to perpetuate the earlier inequities of the eugenics movement. For others, however, the growth of biological psychiatry created a different monster: the fear that biology was the equivalent of an insanity plea writ large. Highly publicized cases of "Twinkie defenses," "abuse excuses," and collaborations between well-paid lawyers and unscrupulous psychiatrists fueled the misconception that biology was an excuse, not an explanation.

Society has been undecided about the line between malice and mental illness for centuries. Clearly, some people are so delusional or incapacitated that punishing them seems worthless at best and inhumane at worst. On the other hand, unless the concept of insanity is rigorously defined, it does present an open invitation to a defendant with deep pockets or a clever defense attorney.

British courts set the modern standard for the definition of insanity. In 1843, a woodturner named Daniel M'Naughten was acquitted of murdering a secretary to Prime Minister Sir Robert Peel on the grounds that his reasoning abilities had deteriorated to the point that he was no longer capable of understanding the "nature and quality of the act he was doing, or, if he did know it . . . he did not know he was doing what was wrong."[49] This right-or-wrong yardstick, a criterion that came to be known as the M'Naughten test, was quickly adopted by the American judicial system as the definitive answer to the question of assessing mental competence.

But legal, medical, and popular interpretations of moral incapacity quickly muddied the judicial waters. Some states, for example, extended the insanity defense to encompass uncontrollable impulses, or "homicidal mania," while defendants pleading mental disease and "temporary insanity" enjoyed an increasingly favorable reception from judges and juries. Eventually the link between violence and mental illness led to a turning point in the definition of competence. In a controversial 1954 decision, Washington, D.C. circuit court judge David Bazelon struck down the M'Naughten rule in favor of a new criterion that he believed better reflected current medical knowledge: "an accused is not criminally responsible if his unlawful act was the product of mental disease or mental defect."[50]

Durham v. the United States had a profound impact on the public's opinion of biological explanations for human aggression. Scandalized newspaper reports documented a surge of successful insanity pleas in some areas; in Bazelon's district, for example, such pleas rose from 0.4 percent at the time of the decision to over 14 percent by 1961.[49] A judicial system that had once punished all but the most obviously deranged now seemed to accept every anxiety or blue mood as a legitimate excuse for murder. *Sick* and *psycho* replaced *immoral* and *evil* as the preferred terms for describing the most violent. Media coverage of high-profile cases, such as that of John Hinckley, Jr., declared insane and therefore acquitted of attempting to kill President Ronald Reagan, reinforced the belief that murderers and rapists routinely exploited the insanity defense to evade punishment.

In fact, recent studies show that fewer than 1 percent of defendants facing felony indictments resort to the insanity defense; of these, no more that one-quarter are successful.[51,52] Both lawyers and psychiatrists report that juries are surprisingly skeptical of arguments attributing violent behavior to mental illness; well-known exceptions, like the jury that acquitted Hinckley, represent, in the words of one forensic psychiatrist, "aberrations that skew the public perception."[52] Unfortunately, these aberrations have convinced the public that biology favors the violent at the expense of victims and that forensic psychiatrists are poised to throw open the floodgates of our prisons.

The exceptions also foster widespread misconception about the potential violence of the mentally ill. A recent study by the University of Pennsylvania's Annenberg School of Communications, for example, discovered that the "psychotic" killer on a mad murder spree represents a common theme in television programming; in fact, 70 percent of mentally ill television characters were also portrayed as violent.[53] This "violent-maniac" stereotype translates into widespread discrimination and distrust, despite the fact that experts estimate that no more than 20 percent of the mentally ill ever commit a violent act.[54]

The crime-fueled victimization of the mentally ill has not been lost on social critics of behavioral biology. They hear a familiar echo of biological determinism inherent in the insanity defense, the well-worn genetic mind-set that causes "millions of people . . . to think that criminals are perhaps born that way; crime is in the blood, the genes, the bones."[49] As a result, biologically inspired legal initiatives have not only enraged conservative Americans but have also paradoxically fueled the worst fears of modern antieugenicists, that biological explanations for violent behavior will be used to isolate, discriminate, and persecute.

A Truce in the War Between Nature and Nurture

The miscalculation, misconduct, and misunderstanding that characterized biologically oriented studies of behavior for over a century created a rupture between nature and nurture that has often seemed as entrenched and as bitter as a civil war. The inevitable consequence of eugenics and psychosurgery, killer instincts and selfish genes, caged animals attacking dolls and murderers bartering for a hospital bed instead of a prison cell has been a profound distrust of behavioral science and the motives of biological researchers.

Criticism of the biological perspective has centered on the determinism that permeated biology after Darwin. If genes rule, people are little more than genetic puppets, their behavior and their moral judgment tethered to the double helix. If they are bad, it can be blamed on an unwelcome mutation, and if they are good, they have a reproductive advantage to enjoy along with a clear conscience.

While this "it's not my fault" view may bring comfort to some offenders, it brings only anguish to victims and their families. As a result, many believe that the biological perspective on aggression has little to offer victims and little to say about the lasting consequences of violence. Biological explanations, they argue, merely allow perpetrators to walk away from their actions or to reverse the charges, blaming the victim and society at large for a lack of understanding.

Explanations for violence based on rigid thinking—whether they exclude the brain or the environment—are easy prey for ideologists. Only a few decades ago, eugenic extremists decried those who opposed discriminatory limits on immigration and harsh laws forbidding interracial marriage as "race criminals." Today social critics like Peter Breggin charge that research on the biology of violence is part of a government-engineered conspiracy to implement "biomedical social control" of minorities, including enforced medication of children deemed to be at risk for committing violent crimes.

The excesses of yesterday's advocates earned the outrage of today's detractors. But are such critics still justified? Does research on the behavioral biology of aggression constitute biomedical social control, or is it an undervalued and misunderstood option for understanding the problem of violence, not just among inner-city offenders and serial killers but in violent husbands, homicidal drivers, abusive parents, and even victims in all walks of American life?

WHILE THE DEBATE over eugenics, instincts, and sociobiology raged, neuroscience underwent a radical transformation, from a gawky scientific adolescent to a vigorous mature discipline. "Twenty-five years of progress"—the slogan celebrating the 1995 silver anniversary of the neurobiological equivalent of the Screen Actor's Guild, the Society for Neuroscience—have not simply brought us more sophisticated skirmishes between the nature and nurture camps, but have significantly altered the way we think about brain structure and function. In a tribute to the sagacity of Flourens, new methods have yielded new data, and fresh interpretations are changing the very foundation of our understanding of the relationship between brain and behavior, gene and environment.

Some observers see these advances as the symbol of a holistic biology that can finally mend the rift. Biology is *not* destiny. The century-old gap between nature and nurture has been bridged by the brain. Critics who charge otherwise are trapped in an outmoded argument that ignores pivotal neuroscience discoveries and a paradigm shift in our understanding of the causes of human behavior. The critics may be fighting the same old battles, but the world itself has changed.

Two

THE VICIOUS CIRCLE

I grew up in an illogical home dominated by a mother who saw everyone around her as an angel or a devil. Gray was not an option in her rigid yet erratic worldview. If someone acted out of character, she simply reclassified them. One day, I was her child. The next, I may have forgotten to hang up a towel, or dropped a glass, or called a friend once too often, and I ceased to exist. Weeks, even months of silence followed; then she would suddenly begin speaking again, as if nothing had happened.

Under the spell of genetic determinism, behavioral research suffered from a similar personality disorder. While gene-centered biologists battled social scientists in this pathological conflict, behavior had to be either nature or nurture, genes or environment, physiology or learning. During the early decades of this century, unbridled enthusiasm for scientific progress deified the "nature" viewpoint, and social behavior, including aggression, was attributed to innate drives and appetites, preprogrammed by heredity. The environment was an afterthought, outside events merely triggers that cued genetically engineered programs in the same way that electrodes planted in the brain could spark elaborate behavioral patterns. But this rigid approach could neither tolerate nor assimilate the challenges posed by the postwar revelation of eugenic atrocities and clinical reports of the failures of psychosurgery. Biological explanations for behavior fell from grace, nurture emerged as the "angelic" position, and behavioral biologists, particularly geneticists, were demonized.

First nature, then nurture. Such simplistic views may have an innate appeal, but they violate the realization that life cannot be accurately portrayed in black and white. A more realistic analysis of behavior demands a more sophisticated perspective, capable of seeing in colors.

Fortunately, neuroscience has outgrown its early methodological and theoretical constraints to welcome what theoretical biologist Stuart Kaufman calls it: "a world of stunning biological complexity."[1] The first truly multicultural biological discipline, it has expanded well beyond the operating theater and the anatomy lab to encompass fields as diverse as veterinary medicine and mathematics, cell biology and biomedical engineering, immunology and computer science, and to exploit state-of-the-art techniques ranging from magnetic resonance imaging to positional cloning. From a core group of about one thousand brain researchers just twenty-five years ago, the community has multiplied to more than twenty-five thousand scientists worldwide, studying behavior not only in the laboratory, but also in the field, on line, and in the clinic.[2] During these two and a half decades of expansion, neurobiology has grown in scope, as well as in size and sophistication, reaching out of the head to embrace the surrounding environment. As a result, studies of integrative processes such as perception, adaptation, and learning have revealed that far from being dictatorial or protectionist, the brain actually maintains an open-door policy, inviting outside influence.

This unified biology does not need to be exiled. Holistic in outlook, it offers a new language and an alternative perspective that not only admits the outside world but suggests how it might influence behavior. Dreadful sights, angry voices, a racial slur, the feel of the trigger, as well as moderating influences—reason, memory, conscience—are not cultural or spiritual ephemera, but depend on the movement of molecules, the integrity of proteins. Conversely, the molecular processes that craft the neural language of chemistry and electricity would falter without the continuous inspiration of current events. Between gene and environment, body and world, there is a brain—and it is the final common pathway of human experience.

Even the most unrepentant assailants, the most cold-blooded murderers, the most sadistic of serial killers were once infants. There was a time when they could barely hold a rattle, much less a gun; when they smiled for Christmas portraits and giggled at peek-a-boo; when they were afraid of fireworks, needed help to feed themselves, and wore shoes no bigger than ring boxes. What happened? What inner or outer factor—parents, schools, genes, morals, abuse, television, neglect, stress, attention deficits, self-esteem, temperament—has the power to transform innocence into violence?

The answer provided by modern neuroscience is "all of the above." Violent behavior, like other complex behaviors, is neither a program nor a reaction but a process. It is not inborn, nor is it made from scratch by culture. It develops.

The "Nurture" of Nature

There are fifty-two cookbooks, three boxes of index cards, two looseleaf binders, and five years of *Better Homes and Gardens Holiday Baking*—a total of nearly twenty thousand recipes—in my kitchen. I've actually taste-tested no more than 10 or 20 percent. And only a fraction of those sampled—about three or four dozen altogether—appear regularly on our dinner table.

Infant cells deciding whether to become heart or brain, stomach, bone, or skin, face a similar decision process. From a full complement of as many as 100,000 genes, they must select the subset that will define the characteristic form and function of a single tissue, the "recipes" for the unique proteins that make a heart cell a heart cell and a brain cell a brain cell.[3] This process of recipe selection, or *differentiation,* has posed a crucial challenge to geneticists, for it implicitly suggests that some entity or force outside the genome acts to limit its allegedly universal power over life. From the chemical cookbooks and genetic recipe boxes, an unseen hand somehow culls a few dozen recipes that define a cellular lifestyle.

Evelyn Fox Keller proposes that early geneticists coped with this nagging tension between genetics and embryology by papering over it, transforming the question, "What regulates gene expression during differentiation?" to "How do genes control development?"[4] She suggests that Watson and Crick's identification of DNA as the medium of genetic transaction, and subsequent work that outlined the gene-to-protein control sequence—DNA makes RNA, which in turn acts as the blueprint for proteins—allowed geneticists to annex embryology and "talk instead about development in the abstract and the genetic programs or instructions that are needed to guide it."[4] The answer to the riddle of development was simple: a nerve cell becomes a nerve cell because its genetic program orders it to do so.

The so-called central dogma of molecular genetics may have answered the question of how the cell starts with genes and ends up with proteins, but it did not explain the puzzle of differentiation. Nor did it succeed in excluding embryologists, who quietly adapted the new genetics to their own ends. The resulting hybrid science—developmental biology—championed a dynamic view of the genome, in which "talk about 'gene action' subtly transmutes into talk about 'gene activation,' with the locus of control shifting from genes themselves to the complex biochemical dynamics . . . of cells in constant communication with each other."[4] Thanks largely to these new embryologists, the scientific community has come to recognize that the environment is not at odds with the genome, but in fact is essential to its function. Developmental biology has demonstrated that from the first moments of its existence, the developing organism is more than a prefabricated structure assembled according to a DNA blueprint; it is an organic promise, utterly reliant on environmental direction to select, control, and regulate gene expression.

The developing nervous system has played a critical role in this discovery. With more different cell types than any other body organ and up to 10^{15} specific cell-cell connections, the human brain presents a challenge that cannot begin to be met simply by the genome, for the number of connections exceeds the total number of instructions by at least ten orders of magnitude.[5] Only an active dialogue between the genome and the external environment can guide the complex decision-making needed to build a fully functional nervous system.[6] As a result, the environment has a profound influence on the brain from the earliest stages of development, and it retains that sway throughout life. Geneticists may have taken little notice, but as early as the 1920s, developmental neurobiologists had already begun to undermine the omnipotence of the genome with their discovery of this symbiotic relationship between genes and the outside world.

The cells that will ultimately form the adaptable and elaborate mammalian brain begin life as part of a thin sheet swaddling the growing embryo.[7,8] In a propitious decision that occurs when the nascent animal is still little more than a dimpled ball of cells, a central ribbon of the tissue sheet secedes from its cellular neighbors, which will ultimately give rise to the overlying muscle and bone; this panel then curls up and fuses to form the *neural tube* that will become brain and spinal cord. The birth of the neural tube is not a display of cellular independence, however; it is a forced choice instigated by the surrounding tissue, the *mesoderm,* as shown in a classic experiment published in 1924 by Nobel Prize–winning embryologist Hans Spemann and his student, Hilde Mangold.[6,7]

Building on earlier studies in which Spemann had perfected the delicate art of prenatal surgery on amphibian eggs, he and Mangold transplanted a sliver of mesoderm from a light-colored newt embryo to the underbelly of an embryo of a dark-colored species. As shown in Figure 2.1, the pale immigrants provided more than local color; they provoked a revolution among the adjacent host cells, convincing them to trade in a life as belly skin and recreate a mirror image "Siamese twin" of the original dark embryo, complete with a second fully formed nervous system. Spemann called the activist mesoderm the *organizer,* a term that openly suggested a conspiracy between the cells recruited to form the neural tube and the prenatal environment.

The discovery of the organizer launched an intensive search for environmental determinants of differentiation right in the shadow of a biology dominated by genetic programs and predetermined schedules. Ironically, the

Figure 2.1

Cells located on the lower rim of a dark-skinned newt embryo normally mature into skin cells located along the belly of the young animal (left panel). "Organizer" cells transplanted from a light-colored embryo (right panel) induce the skin precursor cells to form a second neural tube instead—and, ultimately, an entire second embryo.

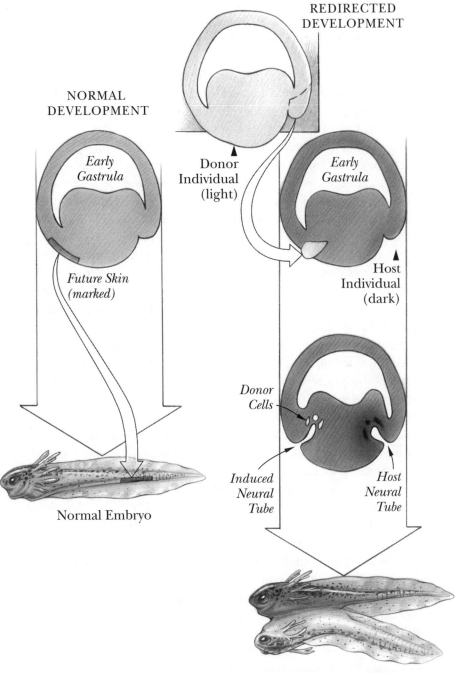

REDIRECTED
DEVELOPMENT

NORMAL
DEVELOPMENT

*Early
Gastrula*

Donor
Individual
(light)

*Early
Gastrula*

*Future Skin
(marked)*

Host
Individual
(dark)

*Donor
Cells*

*Induced
Neural
Tube*

*Host
Neural
Tube*

Normal Embryo

Induced (Double) Embryo

directive issued by Spemann's mesoderm proved one of the most elusive, evading capture until 1993, when Richard Harland of the University of California at Berkeley discovered a protein in the frog (christened, appropriately enough, "noggin")[9] that could convert undifferentiated embryonic cells into neural tissue. But by that time, the pursuit of outside influences on the growing brain had turned up an entire family of parental proteins known as growth, or *trophic,* factors (from the Greek word meaning "to nourish"). As embryologists have learned, these proteins not only guide and nurture developing nerve cells; they provide cradle-to-grave services, ranging from population control to career counseling, genetic engineering to communications management.

MANY CELLS of the nervous system would never survive beyond infancy if it were not for critical growth factors supplied by the extracellular environment. This chemical dependency was first observed in the peripheral nervous system—the neural network that reaches out beyond the brain and spinal cord to innervate the sense organs, muscles, and body tissues—by an Italian histologist transplanted to Washington University in St. Louis, Rita Levi-Montalcini. Levi-Montalcini—a survivor herself, who had conducted her first experiments on neural development in a makeshift laboratory in her bedroom during World War II—had become intrigued by the growth-promoting effects of certain tumors on the clusters of peripheral neurons that chain alongside the backbone.[10] These knots of cells, the *sympathetic ganglia,* are the base camps of the sympathetic neurons that speed up heart rate, breathing, and gastrointestinal function in response to excitement. By excising the tumor and culturing it outside the body, Levi-Montalcini learned that she could concentrate the growth-enhancing elixir and observe its effects on a single ganglion at close range: "Twenty years later I recall, as if it had just happened, the astonishment and wonder of that morning when, for the first time, I saw in the light microscope the outcome of these experiments. Nerve fibers had grown out in twelve hours from the entire periphery of [the] . . . ganglia . . . and had spread out radially around the explants like the rays of the sun."[10]

Levi-Montalcini and her collaborators, Viktor Hamburger (another of Spemann's star pupils) and biochemist Stan Cohen, christened the tumor protein responsible for these neural sunbursts "nerve growth factor"—NGF for short.[11] Sympathetic neurons needed NGF like water. The probing ends of their outgrowing fibers sopped it up like sponges, using receptor proteins on the surface of the cell membrane to capture the growth factor and introduce it to signaling proteins inside the cell. Loss of the factor was catastrophic. When Levi-Montalcini and Cohen injected an antiserum to NGF into newborn mice, starving the infant sympathetic neurons at a critical point in their development, over 95 percent of the cells in the ganglia died.[12] Subsequent investigators identified additional neuronal populations that depended on NGF, including the sensory neurons that mediate pain perception and, perhaps most important, cells in the brain itself.[13-15]

NGF was only the tip of the iceberg, the first example of a family of survival-oriented proteins known as *neurotrophins.* A second member of this collective, *brain-derived neurotrophic factor* (BDNF), isolated in fleetingly small amounts from kilograms of pig brain, bears an uncanny resemblance to its less-secretive cousin; more than half of their constituent amino acids are the same.[16] Sympathetic neurons aren't interested in BDNF. But the sensory neurons that innervate muscles and skin are as dependent on this neurotrophin as their sympathetic counterparts are on NGF.[16] A third member of the family, neurotrophin-3 (NT-3), is the patriarch of the clan. In fact, sympathetic neurons first need NT-3 to elaborate the receptor machinery needed to recognize and use NGF.[17]

The influence of the neurotrophins extends far beyond basic survival. In addition, these proteins guide and instruct the young neurons as they forge an identity, selecting from the hundreds of thousands of gene recipes present in each neuron those that will allow it to do its appointed task in the adult nervous system. For example, the sympathetic neurons under the care and tutelage of NGF are descendants of a founder population of cells that also give birth to sensory neurons, the parasympathetic nervous system, supporting cells known as glia, and the epinephrine-secreting chromaffin cells of the adrenal gland.[6] These precursor cells, collectively known as the *neural crest,* migrate out of the neural tube and stream along invisible pathways to their appointed posts in the body, where some meet and coalesce to form the sympathetic ganglia. Within the emergent ganglia, NGF, in conjunction with the skin cell trophic protein, fibroblast growth factor (FGF), guides the precursors through the selection of genes that commit them to becoming sympathetic neurons.[6] In contrast, precursors that migrate to the infant adrenal gland and find themselves exposed instead to high levels of glucocorticoid hormones select the chromaffin cell gene menu.[6]

Additional signals in the ganglionic environment then assist newly born sympathetic neurons in the selection of a chemical dialect, or neurotransmitter, they will use to communicate when mature, counseling them to turn on the genes that control the synthetic and recycling apparatus appropriate to this transmitter and shutting off those coding for others. In a series of studies conducted in the 1970s, Harvard University researchers Paul Patterson and Linda Chun demonstrated that sympathetic neurons grown in an artificial culture medium in the absence of growth factors grew up to speak with the transmitter norepinephrine, while sister cultures reared in the presence of the growth factor ciliary neurotrophic factor speak a different language; these chemically socialized neurons reject the norepinephrine recipes in favor of genetic instructions for the protein machinery needed to make and use another small-molecule transmitter, acetylcholine.[18]

Young sensory neurons in their ganglia receive similar extracellular guidance. Here, NGF directs a cadre of uncommitted cells to become high-threshold pain-detecting neurons. NGF depletion, or substitution of NT-3, converts the cells from their original destiny to a genetic program coding for touch-receptive cells that ignore painful stimuli but respond instead to low-threshold stimulation of skin hairs.[17]

For every cell, there is only one genome but many life choices, each characterized by the selection of gene recipes that determine that cell's identity. The choices are governed not by genetic commands alone but by a process that relies on chemical mandates issued by the environment.

Genetic Engineering

The car, a 1974 Plymouth Duster, was a bargain at $450, even with a failing battery and a mysterious click. The battery could be replaced—and I knew that my father would be able to discover the cause of the noise.

My father was an accountant with a mechanical mind. Engines told him their secrets. Rebellious machines sized him up, then acquiesced. The click had baffled my local mechanic, the Duster's former owner, and a fellow graduate student who claimed to be a "car junkie." But all my father had to do was listen.

"It's the differential."

"The what?"

"Differential. It's part of your rear wheel mechanism. I added some more oil to that bearing, and maybe you'll be lucky."

The click came and went for nearly two years before my luck ran out. A piercing metallic whine, a snap, and the car dropped like a flatiron to the pavement, the rear axle sheared in half. The guy who towed it said it sounded like the differential had gone.

NOT ALL MECHANICAL problems are as mysterious as my Duster's fatal click. In fact, as my father was fond of noting—having pointed out dozens of switches and plugs to friends and family—most were actually rather simple to solve. "Won't work until it's turned on," he'd have observed.

The gene menus selected during differentiation don't work until they're turned on by environmental signals, such as trophic factors. But how exactly does the outside world get into the brain to do this? Simple recognition—the mere binding of a growth factor to its receptor at the cell surface—isn't enough to alter gene expression in the nucleus, just as simply pushing your key into the ignition isn't enough to start your car. To get moving, you have to turn the key and flip the switch firing the car's electrical system. To motivate an NGF- or BDNF-receptive cell, the growth factor–receptor complex must penetrate the nucleus, to select and "tinker," as Dr. Steven Hyman, director of the National Institute of Mental Health, puts it, with the recipes for proteins that frame outgrowing nerve fibers, synthesize neurotransmitters, bind and translate signals, forge cellular identity.

Nuclear access hinges on flipping switches, from the cell surface right down to the genome. In fact, neurotrophin receptors themselves can be thought of as switches, flipped into the on position by a chemical reaction initiated when receptor and growth factor hook up. The animated receptor in turn trips a progressive cascade of similar reactions, a chemical waterfall that "turns on" critical enzymes and regulatory proteins closer and closer to the genome, ending with a *transcription factor*—the protein key needed to unlock one or more genes (see Figure 2.2).[19,20] Loops and coils on the transcription factor slot into the turns and grooves of the double helix, opening the door for the enzyme RNA polymerase to read and copy the recipe for the protein encoded in that gene. The cascade of switching reactions, a process molecular biologists call "stimulus-transcription coupling," transforms receptor binding—an event that began in the environment—into a genetic event, regulating, adapting, and fine-tuning gene expression.

After binding to cell-surface receptors, the protein cascade activated by NGF or BDNF pumps up the expression of DNA sequences known collectively as *immediate early genes* (IEGs), that code for a family of short-lived transcription factors with exotic acronyms like c-*fos,* c-*jun,* and CREB.[21] The IEG transcription factors might be better termed conscription factors, for they draft into service an army of additional genes critical to neural function. The result is an adaptive modification of the target cell, a genetic response to chemical signals in the extracellular environment that reveals the exquisite plasticity of a system long assumed to be hard-wired.

INTERACTIONS BETWEEN gene and environment are not limited to the charmed interval between conception and birth. The environmentally sensitive nervous system remains malleable across the lifespan, and our perceptions, actions, and experiences tinker ceaselessly with the proteins assigned to carry out the myriad small tasks that collectively fashion our relationship with the world around us.

Molecular tinkering based on recognition and response, like the developmental decisions based on neurotrophic signals, cannot occur until the interior of the neuron learns about events in the outer world. On a cellular level, therefore, perception is essentially a translation process, one that again begins with receptor binding, progresses to activation of a web of "second (and third) messengers," and ends with gene-switching transcription factors.

Stripped of its neighbors and stretched out in isolation, the nerve cell, or neuron, resembles a carrot seedling. The body of the cell, the split kernel of the seed, which houses the genes and carries out the chores of daily living, is capped by a spidery tuft of fibers called *dendrites.* These fibers represent the receiving

Figure 2.2

The binding of chemical messengers, such as neurotransmitters or growth factors, to receptors on the surface of the neuron initiates a cascade of reactions inside the cell that culminates in gene activation.

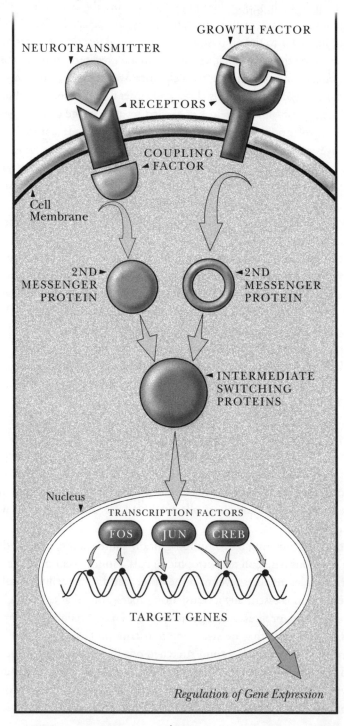

GROWTH FACTOR

NEUROTRANSMITTER

◄ RECEPTORS ►

COUPLING
◄ FACTOR

▲
Cell
Membrane

2ND ►
MESSENGER
PROTEIN

◄ 2ND
MESSENGER
PROTEIN

◄ INTERMEDIATE
SWITCHING
PROTEINS

Nucleus
▼

TRANSCRIPTION FACTORS

FOS   JUN   CREB

TARGET GENES

Regulation of Gene Expression

end of the neuron, the entry point for as many as 150,000 incoming messages from adjoining cells.[22] At the other end, a single fiber trails away from the cell body like an emerging root; this *axon* ferries the signal condensed from those thousands of inputs to the next cells in the communication chain, sometimes over an extraordinary distance (Figure 2.3).

Data, including sensory inputs and other environmental information, are coded in the form of brief surges of current that flow from the body of the nerve cell down the axon. At the terminal end, the arriving signal, like a miniature garage door opener, trips dozens of portals in the cell membrane to release tiny amounts of a neurotransmitter—one of over fifty chemical messengers that include small molecules, like norepinephrine and acetylcholine; proteins and protein fragments known as peptides; steroid hormones; and even water-soluble gasses, such as nitric oxide—into the cleft separating the transmitting cell from its next-door neighbor.[23] The eager finger of a dendrite pokes into the cleft, like a nosy neighbor sniffing for gossip, listening, anticipating. There it waits for messages to seep to its doorstep, inviting capture by specific receptor proteins protruding from the dendritic surface or hovering just below it.

Some of these receptors, including those for steroid hormones (estrogen, progesterone, thyroid hormone, cortisol), are actually transcription factors. They head for the nucleus as soon as they've acquired a hormonal partner, where their finger-like projections grasp a DNA switch known as the *enhancer* region, and pull it to activate "downstream" genes coding for specific structural and functional proteins.[3] Other transmitter-receptor complexes remain fixed in the synaptic membrane; their limited mobility precludes a direct audience with the genome. Transmitter binding to these receptors, like the binding of NGF or BDNF to their respective receptors, alerts second messengers that can initiate a switching cascade culminating in the induction of IEGs and other transcription factors.[19,20] These transcription factors regulate the expression of target genes for transmitter-synthesizing enzymes, receptors, and other neuronal proteins, changes that may underlie such long-term responses to the environment as adaptation, learning, and memory.

Circular reasoning abounds in this chemically mediated coupling between stimulus and gene expression. Among the target genes receiving a wake-up call following neurotransmitter receptor binding are those coding for growth factors, including NGF and BDNF.[24,25] And, in addition to their well-known effects on survival, differentiation, and cellular identity, recent data suggest that these growth factors can also reach back to tinker with the process of neurotransmission, aiding and abetting transmitter release and action.[26] Growth factor influence, in other words, circles from environment to gene and back to the environment, first protecting and guiding infant neurons and then, reawakened by activity at the synapse, strengthening the relationship between communicating cells, to create, in the words of Ira Black, head of the Department

Figure 2.3

The space between adjacent neurons (the synapse) is not a barrier to cell-cell communication. Neurotransmitters released by the "sending" cell deliver the message across the gap to receptors on the "receiving" cell. A transmitter that is no longer needed is broken down by enzymes or taken up by transporter proteins on the postsynaptic cell for recycling.

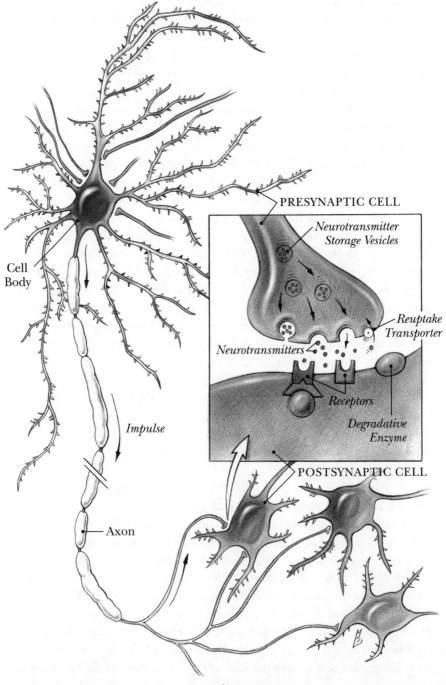

of Neuroscience and Cell Biology at the Robert Wood Johnson Medical School and former president of the Society for Neuroscience, "a doorway from environment to brain to memory."

ENVIRONMENTAL INFLUENCE is not always benign. Physical injury to the brain, for example, can cause persistent damage that even growth factors are hard-pressed to repair. Recurrent electrical stimulation can have equally devastating consequences that culminate in a genetic response. Researchers at the National Institute of Mental Health (NIMH) have studied such electrically induced changes in behavior, biochemistry, and gene expression after *kindling,* an experimental technique in which repeated small pulses of electricity are used to "turn on" cells deep in the temporal lobe of the rat brain.[27,28] Daily intermittent stimulation for a mere second is enough to progressively destabilize the cells surrounding the electrode; a few days of such stimulation, and the irritable neurons fire back. The rat freezes, locked in place. Its snout, mouth, and facial muscles twitch; its forepaws jerk and tremble spastically. If the repetitive stimulation is continued, the excitation reaches a critical threshold, and an uncontrolled surge of electrical activity rampages through the brain as the rat rears, goes rigid, and topples convulsively in a full-blown motor seizure.[29]

In the wake of the seizure, long-lasting changes in responsiveness are "branded" into the activated neurons. Up to a year later, the merest hint of electrical stimulation readily kindles a recurrent seizure.[28-30] If seizures are kindled often enough, they begin to take on a life of their own, erupting spontaneously even in the absence of stimulation. The evolution from electrically dependent to spontaneous seizures is mirrored in a characteristic sequence of changes in the electrical and biochemical properties of the overexcited cells.[28] Some cells increase their production of key neurotransmitters; others reduce their output. Receptor populations fluctuate, and dendrites sprout new knobs and branches. Some unlucky neurons, like the legendary girl in the devil's red shoes, may become agitated to the point of lethal exhaustion. The escalation to electrical storm is accompanied by a spatial wave of gene induction. Beginning close to the electrode site, the disturbances initiated by sensitization percolate into the cell, awakening c-*fos* and other IEGs. As the kindling process evolves, IEG recruitment spreads, engulfing a wider and wider region around the electrode; eventually, with the emergence of generalized seizures, it spreads to the opposite side of the brain as well.

Robert Post, chief of NIMH's Biological Psychiatry Branch and an expert on the dysfunctional relationship between environment and gene evoked by kindling, believes that this IEG induction may "program the cell for the second and third wave of longer-term induction of genes and longer-term modifications of peptides, proteins, and . . . neurotrophic factors . . . that may alter the microanatomical environment."[28] As evidence, he notes that the spreading wave of c-*fos* induction during the progression from twitches to seizures is followed by a second wave of microanatomical changes: the signature alterations in peptide and hormone transmitters, neurotransmitter receptors, and the growth factors NGF and BDNF.[28] These changes, Post argues, may reflect molecular events responsible for the evolution of spontaneous seizures.

During kindling, electrical stimulation induces a familiar protein activation cascade, prodding transcription factors to tinker with the neuronal genome in a way that greatly increases cellular sensitivity to subsequent stimulation. Environment and gene interact, as they do during development, in a dynamic tit-for-tat process that spirals toward a malevolent counterpart of the supportive interchanges that occur during development and the communication-building exercises that accompany ordinary neuronal activity. Good or bad, what we see, hear, learn, experience, and do intersects with the person we already are, woven by the brain into the seamless fabric of a life.

The "Nature" of Nurture, or What's a Gene Good For, Anyway?

The environment is a crucial player in the drama of differentiation. But environmental determinism is no more satisfactory as an explanation for behavior than the gene-dominated view it purports to civilize. Genes may not dictate our actions, but without the gene recipes selected during development, we could never construct, operate, or maintain the delicate machinery of the nervous system, and the unique identity of distinct neuronal populations would be obliterated, their purpose lost in a sea of infinite possibilities.

The sequential decisions about identity and function that progressively narrow the neuron's lifestyle options during differentiation prevent adaptability from degenerating into chaos. By locking in a menu of genes that code for characteristic features, like transmitter choice, the process of development defines the ground rules for postnatal interactions with the world, rules that are not designed to set unreasonable or rigid limits on the capacity to adapt, but to provide a framework for making sense of the cacophony of sounds, sights, and experiences that swirl around us, and to organize mechanisms for learning, remembering, and responding.

Some of these innate rules define the boundaries of perception that qualify our experience of the world. For example, human vision is tethered to a discrete range of wavelengths, so, unlike the owl that calls in my backyard after sunset, we cannot see in the ultraviolet to track small mammals by the fluorescent trails of their scent marks. We cannot compensate with our noses, for our sense of smell pales in comparison to most animal hunters, and our meager supply of scent receptors—less than 1 percent that of the average German shepherd[31]—could not tell the difference between a field mouse and a tree shrew.

Were we actually to catch our prey by sheer human cunning, we would be deaf to its ultra-high-pitched cries of panic; should it snap in self-defense, few of us could retract a hand in time to prevent a nasty bite.

Genes map out this human range of perceptual capabilities. The resulting biological framework demarcates the limits of awareness, framing and coloring our apprehension of the world around us. Individual variations in this framework, like house-to-house differences in floor plan, allow for interpersonal differences in perceptual acuity and responsiveness, for challenging deficits as well as exceptional ability. Some of us see only in shades of gray, most can see well enough to drive, and a select few see well enough to paint.

The gene menu also outlines starting points for action. By birth, the brain has elaborated a species-specific repertoire of elementary response patterns, a behavioral tool kit that includes reflexes, facial expressions, vocal patterns such as crying and cooing, orienting and alerting responses. These simple modules provide the basic operating procedures needed to initiate life, the nucleus of more elaborate behavior patterns that will characterize the mature individual, including complex social behaviors such as aggression and affiliation. The rudiments of emotional behavior are also in place in the form of individual variation in responsiveness, intensity, ability to adapt to change, reactions to novelty—traits that collectively define the characteristic approach to the world known as temperament and that are demonstrably sensitive to genetic influence (see Chapter 8).

Finally, the neural framework constructed from genetic recipes maps out biases as well as boundaries. These physiological filters set priorities for the organism that favor responses keyed to particular classes of environmental stimuli. Like perceptual boundaries, they begin to structure an otherwise overwhelmingly complex world.

Primates, including humans, have socially biased nervous systems, constructed to notice and attend to social cues over other features of the environment. The importance the brain attaches to the social realm is revealed in the complexity and sophistication of the mechanisms devoted to interpersonal communication, to giving and receiving messages from others, a complexity that reaches its apex in human language. Language is so important—as Paul Broca discovered in his landmark studies of aphasic patients—that entire sectors of the human brain are devoted to the management of this critical function. But even in the absence of spoken language, monkeys and apes have developed rich and intricate systems of vocal communications to warn their companions of predators, celebrate the discovery of food, pick fights, solicit help, initiate play, and intimidate subordinates. In studies of African vervet monkeys, for example, Dorothy Cheney and her colleague Richard Seyfarth describe six acoustically distinct calls the monkeys use just to distinguish commonly encountered predators, each of which precipitates a unique response: the *leopard alarm,* which sends vervets on the ground scrambling into the trees; the *eagle alarm,* which alerts the monkeys to look up and run for cover close to the ground; the *snake alarm,* which prompts the animals to stand on their hind legs and scan the surrounding grass; and *minor mammalian predator, unfamiliar human,* and *baboon alarm* calls, all of which cause the monkeys to pause with a cautious, wait-and-see attitude.[32] In addition, vervets emit a surprising variety of grunts, screams, and chutters during social encounters, including contact with members of neighboring groups. Human observers may have difficulty deciphering these subtle variations of "vervetese," but the monkeys themselves readily match voices and identities, tone and intent. Grunts, for example, a common and seemingly nondescript element of vervet conversation, can be grouped into four categories, each matched to a particular social situation: approaching a more dominant individual, approaching a subordinate, observing another monkey move into an open area, and catching sight of a neighboring group. Voice recognition studies, in which Cheney and Seyfarth recorded vocalizations from individual vervet monkeys and played back these calls from a concealed loudspeaker, demonstrated that the animals easily recognized the grunts and chutters of specific individuals—juveniles as well as adults, and even those belonging to other clans—by voice alone.

Faces are as crucial as voices to primates, and the mechanics of facial signaling are no less sophisticated than those underlying conversation. A dense plexus of nerves provides for an unparalleled flexibility and subtlety in the muscles that control facial expression;[33] on the receiving end, individual neurons within the temporal cortex respond to specific facial features, the angle at which the face is viewed, or the direction in which it is gazing.[33] Additional cell groups define and interpret the subtle emotional nuances of expression, creating an intricate network dedicated to detecting, classifying, comprehending, and responding to facial signals. Noting that such signals provide crucial insight into motive and intent and that "the recognition and interpretation of these social cues are extremely important for the smooth functioning of a social group," psychologist Martin Tovee suggests that the sophisticated machinery dedicated to facial expression and recognition—the physical expression of a socially biased gene menu—constitutes a "social semaphore system" that helps primates skillfully navigate the complexities of group behavior.[33]

The primacy of social responses provides additional evidence of an innate neural bias toward the social features of the environment. For example, within a few days of birth, human infants not only recognize and prefer their mother's voice to a stranger's voice but also display a precocious talent for nonverbal communication. Developmental psychologist Tiffany Field and colleagues reported that one-day-old infants not only responded differentially to sad, happy, or surprised faces—suggesting they were capable of discriminating the definitive features of some facial expressions—but also attempted to imitate the experimenter's expression.[34] By four weeks, infants wave their limbs and wiggle their brows to signal their recognition of their mothers; at three months, they skillfully use facial expressions and gestures, as well as vocalizations, to solicit the attention of a caretaker or to register their discontent.[35,36] Even infants who are congenitally blind smile, scowl, gape with surprise at a sudden sound, turn their heads to track voices, reach out for their mothers.

Humans are clearly less single-minded than monkeys, who have, in the words of Cheney and Seyfarth, "a kind of laser-beam intelligence."[32] Nonetheless, we seem to share a similar bias toward the social world, a "predisposition that makes it easier for individuals to understand relations among conspecifics than to understand similar relations among things."[32] This predisposition is the oil that allows the social engine to run smoothly. But it is also a sensitive barometer of the climate of our social relationships; when these relations cloud over or grow stormy, our fine-tuned, socially conscious nervous system cannot simply look away.

SOCIAL RELATIONSHIPS are not the brain's only pressing concern. The human nervous system is also biased toward survival—not merely in the long-term Darwinian sense, but in the day-to-day effort to hold body and soul together. Again, evidence of this bias can be found in the complex innate mechanisms dedicated to perceiving and reacting to events that threaten physical or emotional well-being. As we will see, the brain has dedicated entire circuits to the detection and processing of stimuli relevant to survival, including an "alarm" system that focuses attention on threatening events and that continuously monitors the intimate connections with the body regulating internal equilibrium and a reward system that differentiates positive and negative experiences. Other circuits, capable of a more refined examination of the environmental data, fine-tune the alarm system, overriding it if necessary to prevent self-defense from devolving into dangerous overreaction.

Survival bias, like social bias, makes itself known early in life. With its first squalls, the newborn signals its pursuit of the nourishment and physical contact it needs to survive. The quick recognition of the mother's voice and face, the early attempts to imitate her facial expressions and catch her attention are geared toward winning and keeping her essential interventions. The first responses to these infantile social cues set the stage for an effective communication system between the infant and the family, a development crucial to physical and emotional health. Later, the capacity for self-care (including self-defense) emerges in parallel with an age-appropriate decline in parental intervention. And as adults, our ongoing and vigorous efforts to stay fed, preserve our safety, and form meaningful relationships provide continuing testimony to our inborn and persistent desire to stay alive.

No Safe Place

The lessons of neural development have taught us that complex behaviors like aggression require two inputs: one from the inner world and one from the outer. And the people one Vietnam veteran calls the "hollow men, weighed down by memory," have taught us that aggression requires two participants as well: one who performs the aggressive act and one who suffers the consequences. Recent progress in neurobiology has not only given us a new framework and a new language for understanding the origins of aggression, but has also led to a revolution in our knowledge of the neurobiological consequences of violence. As a result, biology has become more than just a ploy for getting aggressors off the hook; it has drawn victims into the equation.

The pivotal event behind this research effort was the return of thousands of traumatized veterans after the Vietnam War. Perhaps this was only fitting, for war, the ultimate act of human violence, has been our gateway to the biology of trauma for over one hundred years. Writing in 1871, Jacob DaCosta, an attending physician in a Philadelphia military hospital constructed after the Civil War, described veterans incapacitated by chest pain, palpitations, and exhaustion brought on by their experiences on the battlefield. DaCosta hypothesized that the men's symptoms actually originated in the nervous system, not the cardiovascular system, and that "the heart has become irritable from its overexertion and frequent excitement, and that disordered innervation keeps it so."[37] In the 1940s, Abram Kardiner also described a "physioneurosis"—physical symptoms elicited by an emotional reaction to combat—rampant among World War I veterans.[38] Colloquially dubbed "shell shock," afflicted soldiers startled easily, slept poorly, and overreacted to sights, sounds, and events that reminded them of their traumatic experience.

Kardiner's report piqued little interest in the research community, but it did prompt military authorities to draft policies designed to reduce psychological casualties in subsequent conflicts, including screening recruits for signs of emotional instability and reducing exposure to combat by limiting the average tour of duty. Their efforts were resoundingly unsuccessful. According to the National Vietnam Veterans Readjustment Study (NVVRS), a congressionally mandated epidemiological survey of more than three thousand veterans and civilians, up to one-third of Vietnam veterans exhibited symptoms similar to those described by Kardiner at some point following their combat experience.[39] Less comprehensive studies, including a survey of the medical consequences of Vietnam service conducted by the Centers for Disease Control (CDC) and a Veterans Administration study of twin pairs (one of whom served in Vietnam), reported rates of combat trauma symptoms reaching 15 to 16 percent.[40,41] To place these figures in context, compare them to the rates of several common psychiatric disorders in the general population, tabulated by the National Comorbidity Survey. Among a sample of more than eight thousand randomly selected Americans between the ages of fifteen and fifty-four, 17 percent had been depressed at some time in their lives, 14 percent had been diagnosed as alcoholics, and slightly more than 11 percent—about one-third the proportion suffering from trauma-related stress symptoms in the NVVRS survey of veterans—reported a fear (e.g., of spiders, heights, or closed-in spaces) intense enough to be classified as a phobia.[42]

America after Vietnam was not the quiescent America of the 1950s. Unlike those emotionally incapacitated by earlier wars, Vietnam veterans resented efforts to ignore or trivialize the debilitating anxiety that persisted long after their return home. Their activism forced the government, the public, and the mental health community to recognize that the behavioral and emotional devastation caused by combat trauma was not due to a character defect, but to long-lasting physiological changes in the wake of an extraordinary emotional experience. In addition, it opened eyes to similar pathological stress responses in people who had experienced other kinds of trauma, such as natural disasters, domestic violence, physical or sexual abuse, torture, and violent crime. This recognition of the devastating physical and neurobiological consequences of trauma revitalized clinical and basic research on stress disorders and prompted scientists to broaden their perspective on the biological mechanisms of violence to include victims.

By 1980, the American Psychiatric Association formally acknowledged the clinical significance and presentation of trauma-induced symptoms in the form of a new diagnostic entity: *posttraumatic stress disorder* (PTSD). Criteria defining the new disorder appeared for the first time in the third edition of the association's diagnostic handbook, the *Diagnostic and Statistical Manual of Mental Disorders* (DSM-III); the latest update of these criteria (published in 1994 as part of DSM-IV) grouped the "core symptoms" of PTSD into three categories, or symptom clusters, listed in Table 2.1.[43] *Reexperiencing* symptoms include acute physical discomfort when confronted with reminders of the traumatic event, as well as nightmares, intrusive memories, and flashbacks—"waking dreams" reenacting the original trauma. *Avoidance* symptoms represent efforts to avoid activities, emotions, or interactions that are associated with trauma. Symptoms of overreaction, or *hyperarousal*—insomnia, irritability, rage outbursts, exaggerated startle responses—reflect an enduring sense of imminent danger, feelings that one psychologist has called the "constant expectation of harm." Experts now believe that the symptoms of PTSD can follow seeing, as well as actually experiencing, a threatening event and may not appear until months or even years after the trauma.[44]

Table 2.1. Symptoms of Posttraumatic Stress Disorder

Reexperiencing Symptoms

- Physical distress when confronted with trauma reminders
- Emotional distress when confronted with trauma reminders
- Nightmares
- Flashbacks
- Intrusive memories

Avoidance Symptoms

- Efforts to avoid events, places, people, or feelings associated with the trauma
- Amnesia for details of the trauma
- Diminished interest in previously pleasurable activities
- Detachment, emotional numbness

Hyperarousal Symptoms

- Insomnia
- Irritability
- Difficulty concentrating
- Hypervigilance
- Exaggerated startle responses

Adapted from American Psychiatric Association, *Diagnostic and Statistical Manual of Mental Disorders,* 4th ed. (Washington, DC: American Psychiatric Association, 1994).

Research on PTSD and the long-term physiological effects of stress has inspired a renewed commitment to the diagnosis and treatment of stress-related disorders. In addition, this research has uncovered surprising congruencies between the neurobiological mechanisms underlying stress disorders and those underlying aggression. Fear and anger, it seems, share pathways and signaling mechanisms, a commonality that shows the relationship between aggressor and victim to be as fundamental as the indivisible link between gene and environment. Physiologically (as well as legally and morally) we cannot talk about violence without acknowledging recipients as well as perpetrators.

The Vicious Circle

The biology that advocated phrenology, eugenics, and psychosurgery as solutions to the problem of violence saw violent behavior as an isolated defect in a single individual: the violent criminal. But the biology that has charted the exquisite intricacy of neural development, the complexity of vocal communication, the anatomical precision of facial recognition, and the behavioral devastation that is the legacy of trauma knows that such one-dimensional reasoning is hopelessly oversimplistic. Violence cannot be linked to one gene, one brain region, one actor; it cannot be viewed in isolation, and it cannot be detached from history. The product of both nature and nurture, aggressive behavior is an ongoing and collaborative effort between the world of genes and proteins inside the neuron and the constantly changing and occasionally hostile world on the outside. Behavior may be constrained by structural and functional boundaries, but it is never locked in place.

We are not helpless pawns of either our DNA or the environment. From the earliest stages of development, the outside world reaches in to select and activate relevant genes, to shape and weave the tapestry of a unique nervous system. After birth, the genetically opinionated brain reaches out to sample key features of the environment and organize essential social relationships.

Beginning from the baseline level of responsiveness, intensity, and adaptability set during development, each of us strives from birth to make sense of our personal worlds, define the limits of safety, and establish a measure of control over self and others. Developmental psychologist Edward Tronick has captured the essence of this process in a comparison of two very different mother-infant interchanges.

> Imagine two mother-infant pairs playing the game of peek-a-boo. In the first, the infant abruptly turns away from his mother as the game reaches its "peek" of intensity and begins to suck on his thumb and stare into space with a dull facial expression. The mother stops playing and sits back watching her infant. After a few seconds the infant turns back to her with an interested and inviting expression. The mother moves closer, smiles, and says in a high-pitched, exaggerated voice, 'Oh, now you're back!' He smiles in response and vocalizes. As they finish crowing together, the infant reinserts his thumb and looks away. The mother again waits. After a few seconds, the infant turns back to her, and they greet each other with big smiles.
>
> Imagine a second similar situation except that after the infant turns away, she does not look back at her mother. The mother waits but then leans over into the infant's line of vision while clicking her tongue to attract her attention. The infant, however, ignores the mother and continues looking away. Undaunted, the mother persists and moves her head closer to the infant. The infant grimaces and fusses while she pushes at the mother's face. Within seconds she turns even further away from her mother and continues to suck on her thumb.[35]

The infant's first vital interactions with her caretaker set in motion a life-long adaptive process, initiating reactions—a joyous smile, a steady patience, a frustrated tongue click—that will interact with key features of her innate response style to mold personality and behavior. In turn, her reactions to the behavior of others feed back on the brain to alter structure, chemistry, and function, coloring future perception, updating memory, resetting response mechanisms to correspond with the evolving picture of the environment. This readjustment in outlook and responsiveness elicits a complementary adaptation on the part of the caretaker that starts a new cycle of action and reaction. As Tronick, commenting on the hypothetical peek-a-boo games outlined above, observes, "The affective communications of each infant and mother actually change the emotional experience and behavior of the other."[35] Today's effect, in other words, becomes tomorrow's cause.

Our mature social interactions build on the template laid out by these first exchanges between brain and environment. Reciprocal interactions with parents and siblings, teachers and peers, coworkers and spouses, strangers, neighbors, and, ultimately, our own children continue to mold and select reactions, shape neural pathways, switch on genes. Over time, repeated experiences of empathy, encouragement, anxiety, conflict, or outright danger lead to adaptive responses not only at a behavioral level—defining when, where, how, and with whom specific emotional behaviors are permissible, as well as essential—but at the level of cell and synapse, shaping neural activity to meet the demands presented by the environment.

This inclusive biology views violence as a developmental process rather than a genetic or cultural mandate. The process begins with a nervous system biased toward survival and social responsiveness, equipped to respond aggressively or fearfully when the threat to survival is extreme. The constitutional boundaries of such a nervous system are not constant but vary from individual to individual, reflected in differences in perceptual acuity, interactive style, attentiveness, and social orientation. Features of the environment pull out critical threads of this basic fabric to create the raised design that we know as behavior, progressively increasing or decreasing the risk of violence, or provoking and maintaining fear, an interactive process between gene and environment, performed by a brain seeking balance.

The path linking brain to behavior does not follow a straight line. Instead, it circles back and forth between the world on the inside of the brain and the world on the outside. Ideally, the trade-off between physiology and environment leads to a well-adapted, fully functioning human being. But when the interchange has been hostile, unrewarding, or unproductive, the struggle to balance response and demand strains the physical boundaries of the nervous system, and the dynamic interplay between physical and environmental insults degenerates into a vicious circle, spiraling toward violence and rotating compulsively back to fear (see Figure 2.4).

Figure 2.4

Events in the outside world, including social interactions, have lasting effects on the neurobiological processes that underlie behavior. Positive exchanges between the brain and the environment push the individual toward socially acceptable behavior. Negative interactions increase the perception of threat; over time, the process may develop into a "vicious circle" that leads to violence.

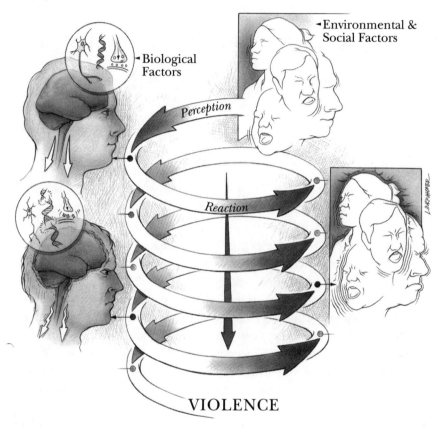

WE BEGIN LIFE aching to breathe, to suckle, to sleep next to a heartbeat. Warmed and fed, we play peek-a-boo with life, teasing and testing faces, voices, objects. For some, like my father, awareness may grow into a studied detachment and an intimacy with engines. But for others, like my mother, anxiety can crystallize into nightmares, simple frustration into aimless rage.

Three

FROM WILDERNESS TO LAB BENCH

ANIMAL MODELS OF AGGRESSION

East Baltimore, late fall, 1981. It's the morning of my Ph.D. qualifying exam, and I'm circling the Johns Hopkins University Medical School, searching for a parking space.

This is not a good neighborhood for street parking. Three days after my arrival in Baltimore, a medical student headed back to his car was shot and killed in broad daylight. Nearly everyone I know has lost a wallet, a purse, a car radio. Only a few weeks ago, a friend was stranded at the lab in the middle of the night after his car was stripped practically to the ground. It had been parked about ten feet away from the guard station shielding the main research building.

This morning I'm late, wearing two-inch heels, and preoccupied with rehearsing intelligent answers to hypothetical exam questions. Finally, in desperation, I opt for one of the last remaining spaces on the roof of the hospital parking garage. This will prove to be a nearly fatal indulgence.

Eight hours later. The air has an unpleasantly cool edge and it's already dark. But I hardly notice. The exam is over, I've passed, and I can finally get some sleep. I don't even see the guy until he has pinned me against my car.

"Get in the goddamn car," he orders. "I have a gun."

I know I should stay out of the car, try to run, to scream. But I can't react. I can't even breathe.

The dim light inside the car accentuates his harsh, contorted face, no older than my own. His clothes stink, with a decaying, subterranean reek that reminds me of the animal room at the laboratory early Monday morning, before the care-takers arrive. He empties my handbag and pockets the wallet. He'll find no more than fifteen, maybe twenty dollars inside. But this guy's after more than just money.

I've slipped my shoes off by now, so that I can run faster when I find the courage to throw open the passenger door. Where I'll run to up there on the roof is anybody's guess. I'm really not thinking too clearly. But how can anyone think when a stranger who might have a gun is only a few inches away, grabbing her breast?

Lights. A car wheels around the corner from the other side of the roof, rolling slowly toward the exit. My stranger pulls back, eyes narrowed. He's calculating rapidly. As soon as the car is gone, he shakes my handbag a second time, then slides back out of the car, the bulge under his coat pointed directly at my face.

"Don't move until I'm gone. Hear that bitch? Don't move."

The roof is silent and empty, and eventually I can move again. A few minutes later a medical resident on his way home discovers me. The stranger with a gun has vanished over the edge of the roof into the night below.

HUMAN AGGRESSION waits for the darkness or hides behind closed doors, carefully shielded from onlookers. As a consequence, unless they're also law enforcement officials or prison psychiatrists, researchers rarely have an opportunity to observe the violent act in progress, much less note what happened in the crucial moments preceding the quarrel. And what they do learn is likely to be one-sided, given that only one of the participants may be alive or available after the fact. For "real-time" data, therefore, scientists have turned to aggressive encounters between animals to learn more about the factors governing the decision to use force, the contributions of the participants, and the choreography of the conflict.

Animal research on aggression can take place indoors or outdoors. The naturalistic approach pioneered by Darwin and the early ethologists goes to the animals. Field observation has the advantage of capturing behavior as it occurs in the wild, unencumbered by cages, feeding schedules, or curious visitors. But it is time-consuming and offers few opportunities for the experimenter to test theories about the origins or consequences of behavior.

Captive studies—in the laboratory, a zoo, or a free-ranging simulation of the natural habitat—offer a level of control and detail not possible in field research.[1] Such studies not only fill in the blanks in the ethologist's notebook, but also set the stage for human studies of aggressive behavior.

Modern behavioral biologists believe that both kinds of studies are critical to a comprehensive understanding of behavior; favoring one to the exclusion of the other constitutes a "false dichotomy."[1] Observations in captivity complement and clarify those recorded in the wild; field studies provide a way to check the biological validity of laboratory models. Together, the two approaches allow a comprehensive analysis of conflict in animal societies that has much to tell us about our own behavior.

Watching and Waiting

Early studies of animal aggression emphasized the individual. Aggressive encounters were viewed as personal responses to an innate drive designed to maximize gain and avoid pain: by defending a home turf, coping with frustration, blowing off instinctive steam. Gradually, however, ethologists have come to recognize that social behaviors, including aggression, are more than preprogrammed reactions to internal or external triggers—they are responses to the behavior of another organism. For solitary species (e.g., the mouse), aggression is a solution to the problem of keeping others away. But for highly social species like primates (including humans), aggression is one solution to the challenges of group living. For these species, aggression is a tool that group members can use to define complex social relationships, a behavioral probe that periodically tests the reactions of others and adjusts the tenor of future encounters accordingly, a mechanism that paradoxically promotes cohesion rather than dispersion. As a result of this new recognition that aggressive behavior is a two-way street, today's behavioral researchers stress the importance of studying aggression—whether in the lab or in the field—from a social perspective that includes the victim as well as the aggressor and bystanders as well as participants.

An emergent ethology that saw aggressive behavior only in terms of territories and rigid dominance hierarchies also viewed animal aggression as a way to ensure survival without resorting to bloodshed. The result was a widespread belief that animal aggression, unlike human violence, is primarily bluff. Animals engaged in an instinctive, stereotyped show of bravado that results in more ruffled feathers than fatal wounds, while we were "the only mammals who kill members of their own species."[1]

Conversely, philosophers and theologians argued that human society, in contrast to animal communities, had replaced biological mechanisms for maintaining the social order with morality and reason. Our status as a unique and separate life form conferred the ability to will our "killer instinct" into submission. In addition, it enshrined us as the only living things capable of compassion, empathy, and reconciliation.

The combination of field research and captive observations championed by ethologists over the past several decades has challenged these traditional beliefs. Researchers have learned that patience is a scientific as well as a personal virtue, and that by watching a particular group of animals longer or more often, they may witness events that short-term observers missed. In addition, the growing emphasis on social context has led to a greater detail in recording interactions, as well as to a reinterpretation of the function and significance of these encounters.

Primatologists have led this theoretical and methodological revolution. Long-term naturalistic studies, such as those initiated by Jane Goodall at Gombe Stream National Park in Tanzania and Dian Fossey in the rain forests of Zaire and Rwanda, have provided tantalizing new insights into the social organization of the great apes, our closest animal relatives. Their patient and arduous work in the field has been complemented by equally painstaking observations of captive primate groups at sites ranging from the Arnhem Zoo in the Netherlands to Cayo Santiago, an island hideaway for monkeys off the east coast of Puerto Rico.

The revelations of these pioneering primatologists did more than entertain readers of *National Geographic.* They narrowed the gap between us and the rest of the animal kingdom. Data collected in the wild and in captivity demonstrate that while we humans can take credit for automatic weapons and nuclear warheads, we invented neither war nor murder. And for all our moral rectitude, our primate cousins can surprise us with their capacity for random acts of kindness.

"ONLY ABOUT ONE in ten students proves to be a good observer," insists Frans de Waal. "The rest are too easily distracted by details. To understand primate behavior you need to see the big picture as well as the details."

De Waal compares watching primates to watching team sports. To understand the game, he explains, you have to focus simultaneously on both the group and the individual team members. Similarly, a skilled observer must learn to track each animal within the context of the evolving relationships between group members. As evidence that some just can't make the grade, de Waal recalls a former student who witnessed a major confrontation in a chimpanzee community but could remember only the physical attributes of a single male, remarking over and over again, "Did you see how big his teeth were?" De Waal still marvels at the student's narrow point of view. "He was distracted by the teeth, of all things. But he indicated to me that observing is very difficult. If you zoom in on one individual, you're going to miss everything that happens."

The Yerkes Regional Primate Center Field Station offers the exceptional student with the requisite "feel" for primate watching a wealth of opportunity to exercise this skill. Set on a 117-acre pocket of Georgia pine forest about one hour north of Atlanta, the field station is home to two thousand monkeys and apes, representing eight different species.[2] The animals roam freely in enclosed "communities," all similar in size and composition to naturally occurring groups in the wild. Like neighboring countries, each enclosure has its own rules and customs, history and personalities. Some are rigidly hierarchical. Others are more like a liberal university campus, valuing tolerance over strict obedience. But within each community, group cohesion is paramount, and the prevailing social conventions operate to preserve and promote critical long-term relationships among the residents.

From the observation window that forms the rear wall of his tower office, de Waal ponders the complexity of such relationships among members of the station's chimpanzee colony. He has been collecting data on chimpanzee behavior for over twenty-five years, including observations of the world's largest captive chimpanzee colony, housed on a two-and-a-half-acre island in the Arnhem Zoo. It was there, at Arnhem, that de Waal first recognized both the dark side and the bright side of primate nature.

He knows that even socially sophisticated apes are capable of murder. In 1980, following the dissolution of a long-standing alliance between the Arnhem colony patriarch, Yeroen, and the ruling alpha male, Nikkie (see Figure 3.1), a third contender, Luit, usurped the top spot in the male hierarchy.[1] Such transitions are always tense. Until the new dominant has firmly cemented his position, demanded and received assurances of loyalty from the losers, threats, noisy confrontations, and outright physical attacks disrupt day-to-day living on a regular basis.

Left to themselves, Luit and his deposed rival, Nikkie, may ultimately have negotiated a truce. But the presence of a third male, Yeroen, thoroughly derailed the peace process. Despite Luit's best efforts to keep the older male and Nikkie separated, Yeroen labored to mend his coalition with the former leader. Weeks of troubled negotiation ended in disaster. Finally reaching an agreement, Nikkie and Yeroen ambushed Luit in his night cage, attacking him so viciously that he died of his injuries the next day.

De Waal points out that Luit's death disabused the Arnhem researchers of their "romantic notions" about the limits of chimpanzee rage. However, it confirmed reports of simian brutality documented by Jane Goodall nearly a decade earlier. In 1970, the previously stable group that formed the basis of her observations suddenly split, with seven males, three females, and their infants retreating to form a separate community in the southern end of the home range.[3]

Social cohesion is so important, to humans as well as to chimpanzees, that secession rarely turns out peacefully. Border skirmishes escalated into civil war when males of the northern community cornered a rebel male, beating, kicking, and biting him for over twenty minutes. Neither Goodall nor any of her assistants saw the victim alive again. Over a three-year period, northern male war parties continued to track and savage chimpanzees from the southern community, ultimately killing five of the original seven males (the two others died of natural causes) and at least one female. Goodall, who had watched the

Figure 3.1

Primate assassins: Nikkie and Yeroen.

Reprinted by permission of the publisher. From *Peacemaking Among Primates*, by Frans de Waal (Cambridge, MA: Harvard University Press). Copyright ©1989 by Frans B. M. de Waal.

Gombe chimps for seventeen years, witnessed for the first time the systematic extermination of one chimpanzee group by another.

More recently, Lee White, a biologist working with the Wildlife Conservation Society in Gabon, has collected evidence of another devastating war between rival chimpanzee communities.[4] This war, White suggests, may have been started unintentionally by humans. Modern logging operations in Gabon deploy a three- to six-mile-wide column of men and machines that invade a resident clan's particular patch of forest, steadily backing the group farther and farther up the boundary line that separates their territory from that of the neighboring clan. When they cross the line, the natives attack. White has observed border clashes that resulted in the death of as many as 80 percent of the invading refugees and believes that such encounters have reduced Gabon's chimp population over the last few years by nearly 40 percent.

The examples of murder and warfare documented by Goodall, de Waal, and White have forced behavioral researchers to acknowledge the lethal potential of animal aggression. But murder is not the only surprising revelation of observant primatologists. Attack, it seems, is only half of the aggressive equation. After the anger has subsided, chimpanzees and other primates do not simply retreat to separate corners; they actively seek to mend the breach. Slaps give way to gentle pats, screams to soft pants of contrition:

> On this occasion, Nikkie, the leader of the group, has slapped Hennie during a passing charge. Hennie, a young adult female of nine years, sits apart for a while feeling with her hand the spot on her back where Nikkie hit her. . . . More than fifteen minutes later Hennie slowly gets up and walks straight to a group that includes Nikkie and the oldest female, Mama. Hennie approaches Nikkie, greeting him with a series of soft pant grunts. Then she stretches out her arm to offer Nikkie the back of her hand for a kiss. . . . This contact is followed by a mouth-to-mouth kiss.[1]

Reconciliations such as these are the foundation of de Waal's ongoing research on social connectedness. By settling the score so consistently, monkeys and apes defy the contention of earlier ethologists that the only result of aggression is dispersal. Rather, the "fight, then make up" sequence emphasizes the importance of belonging; of testing, but not severing, crucial relationships. Conflicts of interest, in de Waal's view, may be an inevitable part of group living, but reconciliation provides a way of preventing such disagreements from destroying the group.[1]

Unresolved conflicts don't go away. They fester, ready to erupt in new clashes that further erode the relationship between combatants. Because victims have the most to lose—an enemy at large can decide to pummel them again at a moment's notice—they're often the first to bury the hatchet. But it's in the aggressor's best interest to accept the apology, particularly if the object of his wrath is a potential ally or a member of his family. By forgiving and forgetting, both winners and losers come out ahead. The risk of future aggression goes down, and a critical bond is reaffirmed.[5-8]

Animals, especially primates, and humans are more alike than we may want to admit. Examples of murder, treachery, and warfare in primate communities demonstrate that we are not the only animals familiar with the dark side. However, expressions of empathy and reconciliation may be the animal prelude to human morality. For both animals and humans, aggression may get out of control, but sympathy may also transcend all expectations.

Moving Aggression Indoors

Primates aren't the only species with complex social structures that govern and regulate aggression. Other investigators have sought to adapt the new socially conscious ethology to the study of other mammals, especially rodents. Like the primatologists, they also emphasized the importance of extended observations and the analysis of aggressive behavior within a social context.

At the mouth of the Delaware River, which separates Pennsylvania and New Jersey, the wooded civility of suburban communities and river islands farther north fades into a sullen, metallic grittiness. Here, after dark, in a decaying spot on the riverbank, another pair of contestants prepares for mortal combat.

In the clammy blackness, there is a quick scuffle, a thousand tiny clicks, nail against metal, the determined scrape of teeth. The floor seethes with rats. There are at least fifty, perhaps as many as a hundred, all related, bound by blood and a common purpose: keeping their warehouse to themselves.

A dominant, or alpha, male reserves the right of first refusal in any confrontation.[9] Earlier in the evening, one of this colony's alpha males discovered an unmistakable sign of a need for his fighting skills: the scent of a strange male, an enemy secretly deployed by one of the delivery trucks. Strangers spell trouble; if one breaches the defenses, dozens more may follow. An immediate show of force is essential.

The alpha male tracks the intruder, finally cornering him behind a stack of wooden pallets.[10,11] Cautiously, he sizes up his opponent, circling, sniffing, steadily winding a tighter and tighter spiral of intimidation until the tip of his quivering nose brushes the strange tail. The sudden contact sparks the attacker, and he bristles with rage. Teeth flash and find their target, sending the invading rat racing for cover. Lucky enough to find a rat-sized hole in the floor or a wall, he may escape with minimal damage. Trapped, with no escape, he must stand and defend himself, rearing and boxing with his forepaws to ward off further attacks. But the alpha male is relentless. Teeth chattering, he edges toward his opponent and throws a body block, wheeling and lunging in an attempt to land another bite. The hapless victim sinks onto his back, the victorious alpha male trampling him underfoot, rooting and shoving until he succeeds in exposing the back for one final snap before the defeated intruder finally wrenches himself away and squeezes through a space between the door and the night.

You can take the rat out of the wild, but you can't take the wild out of the rat. Even when relocated to the civilized quarters of the laboratory, the common brown, or Norway, rat retains its animosity toward strangers. Scottish behavioral biologist S. A. Barnett was one of the first to observe and categorize social behavior in these combative rodents, both in the wild and among the members of groups trapped and housed together in large cages. Based on detailed observations of rat colonies, Barnett constructed a "criminal profile" of rat aggression, consisting of characteristic postures, sounds, and maneuvers: the *threat posture,* in which "the flank is presented to the opponent: the legs are fully extended and the back is arched"; *boxing,* in which "two animals face each other on their hind legs and each pats the other with the forepaws"; and *attack,* in which "the animal leaps at its opponent with rapid adductions of its forelimbs and . . . bites."[12,13]

The old animal models of aggression, based on pain or frustration, bore little resemblance to a typical aggressive encounter in the wild. But Barnett's captive rat colonies were different. Encounters between an established resident and an intruder offered researchers a paradigm based on naturally occurring behavioral responses rather than psychological theory. Laboratory researchers, like their colleagues in the field, could finally study both aggressor and victim, as well as the intricate choreography of their dangerous dance.

British behavioral researchers Michael Chance, Ewen Grant, John Mackintosh, and Paul Silverman of the University of Birmingham extended and refined Barnett's observations to develop "ethological" models for studying rodent social behavior in the laboratory.[14-18] These models, unlike the pain-induced fighting that dominated aggression research at the time, were based on a detailed description of normal behavior known as an *ethogram*. The Birmingham group created ethograms for four commonly used species of laboratory rodents—rats, mice, guinea pigs, and golden hamsters—painstakingly recording and categorizing solitary, social, and aggressive behaviors to construct a catalog of the full behavioral repertoire of that species.[15-17] They then used these behavioral templates to study the effects of experimental manipulations, including drugs and housing conditions, on offensive and defensive behaviors by tracking minute-by-minute changes in the activities of both aggressor and victim.

A laboratory-adapted ethology improved the resolution as well as the credibility of aggression research. The extended observations that could be carried out in the laboratory demonstrated that aggressive encounters are not random combinations of behaviors; they follow a characteristic sequence and pattern. With the advent of video monitoring and computer-assisted analysis of observational data, researchers could describe these patterns with an exquisite detail that captured the full complexity of the encounter, time, position, sequence, and intensity.

In rats, for example, Tufts University behavioral biologist Klaus Miczek and his colleagues have shown that a typical confrontation between a resident male and an intruder follows a sequence: pursuit, threat, attack, and aggressive posture. Two of these elements are shown in Figure 3.2.[19,20]

Aggressive activity tends to occur in bursts, followed by quiescent intervals with little social contact. Over 90 percent of aggressive behaviors during the encounter were separated by less than seven seconds; only 10 percent occurred during the more leisurely intervals between these explosive periods.[19,20] Aggression in mice and other rodents follows a similar burst-gap pattern, although the duration of both bursts and gaps differs for each species. Even human aggressors—the abusive spouse who rages at the end of the work week, the boss who terrorizes his staff before a critical deadline, or the serial killer who strikes, then vanishes—tend to act out intermittently rather than continuously.

Ethological models preserve the integrity and social context of behavior

Figure 3.2

(A) Aggressive posture by the dominant rat arched over the subordinate rat, which is in a submissive-supine posture. (B) Attack bite directed to the back of the subordinate.

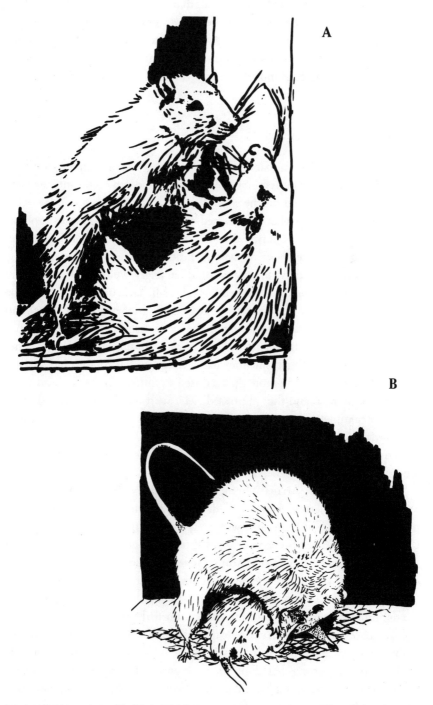

Reprinted with permission from K. A. Miczek, Intraspecies aggression in rats: Effects of *d*-amphetamine and chlordiazepoxide, *Psychopharmacologia 39* (1974): 275–301.

without sacrificing accuracy. Because these models are based on naturally occurring behavior, they provide more detail than simplistic paradigms based on
food rewards or electric shock, and they add a new quantitative dimension
based on the organization of an aggressive encounter, as well as the rate and intensity of the component behaviors. Like de Waal's students, laboratory researchers using ethological models are limited only by their own observational
skill and experience.

Watching Your Back

Rose Scott stabbed her husband, Lawrence, with a butcher knife, slashing a
four-inch gash through his femoral artery; he bled to death on the kitchen floor
before paramedics arrived on the scene.[21] Jerome Handy's common-law wife,
Ariana Thompson, closed her eyes, turned her head, and fired six shots at point-
blank, landing three in Handy.[22] Brenda Clubine clubbed her husband, a re-
tired Los Angeles police detective, over the head with an empty wine bottle.[23]

Three brutal murders committed by three angry women. Or were they?
After his father's murder, Lawrence Scott's son Michael painted a picture of a
long-suffering spouse, weakened by cancer, finally done in by his cruel and
quarrelsome wife. But police arriving on the scene described a bruised and terrified woman bleeding from cuts and abrasions, including a stab wound in her
own leg. It was not the first time they had seen her in that condition. Local authorities had intervened in more than a dozen arguments in the year and a half
prior to the murder, and the "oppressed" Mr. Scott had been convicted of assault after pummeling his ninety-pound wife with a beer can. On the night of
the murder, he violated a protection order to force his way back into the home
he had shared with Rose, beat her savagely enough to crack a rib, and was attempting to strangle her when she grabbed the knife.

Like Rose Scott, Ariana Thompson resorted to violence after legal attempts to control Jerome Handy failed. He too violated a standing protection
order, boasting, "When I'm done with you, you're gonna need an ambulance."
Clubine's husband had broken her jaw, stabbed her, and ripped the skin off her
face during repeated attacks that nearly drove her to suicide.

Each year, approximately seven hundred battered women, like Scott,
Thompson, and Clubine, take the law into their own hands, killing the husbands and boyfriends who have hit, kicked, burned, and shot them.[24] Criminologist Lawrence Friedman suggests that "some form of self-defense underlies
many—perhaps most—of the cases where a wife kills her husband."[25] He cites
a 1978 survey reporting that all but one of a group of thirty women convicted
of murder or manslaughter had killed abusive partners.[26] Far from being cold-
blooded, calculating assassins, women who shoot back see the murder as a last
desperate measure to save their own lives. Their fears are well founded. For

every woman who kills her abuser, three die at his hands;[24] overall, at least two-thirds of all murdered women are killed by abusive partners.[27]

One woman, filing a petition for clemency, said she'd asked herself repeatedly if she'd been wrong, if she'd really needed to shoot her way out. And she concluded, "I still can't see no other way of getting out alive."[23]

UNLESS YOU HAVE the whole story, you can be dead wrong about violence. For example, until researchers turned to ethological models to study aggression, they assumed that if teeth were bared and claws unsheathed, it spelled attack. Psychologists Robert and Caroline Blanchard put this theory to the test, using the "resident-intruder" paradigm developed by Chance and his Birmingham colleagues to demonstrate that the pain- and frustration-induced models that had dominated laboratory research on aggression were flawed from a theoretical, as well as a practical, point of view.

The Blanchards suspected that while electric shock certainly produced a consistent response, it was far from clear whether this was the response an angry rat might make under less stressful circumstances. The researchers questioned whether the rearing, boxing animals were really "fighting" at all. To uncover the true nature of pain-induced aggression, they compared the behavioral response elicited by foot shock with "aggressive behaviors occurring in the real world."[28]

Using the colony model pioneered by Barnett and Chance, the Blanchards organized rats into small groups containing adult males and females, as well as young pups, each housed in a meter-square enclosure.[10,11,28] Every colony quickly developed into a tightly knit neighborhood, ruled by an alpha male who not only maintained law and order within the group but also policed the outer boundaries.

An intruder unlucky enough to be dropped into an established colony could count on a reception from the alpha as hostile as any he might encounter in the wild. And he reacted the same way as his wild cousin, first running away and then, when escape over the perimeter fence proved impossible, by rearing and boxing or rolling over on his back.[10,11,28,29] Earlier investigators might have interpreted the encounter as a classic threat display checked by the interloper's submissive gestures. However, after hours of reviewing videotapes of such alpha-intruder encounters, the Blanchards noted a problem with this explanation: the so-called submission signals didn't end the alpha rat's attack.[28] But boxing, standing up, and rolling over did protect the intruder from injury. Wound counts after the encounter demonstrated that heads, throats, paws, or even tails were rarely bitten, while as many as 90 percent of bite wounds were targeted to a single site: the intruder's back.[28] As long as the intruding male successfully kept his back to the wall or to the floor, he could protect himself, even if he couldn't escape or convince his tormentor to call it a day.

Occasionally a desperate intruder also bit. But his retaliatory bites were not directed to the oppressor's back. Instead, the canny victim aimed his self-defensive response at the attacking alpha male's head.[28] Revenge served to focus the alpha male, emphasizing the hazardous consequences of an attack outside the target zone.

Robert and Caroline Blanchard reasoned that for the alpha male, the object of the encounter is to drive away the rival, not to kill him. So bites are targeted to the back, a site where the wound will be punitive rather than lethal. The intruder's goal consequently is to protect his back, first by running or, if trapped, by rearing and rolling over to shield the target. This "back-attack" strategy isn't limited to the artificial environment of the laboratory colony. By mapping the site of bite wounds in wild rats trapped in the neighboring sugar cane fields, the Blanchards showed that back stabbing is a key feature of naturally occurring quarrels as well.[28]

And what about pain-induced aggression? Like the intruder trapped in an inescapable enclosure with an enraged alpha male, a rat confined to an electrified cage can't escape the painful shocks. As a result, like the cornered intruder, it tries to defend its back, and its "aggressive"behavior—boxing, rearing, and freezing—resembles the defensive posture of a frightened subordinate, not the offensive attack pattern of an angry alpha.[28,29] In fact, an experienced alpha, plucked from his home colony and paired with an opponent in the electrified arena rather than his familiar enclosure, switches attack strategies, shifting from bites aimed at the intruder's back to the defensive pattern of rearing, boxing, and protective bites aimed at the other rat's head.[28]

The Blanchards concluded that pain-induced "aggression" was actually self-defense. In fact, the behavioral response of a rat facing an opponent over an electrified grid was nearly indistinguishable from its response when cornered by a cat or other predator. Azrin, Hutchinson, and Ulrich's shocked rats—the most widely used technique for studying aggression in the laboratory for over a decade—were driven by neither anger nor instinct but by fear, the same fear that prompts survivors of violent crime to adopt dogs and buy guns. The "resident-intruder" model used by the Blanchards, on the other hand, did not confuse attack and defense. As a result, the researchers could analyze an aggressive encounter in the laboratory accurately and relate it in a meaningful way to the behavior of the animals in their natural environment.

The intertwined dance of alpha and intruder reveals the social complexity of the aggressive encounter. Each action on the part of one member of the pair elicits an equal and opposite reaction from the other. The victim rears to protect his back, and the attacker shoves him to one side to expose the preferred target. The intruder, anticipating this sneak attack, pivots to meet the charge head on. When he lies on his back, the attacker pushes and roots, trying to turn him over. Too close, too little attention to detail, and the intruder snaps at the alpha's head, buying enough time for an escape. More than the prepro-grammed responses, or fixed-action patterns believed by early ethologists to characterize aggressive behavior, these "call and response" sequences suggest that fighting in animals engages a complex and dynamic repertoire of comple-mentary attack and defense strategies. The resident-intruder model represents a rational strategy for examining this double-sided nature of aggression.

Modeling the Consequences of Violence: Fear Conditioning

I come back to life in stages after my assault at gunpoint. At first I am numb and confused. Then I downplay the event. The entire incident encompassed a mere five or ten minutes of my life. There's no blood, no shattered bones, no gunshot wounds, so it's easy to pretend I'd never really been in any danger. Maybe my attacker didn't even have a gun after all. I joke about looking on the bright side, about how I'd satisfied my personal quota of street violence until the end of time. That's life in the city. You can't live in fear or you'd go crazy. But I don't park in the garage again.

DOES FEAR EVER go away? Or does it sink like a shard of rusty metal, wait-ing for some storm or shifting mental current to expose its lethal edge?

TWO YEARS PASS. I've just moved from Baltimore to London, a city with few parking garages and no handguns. I don't need a car, and I can go out at night with impunity.

It's already dark when I step off the bus, a half-dozen blocks from my flat. The bus pulls away. In the quiet night, I'm suddenly aware that a man is fol-lowing me, that he's calling out to me. I walk faster. He walks faster. I cross the street. He follows, shouting. I start to run, but it's an intersection, a speeding truck, I have to wait. The man is right behind me. His voice barely penetrates the fog of my terror.

"I say there, can you tell me where I might find Belsize Park Road?"

He's speaking in that exaggerated tone people reserve for the dimwitted. Then it dawns on me. Directions. This person doesn't want to kill me. He wants directions.

Fear. Still alive, still sharp, tripped off by the cold darkness, a man's voice behind me. Fear, roused from torpor a second time about a year ago, when I read that most assault victims who enter a car with their assailant never make it out alive.

I want to believe that armed robbery in my sheltered suburban commu-nity is restricted to shoplifting teenagers bullying the night clerk at the local convenience store. But I still park under streetlights and return to my car with my keys laced through my fingers like daggers.

To New York University's Joseph LeDoux, an expert on how we learn and remember fear, my night terror is not only understandable but expected. As he and others have documented, the brain is exquisitely primed to link fear and memory. They study this pairing in an emotional spin on Pavlov known as "fear conditioning."[30] An unsuspecting rat is introduced to a dimly lit cage equipped with a speaker and floored with a wire grid. During the first session, an unfamiliar but innocuous tone beeps into the chamber every minute or so.[31] Mildly curious, the rat pads around the test chamber, sniffing in the corners, then settles down to groom its fur. Its second visit to the training cage is less benign. When the sound comes on, so does a mild electric current, delivering a brief and unpleasant shock to the rat's feet. The alarmed rat crouches and trembles, its cardiovascular system ratcheting into overdrive (see Figure 3.3).

After only a few such pairings between the tone and the shock—sometimes following no more than a single trial—the connection between the beep and the tingling pinch is firmly cemented.[30,32] Now the sound alone is enough to evoke fright: the rat freezes in place, its heart rate accelerates, and its blood pressure soars to three or even four times the value observed during the initial shock-free session.[31] It startles at the slightest sound.[30] Like the self-defensive rearing and rolling of an intruder fending off an angry alpha male, the threat implied by the tone evokes a characteristic sequence of protective behavioral and physiological responses. The rat, in other words, doesn't learn *how* to be afraid but *when*. "Fear conditioning," explains LeDoux, "opens up this channel of . . . responsivity to new environmental events."[32]

The associations between fear and environment forged during the conditioning procedure are extremely strong; many researchers, including LeDoux, believe they may be permanent.[31] "At best," he suggests, "we can hope only to keep them under wraps."[30] A period of safety illustrates this lingering power of conditioned fear. If the tone sounds many, many times without the shock, the intensity of the rat's fearful response slowly decays; neurobiologists call this deceptive calm *extinction*. But the connection between fear and sound hasn't dissolved; it has merely retreated underground. One shock—or even the rat equivalent of an overwrought Monday morning—and the fear circuit is reactivated.[31,33] Like me trembling on a London street corner, the rat cowers in a corner of the cage, clearly anticipating the long-ago shock. The brain has not erased the fearful memory; it has simply created an overdrive mechanism to circumvent it, a fragile detour that can be blasted away by stress or familiar environmental cues.

Fear conditioning occurs in fruit flies, snails, and reptiles, as well as in rats and humans.[30] In fact, the process is so pervasive, so simple, and so accessible that it has become not only the most widely used technique for evaluating the neural impact of a frightening event, but has also evolved into the gold standard method for studying the anatomical framework and physiological mechanisms

Figure 3.3

A rat that experiences an electric shock at the same time it hears a tone responds fearfully—and is quickly conditioned to fear the sound that accompanied the shock. When the rat hears the tone again—even without the shock—it freezes in fear.

SOUND FOLLOWED BY FOOT SHOCK

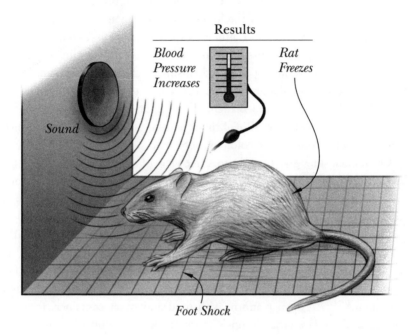

SOUND WITH NO FOOT SHOCK

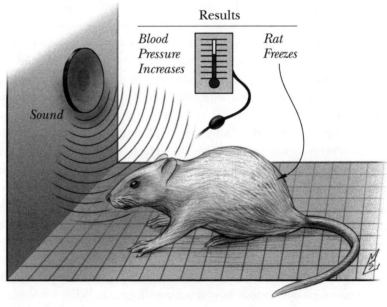

of emotion. Easily elicited and readily related to naturally occurring behavior in a threatening situation, fear conditioning has been the foundation for significant advances in our understanding of how the brain translates events in the outside world into emotional memories of long-standing and profound significance.

Fear conditioning offers a powerful model of the process embedding an alarming event in memory, of the long-term consequences of a few fleeting moments of terror. But not all traumatic experiences are over so quickly. The defensive behavior patterns first described by Robert and Caroline Blanchard provide a model for studying the mechanisms of chronic threat. Since their earlier work using rat colonies to differentiate characteristic offensive and defensive behavior patterns, the Blanchards have refined the colony model to concentrate on the defensive side of the equation, especially the consequences of chronic social stress among colony subordinates.

Open enclosures give investigators free view of aggressive encounters. But rats are burrowing animals, completely revealing their true personalities only under cover. Given an enclosure filled with about a foot and a half of soil, a rat colony excavates a network of interconnected tunnels and burrows within forty-eight hours.[34] The rats are in their element, but the researchers are in the dark.

To effect a compromise between the rats' urge to retreat and the humans' desire to watch, the Blanchards have devised a model burrow system, shown in Figure 3.4, that simultaneously provides shelter and permits continuous observation.[35–38] An open platform with food and water containers forms an above-ground "surface area." Tunnels made of Plexiglas tubing lead out of the platform and connect to a series of four small chambers—the "burrows." Although the surface platform is lit for a standard twelve-hour day, the burrow-tunnel network is illuminated only by infrared light, creating the illusion of constant, subterranean darkness. As a result, rats housed in the visible burrow system (VBS) behave as they would underground, sight unseen, while the scientists can continuously monitor their behavior surreptitiously using low-light videocameras.

With a house as well as property to defend, the alpha male takes challenges to his authority seriously.[36] A strange male in the burrows receives a far more vigorous thrashing than an intruder trespassing in an open enclosure. Subordinates come under increased scrutiny as well. During the early days after moving from an open enclosure to the VBS, the alpha subjects his underlings to the same brutal treatment he would an invader, often singling out one lower-ranking male—typically the one judged most likely to attempt an overthrow—for extra punishment. As a result of this harshly imposed hierarchy, subordinates lead lives of tremendous anxiety. Over time, they lose weight, sleep less, avoid the colony females, and become increasingly defensive. Ultimately they die sooner—not directly, from battle wounds, but gradually, presumably from the exhausting toll imposed by chronic stress. In one study, seven of twelve subordinate males died within four months of moving to the VBS, while nine of

Figure 3.4

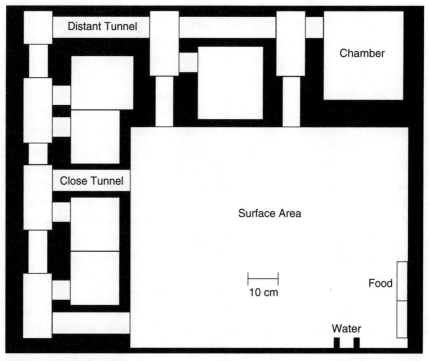

Adapted with permission from C. Monder, R. R. Sakai, Y. Miroff, et al., Reciprocal changes in plasma corticosterone and testosterone in male rats maintained in a Visible Burrow System: Evidence for a mediating role of testicular 11β-hydroxysteroid dehydrogenase, *Endocrinology 134* (1994): 1193–1198. © The Endocrine Society.

twelve subordinates in a control colony housed in the open-enclosure environment lived as long as the alpha male.[36]

Like fear conditioning, living in the VBS induces profound and persistent changes in behavioral responses and the sensitivity to perceived threat. However, like open-enclosure models, it also permits the study of emotional behavior within a social context. Ethological models of chronic threat, like the VBS, provide aggression researchers a second powerful technique for quantifying the physiological and behavioral consequences of being a victim.

Modeling PTSD poses additional challenges to behavioral researchers. More than an exaggerated form of fear, PTSD is a distinct clinical syndrome with explicitly defined psychological and physical symptoms. An accurate animal model of PTSD therefore must do more than subject the animal to unusual or unavoidable stress.[39] The prospective model must replicate the so-called core symptoms of PTSD—involuntary reexperiencing of the trauma, avoidance behaviors that circumvent or block out reminders of the traumatic event, and excessive vigilance—even if exposure to the experimental stressor lasts no more than a few minutes. Poststress symptoms should persist or even worsen over time, reflecting the natural history of the clinical syndrome, and increasing levels of stress should elicit increasingly severe behavioral and physiological symptoms. Finally, some animals should fail to respond to the stressful stimulus, just as some people who experience a severe trauma do not go on to develop PTSD.

The animal models that come closest to meeting these criteria have much in common with kindling—the progressive induction of seizures by intermittent pulses of electricity. Both kindled seizures and the flashbacks and intrusive memories of PTSD are recurrent spontaneous behaviors that reflect excessive reactivity to environmental cues, a hyperreactivity that progresses steadily over time. NIMH's Robert Post, master of the kindling phenomenon, comments, "[Kindled] seizures are paroxysmal discharges of motor circuits. Flashbacks are paroxysmal discharges of memory circuits."[40] The similarity, he goes on, suggests important parallels between PTSD and progressive sensitization that might be exploited in the development of valid laboratory models.

Behavioral sensitization, a neuropsychological parallel to kindling, uses drugs or experimental stressors, such as immobilizing restraint, to amplify responses to subsequent stress.[40–42] Animals treated with a single large dose of cocaine, for example, demonstrate a brief, robust surge in activity—and an equally speedy response the next day to a dose only one-quarter that of the first. The heightened sensitivity to the drug can be reproduced for many months, every time the animal receives a low dose in the original test cage. Like PTSD, cocaine-induced sensitization can occur after a single exposure to a provocative event, which bonds behavioral responses to environmental and physiological cues in a long-lasting relationship.

Brief exposure to stress can also induce persistent changes in activity and responsiveness. In one such procedure, known as time-dependent sensitization, researchers at the UCLA Trauma Psychiatry Program introduced mice into one chamber of a two-room test cage.[43] Ten seconds later, a light flashed on in the antechamber, and at the same time, the door between the two compartments was raised to reveal an inviting dark escape. The mice raced for cover—and promptly received an electric shock.

For the next six weeks, the mice were returned to the lighted antechamber for one minute every week. The door stayed closed, and the animals received no shocks, only periodic reminders of their traumatic experience. And with each reminder, the stress-sensitized mice became more and more fearful. They jumped higher and faster at a sudden noise. Placed in a maze, some crouched and froze, while others raced wildly from arm to arm. Twenty-five of the original thirty animals never completed the experiment; they had killed each other in episodes of vicious fighting.

Sensitization models for PTSD emphasize both permanence and a progressive reactivity to the outside world. Whether sparked by electricity, drugs, or an experimenter's hand, these models reflect a condition that, like the violence that triggers it, is the cumulative result of a dynamic interaction between the traumatized brain and a provocative environment.

What's in a Name?

By early April, the marshy islands splitting the lower Delaware River, still edged with frost, ring with birdsong. Waves of robins have begun to course over the river valley, joining the hardy jays and chickadees that maintained their homesteads during the off-season. A trio of male cardinals, bright as lingering berries against the downy wisps of emerging leaves, insult one another up and down the riverbank. Their scolding merges with the steady rasp of branches and the distant churning of the river, spring background music to accompany the steady task of thawing and unfolding.

A flock of starlings combs through a frost-crisped cornfield on the far bank of the river, unaware of the dark form looping and wheeling steadily upstream. The cardinals panic—they recognize the shape and flight pattern as that of a large red-tailed hawk—but their alarm shrieks reverberate toward the island's interior, and the starlings barely pause. The hungry hawk banks to home in on the warning cries. Although he typically turns up his beak at birds, preferring succulent small mammals such as mice and rabbits, the meager pickings in these final weeks before the summer population surge have left him less finicky.

A well-fed starling will fill the bill. Circling around his lunch selection in a slowly closing death spiral, the red-tail suddenly plunges through the icy air to hit the smaller bird like a warm-blooded harpoon, slamming it into the frozen ground with a force that scatters feathers and flockmates. Effortlessly, he airlifts his catch to the top of a nearby telephone pole, edges out and hoists the little carcass over the phone line, then casually savors his meal in full view of the survivors.

Bloody and lethal, yes. But is killing for food aggression? Is a rat defending its burrow engaging in the same behavior as a troop of chimpanzees waging war against their neighbors? What about fending off a predator?

In a 1969 review, S. A. Barnett noted, "The use of a noun such as 'aggression' for a variety of activities has sometimes an unfortunate effect on the user: it leads him to imply, and even to believe, that there is a single thing, or at least a homogeneous class of events, to match it."[12] When we lump together predation and self-protection, revenge and impulsive rage, status seeking and defensive parenting, we risk repeating the mistakes of the past, oversimplifying a complex problem. In addition, we compromise our ability to relate animal aggression to human violence. Only by understanding that aggression is not one behavioral pattern but several can we effectively compare and contrast the multiple faces of aggressive behavior.

Psychologist Kenneth Moyer was one of the first to propose guidelines for breaking aggression down into its component parts.[44,45] Moyer's map, or *typology*, defined eight distinct types of aggression based on the various types of stimuli he believed sparked aggressive behavior. Table 3.1 outlines his classification scheme.

Behavioral researchers began finding fault with this scheme almost immediately. Some, such as Barnett, objected to the definition of predation as a form of aggression, arguing that "intolerance within a species is quite independent of predatory behavior."[12] Others complained that the categories weren't clear, that some behavior patterns seemed to fit more than one definition.[46] Was the offensive aggressive pattern of an alpha male rat intermale aggression, irritable aggression, territorial aggression, sex-related aggression—or a combination of

Table 3.1. Two Classification Schemes for Aggression

CAUSAL "Aggression is a response to a stimulus" (Moyer, 1978)	
TYPE	STIMULUS SITUATION
Predatory aggression	Presence of prey
Intermale aggression	Presence of strange male
Fear-induced aggression	Response to attack when escape is impossible
Irritable aggression	External source of pain, frustration, or annoyance
Territorial defense	Presence of an intruder
Maternal aggression	Threat to young
Instrumental aggression	Cues for behavior rewarded in the past
Sex-related aggression	Sexual stimuli

FUNCTIONAL "Aggression is a solution to a problem" (Archer, 1988)	
TYPE	PROBLEM
Protective	Threat of physical attack
Parental	Threat to young
Competitive	Threat to status or adequate share of resources

all four? Many disagreed with Moyer's decision to base the system on trigger stimuli, and from a practical point of view, it was just too complicated.

Subsequent researchers worked to develop a simpler and less ambiguous typology. In the early 1970s, neurochemist Don Reis merged several of the original eight groups to derive just two categories: predatory aggression and affective (i.e., emotional) aggression; the affective category comprised both the offensive and defensive behavior patterns described by the Blanchards.[47] A decade later, British behavioral biologist John Archer proposed a totally different way of looking at the problem of how to categorize aggression, in the form of an organizational scheme based on function rather than cause, on the ways in which aggressive behavior furthered survival rather than the ways in which it could be provoked.[46] Archer's scheme (also outlined in Table 3.1) was designed to "provide a useful functional classification of aggression in terms of the particular problem addressed by that type of aggression."[46] It divided aggression into three broad functional categories, each including both offensive and defensive variations and designed to meet a different sort of threat.

Protective aggression is directed toward countering a threat to survival. Noting that to be alive means to be vulnerable to physical attack, Archer pointed out that simple organisms solve the problem of self-defense by digging in or by running away. The aggressive responses mounted by more advanced animals, however, permit the defender to take a stand on site. Pain being one of the more obvious signs of physical danger, Archer, like Blanchard and Blanchard, considered the pain-induced "aggression" that motivated so much of the early laboratory work on aggression one form of protective aggression.

Parental aggression, an extension of protective aggression, is designed to ward off a threat to the young. In fact, protective aggression takes up where parental aggression leaves off; the skill and ability of young animals to defend themselves coincide precisely with the gradual decline in their parents' interest in doing the job for them. Although parental aggression is often equated with "maternal aggression," keeping the children out of danger isn't a job only for mothers, and it isn't as simple as keeping them out of the clutches of predators. Parents of some species must also guard against threats from members of their own kind. A male lion taking charge of a new pride, for example, often begins his reign by slaughtering all of the cubs sired by his predecessor.[48] In these cases, where other adults are known to kill infants, parental defenses are often especially fierce.

The third type, *competitive aggression,* is dedicated to meeting threats to comfort, well-being, or status; to getting and keeping precious resources, including food, sex, space, and peace of mind. It encompasses frustration-induced aggression—a reaction to the failure to satisfy a basic need—and Moyer's instrumental aggression, where winning in the past has taught the value of responding to a given situation aggressively. Competitive aggression, according to Archer, is most likely to occur not when a contested resource is woefully insufficient but when supplies are just limited enough to mean someone must go without. And while it may rely on the same weapons as protective aggression—claws, teeth, voice—used in similar ways, competitive aggression differs from self-defense in one very crucial way: it is less likely to be lethal. Being on top may have its advantages, but staying alive is more important than staying ahead.

Predatory aggression was notably absent in Archer's system. Unlike Moyer and Reis, he considered the motivational and physiological differences between hunting (a response to hunger) and other forms of bloodshed (responses to threat) so profound that predation should not be counted as aggression, but rather as an entirely separate and unique form of behavior.

Of Mice and Men: Answers to Human Questions from the Animal Kingdom

Humans face many of the same problems as animals. If they want to eat, they must compete effectively for scarce resources, whether that means fresh game or new jobs. If they want respect, they must establish and maintain their position in the community. If their genes are to be passed on successfully to the next generation, they must protect and shepherd their children—and stay alive until those children are self-sufficient.

Human aggression, like its animal counterpart, was intended to represent one solution to these problems, to counter legitimate threats to the individual and society. As a result, Archer's classification scheme has clear parallels in human behavior. We also act forcefully to defend ourselves and our children, as well as to guarantee our share of finite resources. And our language reveals that we are familiar with the patterns of offensive and defensive behavior so elegantly outlined by Robert and Caroline Blanchard. How often, for example, have you "looked over your shoulder" to watch out for someone who might "stab you in the back"?

Adaptive aggression stays within bounds because the aggressive individual gauges the potential threat accurately and tailors the intensity of his or her response accordingly. But if threat perception becomes distorted or if the correspondence between perception and response breaks down, aggression becomes maladaptive. For the individual, the cost of such extreme behavior—social rejection, damage to vital relationships, job loss, exclusion from opportunity, prison, even death—outweighs any short-term benefits; for society, the harm to the social structure exceeds the capacity for tolerance.

Aggression directed against the wrong person, at the wrong time, in the wrong place, for the wrong reason, or with the wrong level of intensity is no longer protective or competitive but violent. Threats are seen where none exists or ignored when they ought to sound a wake-up call. Shadows have become enemies—or the impending risk of punishment has shrunk to a mere pinprick of annoyance.

Society defines the boundary between legitimate threat, which can be met with a controlled amount of force, and illegitimate threat, which provokes unacceptable violence. But the boundary is not a straight line. Much of our confusion about what is and isn't violent behavior comes about because tolerable and unacceptable behavior are separated by a gray area, where the definition varies according to culture, time, and opinion, and one society's manly response to an insult is another society's overreaction.

When diverse viewpoints begin to converge, aggressive behavior clearly falls on the maladaptive violent side of the division. And some behavior is so brutal, so irrational, or so far beyond experience and comprehension that everyone would agree that it is not simply misguided but blatantly pathological.

For primates, predators once again define the far edge of aggression. But these predators aren't hunting for other species. They're preying on their own.

HUMANS ARE NOT the only apes capable of waging war. Field research also suggests that we are not the only apes capable of serial murder. Jane Goodall describes a female chimpanzee (ironically named Passion) who kidnapped, killed, and ate at least three infants born to fellow members of her group, attempted to abduct two others, and was implicated in the disappearance of several more.[3] In fact, during the peak years of Passion's killing spree (1974–1976), only one infant born in the northern Gombe community survived. Perhaps worst of all, Passion recruited and trained her own daughter, Pom, to follow in her murderous footsteps, ensuring that the threat of killing and cannibalism would shadow the group even after she was gone.

Male chimpanzees, tolerant of their own youngsters, have little regard for the children of strangers. Females with infants unlucky enough to bump into a border patrol of males from a neighboring community are often pursued and attacked, their babies snatched and killed.[3] But Passion and Pom were the first females known to attack other females and their infants, individuals with whom they shared a long-standing network of social connections. For these simian serial killers, the social virtues of empathy, compassion, and friendship held as little meaning as they do for their human counterparts.

Passion was one of the most unusual chimpanzees encountered by Goodall, who even describes her as having "a strange, mad look."[49] Her behavior was unique enough, extreme enough, to qualify as pathological, yet it did not disqualify her from being a chimpanzee. Similarly, the bizarre, excessive violence of serial sexual killers, the psychopaths who strike and kill with little remorse, cannot disqualify them from being human. Monstrous as they may seem, they are only a living warning of how far behavior can drift over time, how distorted threat can become when the reality check provided by social connectedness vanishes completely. Both human and animal typologies may need to be expanded to include the category of "sociopath."

Forensic psychiatrist Robert Simon suggests that the openly pathological behaviors of sexual predators and serial murderers are only the most extreme example of antisocial impulses and fantasies common to everyone.[50] "Normal" people, Simon points out, almost always score well above zero on psychological tests of psychopathology. And sadistic sexual fantasies, small acts of revenge, or intimidating power plays—behaviors found in more than a few seemingly average, law-abiding people—mark the midpoint of a behavioral continuum that begins in monastic perfection and ends in lurid headlines. As Simon concludes, "Bad men do what good men dream."

A SEASONED NEUROANATOMIST, asked recently to name the most unexpected result of his long and prolific career, replied, "The similarity between the human and rat brain." Nerve cells of rats and chimps, starlings and hawks, mice and men look, act, communicate, develop, and respond in a comparable fashion. The basic processes of perception, recognition, and reaction are also conserved, although the detail and sophistication of these processes become progressively more complex. Direct anthropomorphizing from animal behavior to human behavior, no matter how tempting, is a risky business. But the common features of both brain and behavior across species suggest that the underlying neural mechanisms have much in common.

We watch, we touch, we measure. Animals mirror and model not only what is worst in us but also what is best, our darkest rages and our moments of deepest compassion. Perhaps the real question, after all, is not what we can learn from animal aggression but why we are not better listeners.

Four

ALL THE RIGHT CONNECTIONS

In our cities, the geography of violence has a primal urgency. Rumor, bitter experience, and angry encounters hone city dwellers' sense of direction. As a result, they not only know the right from the wrong side of the tracks, but precisely which streets define the tracks, how that boundary shifts from day to night, where you can linger, and where you'd better keep moving.

Mapping this geography has been a cottage industry in Philadelphia since the late 1960s, when a local landscape architect, university professor, and self-styled "urban environmentalist," Ian McHarg, launched a "modest attempt to bring together the information plant and animal ecologists have developed and apply it to the problems of the city."[1] An early proponent of the dynamic relationship between behavior and environment, McHarg argued that the physical, mental, and social health of a community were intimately tied to the quality of the surrounding environment. Consequently, he reasoned, the distribution of physical disease and social disorder defined "environments of pathology," a geography of local disruptions that ought to inform and guide future planning efforts. To identify such pathological environments within Philadelphia, McHarg and a small group of graduate students from the University of Pennsylvania's Department of Landscape Architecture and City Planning constructed maps of the incidence of violent crime, suicide, drug addiction, alcoholism, juvenile delinquency, and infant mortality in the 200-odd communities that make up this "big city of small neighborhoods."[2] Their results, recorded in McHarg's classic text, *Design with Nature,* describe a swath of destruction that begins about four miles northeast of city hall, where the Betsy Ross Bridge spans the Delaware, and sweeps southwestward across the Schuylkill River to the city line (Figure 4.1).[1] To no one's surprise, this stretch encompasses the city's most notorious and impoverished neighborhoods. But it also includes the very center of the city, the Philadelphia of William Penn, Independence Hall, and the Liberty

Figure 4.1

Environments of pathology in Philadelphia. Data on nine types of violent crime and social pathology are shown. *Black* = highest incidence; *gray* = moderate incidence; *light gray* = lowest incidence. The summary incidence map is a composite of the other nine.

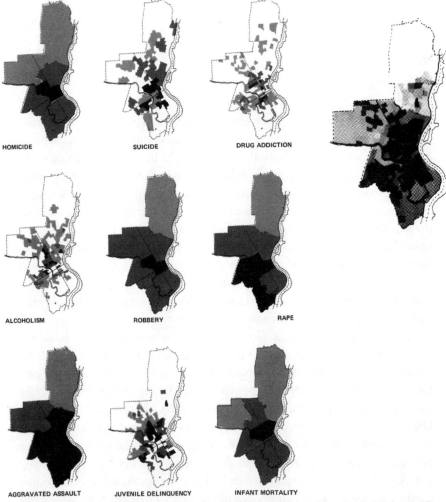

Bell, of Society Hill and Washington Square and the historic ethnic communities that encircle it. McHarg emphasized that "the heart of the city is the heart of pathology."[1]

Ironically, the path of violence also engulfs most of the city's major colleges, universities, and medical centers, including the nation's first institute of higher learning, the University of Pennsylvania. Anchoring a westward belt of communities in need of a social overhaul, its parklike lawns and graceful Gothic architecture cloak the fact that guaranteeing the safety of Penn students costs the university $15 million a year and requires a ninety-eight-member police force.[3,4] Penn's concern with security has a long history. Violence has stalked the Penn campus since the 1870s, when concern over the progressive deterioration of the neighborhood surrounding the school's original site at Fourth and Arch streets prompted university officials to pull up stakes and migrate westward.[5]

Mobility, however, is part of the problem as well as the solution. West Philadelphia's convenient location between I-95 and Schuylkill Expressway expedites quick getaways; an assailant with a car can be well on his way out of the city in minutes. McHarg's jigsaw puzzle maps, in other words, tell only part of the story. The geographical substrate of violence is not confined to troubled neighborhoods—it spreads out to include the network of superhighways, city streets, and back alleys that link hot spot and safe haven, crime scene and hideout, city and suburb.

Like Ian McHarg, neurobiologists specializing in emotion have sought to map the intersections between behavior and environment. And just as McHarg and his students located "environments of pathology" in and around Philadelphia's centers of higher learning, as well as in some of its oldest neighborhoods, anatomical urban planners have tracked aggression and fear to sites of thought and reason, as well as to ancestral regions in the dark core of the brain. They've also taken mapmaking one step further. Using increasingly precise labeling and imaging techniques, modern neurocartographers have begun to add the one essential detail omitted by the University of Pennsylvania team: a road map of the pathways that connect neighborhood to neighborhood.

Functional Localization Revisited: Is the Brain More Than the Sum of Its Parts?

Functional localization—Fritsch and Hitzig's idea that specific behaviors could be mapped to circumscribed areas of the brain—found its consummate champion in the Scottish neurologist Sir David Ferrier (1843–1928).[6] Ferrier extended and refined the work of the German researchers, combining stimulation and lesion procedures in birds, cats, dogs, and primates to create a definitive map of the cortical sites governing movement.[7] In a daring public relations move designed to settle the quarrel between globalists and localizationists once and for all, he introduced a monkey with a purported lesion of the motor cortex to a crowd of more than 120,000 spectators at the Seventh International Medical Congress in 1881.[8] After demonstrating, as a physician might have done with a human patient, the unique deficit created by the lesion and predicting the location of the damage responsible for these symptoms, Ferrier permitted the animal's brain to be publicly dissected by a blue-ribbon panel of his most distinguished contemporaries. They discovered a triangular crater in the motor cortex exactly where Ferrier had insisted it would be found.

Ferrier's monkey won the day for the localizationists but did not instigate a wholesale conversion. In fact, Ferrier's closest friend and colleague, fellow neurologist John Hughlings Jackson (1835–1911), was one of those who maintained reservations about the strict compartmentalization of brain function.[8] A clinical researcher whose precision and fervor on the neurology ward matched Ferrier's in the laboratory, Jackson's own work on localized epileptic seizures had demonstrated correlations between seizure symptoms and areas of focal tissue damage, associations he could have interpreted as a one-to-one correspondence between function and location. Jackson, however, cautioned that the coexistence of an "environment of pathology" and a clinical symptom did not necessarily add up to such a straightforward association. For example, Jackson argued, the injured region might not act as a "center" but as one member of a functional team, overseeing the activity in a sister neighborhood. It was the loss of this executive function that unmasked the symptomatic behavior "downstream."[8]

Jackson's suspicions suggested that the holism so passionately defended by Flourens and the other globalists should not be consigned to the theoretical scrap heap too quickly; the brain, while certainly not an amorphous whole, might indeed be more than the sum of its parts. New anatomical methods developed in the 1960s and 1970s that allowed researchers to unravel the complex highway systems of ingoing and outgoing fibers stringing together multiple brain regions in pathways and systems provided additional evidence that a brain divided into a matrix of centers was a gross oversimplification. Instead, modern neuroanatomists, such as Steve Arnold of the University of Pennsylvania, suggest that the true anatomical substrate for behavior is "a specific, interacting network of brain regions that act together to produce a particular experience or action, whether it's a visual experience or an emotion."

Arnold adds, however, that "the brain is infinitely more complex and organic than one wire connected to another wire." The prevailing metaphor of a brain composed of networks may be the intricate circuitry of the computer, but the functioning nervous system is more like a teeming, dynamic modern city, a collective of interconnected communities of neurons, each characterized by a distinct architecture, style, and local population. Like the urban campuses educating the city of Philadelphia, these affiliated cellular neighborhoods of the brain—Steve Arnold calls them "key nodes"—are linked by a network of axonal freeways and side streets that facilitate discussion, information trading, and gossip. And just as the incessant flurry of seminars, phone calls, computer mail, and research collaborations collectively fashions intellectual life within the city, the ongoing interaction of internal and external stimuli—the sight of a gun, the incessant wail of a baby, a racing heart, an irresistible craving for drugs—activates interrelated neural neighborhoods dedicated to emotion, neighborhoods that work together to coordinate the fateful decision to attack or retreat. In this expanded model of the relationship between structure and function, an event that occurs in one neighborhood has the potential to affect all, and injuries that "break the link" at different points often trigger similar behavioral consequences.

"Once you have a network model—as opposed to 'one-region–one-function models'—it's like a wheel," suggests Yale University psychiatrist John Krystal, director of clinical psychopharmacology at the National Center for Post-Traumatic Stress Disorder. "You can start at any point in the system and get to any other point in one or two synapses." This wheel of key nodes and interconnecting fiber pathways that coordinate sensation, memory, intuition, and response has replaced Fritsch and Hitzig's centers as the basic architectural unit of the nervous system.

The Neighborhoods of Emotion

The surface of the cerebral cortex, with its folds and grooves, is a familiar landscape. But the brain is more than cortex, and mapping the geography of emotion also requires a trip below the visible surface, to neighborhoods where the relationship between the neuron and the environment is more direct, responses swifter and more forceful. Landmarks and a road map are essential.

For the new visitor, a brief walking tour of the brain's interior can be instructive. Figure 4.2 serves as the road map. Such a tour might begin at the hinge between the back of the head and the neck, where the most posterior portion of the brain, the *brain stem,* meets the spinal cord. This is the historic area, home to ancient neighborhoods bearing majestic Latin names and charged with the organization and management of the enduring functions of daily life: breathing, heart rate, digestion, sleep, consciousness. When a hand brushes your face or you bite your lip while composing a difficult paragraph, sensory nerve endings in your skin, mouth, and neck relay this information to the brain stem, just as nerve endings in your hands, legs, and joints carry similar messages to the spinal cord. Directives flowing in the opposite direction operate the muscles you use to speak, chew, roll your eyes, turn your head. An archipelago of cellular islands, the *reticular formation,* continually fine-tunes your level of alertness to keep you awake by day and asleep at night. Specialized islands, dedicated to the complex task of coordinating posture in a top-heavy,

Figure 4.2

A cross-sectional, or midsagittal, view of the human brain. The inset shows the nuclei of the hypothalamus.

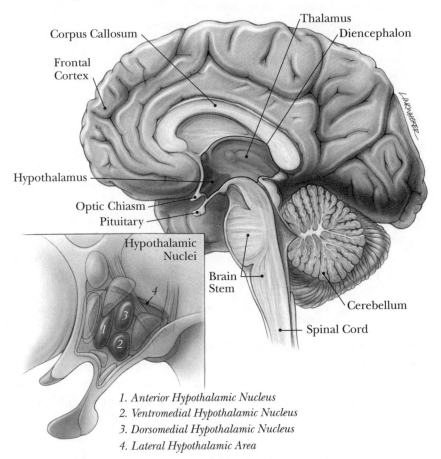

1. *Anterior Hypothalamic Nucleus*
2. *Ventromedial Hypothalamic Nucleus*
3. *Dorsomedial Hypothalamic Nucleus*
4. *Lateral Hypothalamic Area*

two-footed animal, help you stand, maintain your balance as you push open a door or pick up a book, and move your head without a dizzying oscillation of your visual field. Even globalists like Flourens acknowledged the functional specialization of these essential old sectors, recognizing that injury to them had obvious—and often disastrous—consequences.

The tour climbs upward, then turns back into the curved cup at the rear of the skull. The treelike lobes of the *cerebellum* bush up to fill the cup, cascading over the sides of the brain stem like a fleshy garden. If the city of the brain had an athletic department, it would be based here. Impulses streaming over the cerebellar peduncles, two thick fiber cables that hook the cerebellum to the brain stem, mingle with local signals coursing the lobes of the cerebellum to integrate posture, balance, timing, and coordination during the muscle sequences used to walk, to dance, to fight. Recent studies have suggested that the cerebellum—with these elaborate connections to motor response systems—may also "back up" the cerebral cortex, especially on complex tasks involving timing, anticipation, or learning and remembering motor skills.[9–11]

Take a U-turn forward, toward the forehead, and into the center of the brain. Neuroanatomists call this the *diencephalon,* or "between-brain," a reference to its strategic location between the age-old neighborhoods of the brain stem and the high-rent districts of the cerebral cortex. One of the more celebrated residents of the diencephalon is the *thalamus,* a pair of egg-shaped lobes straddling the third ventricle. The name *thalamus* means "inner chamber," specifically, a bridal chamber, a dark and mysterious place of union between streams of incoming information. Here clans of related cells collect, sort, and dispatch the various types of incoming sensory data—sights, sounds, textures—headed northward to the cortex and, conversely, relay outgoing cortical responses to specific muscle groups in the body.

A wall of cortex surrounds the thalamus like a cave, floored by a footbridge of fibers known as the *corpus callosum.* A second arch of tissue, the *hippocampus* (from the Greek for "sea horse," which it roughly resembles; see Figure 4.4), curves between this fiber bridge and the diencephalon, split during brain development from its birthplace next to the overlying cortex by the growing axons weaving the corpus callosum. Overhead, the archway hisses softly with the sound of water drifting through the lateral ventricles, fluid-filled cavities that cut through the interior of the temporal lobes. On either side lies the temporal cortex; in the foreground, the frontal cortex curves around a diffuse spray of fibers, the fornix, that forms the exit road out of the hippocampus.

This is the heart of the brain's emotional network.

One of the earliest descriptions of the subcortical loop encircling the thalamus can be found in the seventeenth-century neuroanatomy text *Cerebri anatome,* notable not only for its primitive stab at functional localization, but also for its lavish and detailed illustrations by a young artist soon to be far better known for his work on the architecture of cities than the architecture of the brain, Christopher Wren.[8,12] But it was Paul Broca, the man who laid the clinical groundwork for functional localization, who is usually credited with the discovery of the neural traffic circle that plays such a critical role in emotion. In 1878, he described a ring, or limbus, of temporal cortex surrounding the underlying thalamus, naming it *le grand lobe limbique*—the limbic lobe.[13] A transitional area between the less differentiated subcortical structures and the ordered precision of the cortex, with its six architecturally distinct layers, Broca's ring, shown in Figure 4.3, included the curved ridges of the parahippocampal gyrus, the cingulate gyrus, and the subcallosal gyrus, as well as the hippocampus and the narrow strip of cortex rolled into its center, known as the dentate gyrus.

Figure 4.3

Le grand lobe limbique (shaded area), as defined by Broca. His original description also included the hippocampus and the dentate gyrus, which lie within the parahippocampal gyrus.

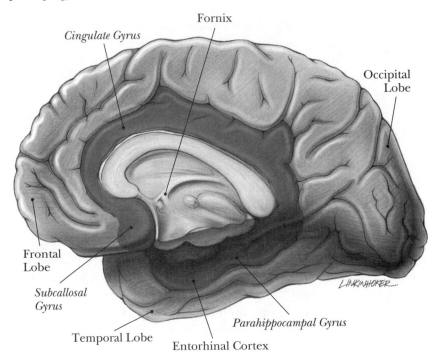

The neighborhood grew. In 1937, Cornell University neurologist James Papez described an "ensemble of structures" that constituted "the anatomic basis of the emotions."[14] Papez's circuit included the cingulate gyrus and the hippocampus—structures that were part of Broca's limbic lobe—as well as the anterior thalamus and a second structure located in the diencephalon, the hypothalamus (Figure 4.4a). Papez and other investigators also discovered a web of connections that linked the so-called limbic lobe and many of the subcortical cell groups that also orbit the thalamus. Recognizing the practical significance of this highway system, Paul MacLean in 1952 extended Papez's circuit to include these associated structures (Figure 4.4b), renaming it the "limbic system."[15,16]

More recent investigations have added a suburban extension to the limbic circuit. The tour ends in a day trip to this swath of cortical mantle, the prefrontal cortex, that loops down from the curtain of fibers fanning out at the tip of the fornix, then sweeps around the front of the brain into the forehead. This newest member of the limbic community association maintains extensive ties

Figure 4.4

The emotional circuit of Papez (A) and the limbic system of MacLean (B)

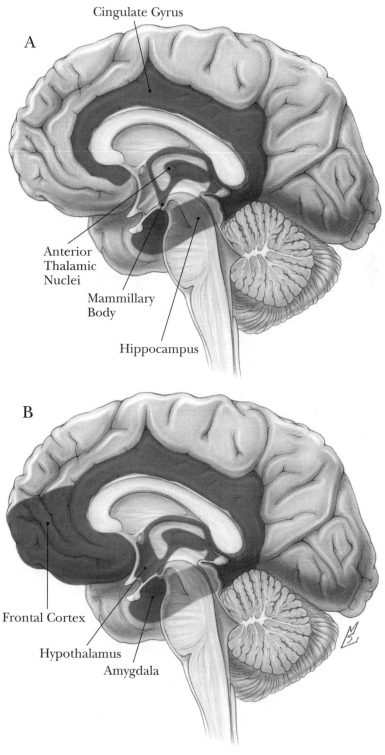

with its downstream neighbors, reciprocal connections that facilitate the exchange of news and gossip between the prefrontal region and the original limbic system structures in the center of the brain.

Steve Arnold suggests that these pathways represent the "integration of cortical and subcortical regions, a way to have cortical information influence autonomic (i.e., sympathetic nervous system) function. And autonomic function is really one manifestation of emotion—you feel butterflies in your stomach, you break out in a sweat, your muscles tense, your blood pressure goes up. In other words, the connections form a point of intersection between emotion and cognition."

Philip Bard, a student of the great physiologist Walter Cannon, uncovered the first evidence linking limbic structures to emotional behavior in the 1920s.[8,17] Starting from the observation that removal of the cerebral cortex in cats turned ordinary house pets into enraged tigers—a phenomenon dubbed "sham rage"—Bard sought to identify the brain area or areas that mediated this aggressive behavior. He severed the brain in several cats, working progressively from front to back, until he discovered one cut that eliminated the sham rage reaction—and that spared a crucial limbic structure, the hypothalamus. Building on Bard's discovery, Papez theorized that his limbic circuit, based on connections between the hypothalamus and the cortex, constituted "a harmonious mechanism which may elaborate the functions of central emotion, as well as participate in emotional expression."[14]

Succeeding decades would support this theory. Like the ring of survival-oriented neighborhoods surrounding the heart of Philadelphia, Broca's limbus of neural neighborhoods has been repeatedly linked to the interplay between internal and external worlds that underlies both aggression and fear. An anatomical correlate of the circular relationship between brain and environment, the intricate neural highway system connecting the key nodes of the limbic system replaces the old idea of an "aggression center" or a "fear center" with an ensemble of interactive regions that link the perception of threat with a potentially life-saving response.

Cat Fights

"If you want to understand behavior, you have to understand anatomy. It's as simple as that." This is Allan Siegel's verdict on functional localization. Siegel, a neuroscientist at the University of Medicine and Dentistry of New Jersey, has spent over thirty years understanding the anatomy of the key node in Papez's circuit: the hypothalamus. His detailed maps of the behavioral geography of this busy neural neighborhood, including its links to both sensory and response networks, provide a compelling example of how centers have evolved into circuits integrating emotion and cognition, perception and action, brain, body, and environment.

Tucked like so many little pebbles into a V-shaped pocket wedged between the thalamus and the base of the brain (see the inset in Figure 4.2), the nuclei of the hypothalamus collectively form "only a trifling part of the whole brain."[18] Yet by virtue of its management responsibility for both the autonomic nervous system—the peripheral neural network orchestrating the fight-or-flight responses of the body—and the endocrine system, this small corner of the diencephalon has played a big role in the neuroanatomy of aggression since Bard's experiments on sham rage. But mapping the emotional circuitry of the hypothalamus would have been impossible without a tool that became compass and sextant to both the neuroanatomist and the neurosurgeon, transforming the descent into the brain from a wandering in the dark to a purposeful expedition.

A British neurosurgeon, Victor Alexander Haden Horsely (1857–1916), invented this better mousetrap.[7,19] Renowned for his dexterity in the operating theater (and his zealous support of the temperance movement in the real world), Horsely was especially fascinated by the problem of epilepsy, the neurological Holy Grail of the late nineteenth century.[7] He recognized how seriously the lack of precision in stimulation and ablation studies compromised a clearer understanding of the mechanisms underlying seizures, and he sought to impose order by taking advantage of the constraints imposed on the brain by the bony cage of the skull. Together with renowned anatomist Robert Henry Clark, Horsely designed a metal brace that immobilized an anesthetized animal's head. Because—thanks to the skull—spatial relationships within the brain are fairly consistent between adult individuals of the same species, a three-dimensional coordinate system could then be used to pinpoint specific sites in the interior of the brain. Once the head was locked into Horsely and Clarke's "stereotaxic" apparatus, cutting and drilling during surgery could not dislodge or tilt the brain out of the plane of the coordinate system, meaning that the investigator could arrive at the intended destination without overshooting or undershooting the mark.

Thanks to stereotaxic surgery, precision replaced the guesswork that had previously bedeviled neuroanatomical experiments, permitting meaningful animal-to-animal comparisons. For the first time, experimenters utilizing ablation to study the selective loss of function could excise the same fragment of brain in each subject, and those who preferred electrical stimulation now had a road map to guide the accurate placement of electrodes. Coupled with advances in microscopy, particularly methods for visualizing fiber pathways, the information gleaned from stereotaxically guided lesion and stimulation studies allowed anatomists to construct maps of the architecture and traffic patterns of the brain with a level of detail and precision that would have astounded Gall and Flourens.

"I CAN REMEMBER so well the exact spot in my garden where I decided in favor of science."[20] The foundation for Siegel's work on the hypothalamus was this fateful decision by a Swiss ophthalmologist, Walter Rudolph Hess, who, in 1912, at the age of thirty-one, renounced a successful medical practice to accept a financially precarious post as a teaching assistant at the University of Zurich.[20,21]

Hess was fascinated by the "function and regulation of the internal organs."[20] For him, the fundamental question facing physiologists was "how a group of organs is used to create and maintain inner conditions in such a way that every element of the collectivity . . . encounters the necessary prerequisites not only for subsistence, but also for carrying out certain functions," a process that he recognized was not static, but "in reality, a system of antagonistic forces resulting in a dynamic equilibrium."[20] His search for the control center of this dynamic equilibrium, which Cannon had named *homeostasis,* led him to the brain and to the largely uncharted territory of the Papez circuit. The explorations culminated not only, as he had hoped, in the location of sites that influence basic bodily functions, such as heart rate, respiration, and bladder control, but also in the unexpected discovery of something equally profound: neighborhoods that were home to aggression.

Hess took the art of brain stimulation to new heights. Using Horsely and Clark's stereotaxic procedure, he threaded delicate wire electrodes into dozens of pinpoint sites in the diencephalon of the cat, then cemented the probes in place to attach them permanently to the cat's skull. After the animal recovered, Hess wired each electrode to a long strand of thin cable, permitting him to stimulate the discrete point surrounding the tip of the electrode and observe the resulting behavior in a free-moving, unanesthetized, unrestrained animal.

To Hess's astonishment, some of the cats did not react to brain stimulation with a mere twitch of a leg, a blink of an eye, or an accelerated heart rate, but with the complex sequence of behaviors "we obtain under physiological conditions when the cat is confronted by a dog and, in this threatening situation, seeks to defend itself."[20] These cats crouched and hissed, teeth bared and pupils dilated. Their fur stood on end. They arched their backs, flattened their ears, and raised their puffed-out tails, poised to attack.

When the brains of the agitated cats were examined afterward, Hess confirmed that the tips of the electrodes clustered in the hypothalamus. But he immediately recognized that the complexity of the behavior elicited by hypothalamic stimulation implied that this anatomical structure was more of a meeting place than an active dictator. The hypothalamus, he reasoned, acted like a "bridge thrown over a gap . . . between . . . physiology and psychophysiology,"[21] a conference center where sensation and thought connected to spark the autonomic, hormonal, and muscular elements of the emotional response. The nexus of visceral connections that coordinated homeostasis, the dynamic physiologic equilibrium essential to the internal survival of the body, also played a crucial role in coordinating the balance between environment and behavior that governed survival in the external world, including the self-defensive response patterns evoked by threat.

Inspired by Hess's pivotal experiments, other scientists also learned that they could provoke aggressive responses by stimulating the hypothalamus. Yale researcher John Flynn, for example, discovered that hypothalamic stimulation could spark not only the defensive aggression so well characterized by Hess, but also predatory aggression—that troublesome stalk-and-kill sequence that strikes terror in the hearts of mice, rats, and birds and dissent in the ranks of ethologists.[22,23] In agreement with classification schemes that viewed predation and self-defense as two distinct behavior patterns, Flynn observed that the "affective defense" pattern Hess described was elicited by stimulation of the midline region, or medial hypothalamus, while predatory aggression appeared after stimulation of the lateral hypothalamus (see Figure 4.2). Siegel, a student of Flynn, has extended the work still further. Using a painstaking combination of electrical stimulation and modern neuroanatomical tracing techniques, he has not only revisited Hess and Flynn's neighborhoods in even greater detail, but has also begun to unravel the circuits ferrying information into and out of the hypothalamus along the neural highways that coordinate aggression.

I VISITED SIEGEL on a blistering summer day. The temperature had already soared into the high nineties by the time I hit the Newark train station, and the humidity made it feel at least ten degrees hotter. People were limp, surly, disoriented.

Siegel, however, was unruffled; in fact, he seemed barely aware of the heat. He fired off details of his odyssey across the hypothalamic landscape over the past three decades like a toll collector rattling off directions on the New Jersey Turnpike, rapidly sketching circuits and loops in the air as he spoke. "The cat is the best model of aggression in the world," he insisted. "It earns its living by catching and killing a rat. It defends its turf with defensive rites. And you can elicit these responses quite easily from the hypothalamus many, many times."

Siegel went on to describe his unique approach to studying this perfect brain. "We started by identifying the regions that produce the [aggressive] behavior, then those that modify it. The next question was, 'What is the circuitry for both of these systems?' We learned that there are two kinds of circuits—one that creates the behavior and one that speeds it up or puts a brake on it."

The work required meticulous patience, an artisan's skill. In the first stage, Siegel, working first with Flynn and then in his own lab, traced the routes leading to and from the sites where electrical stimulation provoked affective defense and compared these pathways to the routes favored by predatory attack. They implanted dozens of tiny guide tubes throughout the hypothalamus, permitting them to verify the response at that site by passing an electrode through the guide tube. When a site proved positive (electrical stimulation elicited either the defensive behavior or mouse catching), the roads leading into that point from other brain regions were tagged by replacing the electrode with a fine-gauge needle and injecting an enzyme called horseradish peroxidase.[24-26] The enzyme is swallowed by the tips of incoming fibers and shuttled back up the axons to their site of origin. When the labeled brain is later removed, sliced into thin sections, and soaked in a broth containing the "reporter" chemical diaminobenzidine, fibers impregnated with the enzyme bind and break down this chemical to generate a trail of golden-brown crystals readily traced back to the starting point. These points represent entry nodes in the circuit, the stations that relay sensory and cognitive information to the hypothalamus.

The roads out of the hypothalamus were marked by injecting a radioactive, rather than enzymatic, tracer compound.[24-27] The labeled molecules are used as building blocks in the construction of neural proteins; when the proteins are shipped out of cell bodies in the hypothalamus and down the axon, the tracer flows downstream too. Energy emitted from the radioactive proteins strung along the outgoing fibers can react with a photographic film; the developed film reveals a chain of microscopic street lights illuminating the path to outposts under hypothalamic command.

In the next phase, Siegel evaluated the influence of the neighborhoods linked to the hypothalamus. He and his coworkers implanted two guide tubes— the first in the hypothalamus and the second in one of the connecting nodes. By electrically stimulating both sites, the researchers could determine if simultaneous activation facilitates or suppresses the aggressive response, that is, if the connection between the two "speeds up the behavior or puts a brake on it."

Combine a cat, a rat, and a minute spark in the lateral hypothalamus. The cat stalks the rat, just as it might in a barnyard or alley, ultimately pouncing to deliver a stunning slap and a killing bite to the neck. The entire body focuses on the attack, and sensory acuity skyrockets. The slightest touch near the mouth triggers reflexive biting; the eyes track a moving object with maximal efficiency.

Anatomically, Siegel concluded, this behavior pattern made perfect sense.[24-27] Stimulation of the lateral hypothalamus triggers the search for prey. Incoming pathways from some limbic system nodes—the ventral or lower surface of the hippocampus; the lateral edge of the elliptical cell group known as the amygdala; and the septum, the curtain of cells marking the foremost edge of the fornix—pump up predatory attack, while inputs from complementary limbic sites, including the dorsal (upper) surface of the hippocampus, the core of the amygdala, and the medial cell group of the septum, damp down the attack sequence.[24,25] Stimulation of these limbic nodes also expands or contracts the facial skin surface area that responds to touch with a reflexive bite, suggesting that one way in which they may modulate predatory aggression is by fine-tuning sensitivity to environmental stimuli.

In addition, the lateral hypothalamus receives incoming messages from cortical nodes in the limbic circuit, especially the prefrontal cortex and cingulate gyrus.[24-26] Because these cortical inputs also suppress vital functions such as feeding, Siegel believes that in contrast to the selective effects of inputs from the amygdala and the hippocampus, the cortex may act in a more general fashion to regulate the activity of the hypothalamus.

Outgoing pathways have a similar elegant logic, issuing hypothalamic commands to brain stem centers that control jaw opening and closing, visual tracking, and alertness.[24] Cells of the lateral hypothalamus also activate fibers that project to the stomach and intestinal tract, stimulating gastrointestinal blood flow in anticipation of a good meal.[24]

Defensive aggression follows different pathways.[24-27] Signals ignited by electrical stimulation of the medial hypothalamus are either fast-forwarded to cells in the midbrain or detoured through the anterior hypothalamus to pick up additional information from other limbic nodes and the endocrine system before being routed downstream. The two inputs converge in the periaqueductal gray area (PAG), a midbrain cell group surrounding the narrow conduit that joins the third and fourth ventricles. The PAG coordinates the information from the hypothalamus with directives from sensory processing areas of the cerebral cortex, then transmits its conclusions downstream to clusters of cells overseeing visceral responses and the muscle groups needed to coordinate a self-defensive attack—sites that orchestrate heart rate and blood pressure, jaw opening, the slash of a paw, hissing and growling.[24,26]

Ethologists may disagree about whether predation ought to be considered aggression or a unique form of behavior, but Siegel's work confirms that anatomically, killing to eat *is* in a class by itself. And it illustrates that behind observable behavior patterns are logical neural mechanisms.

Research like Siegel's—time-consuming, delicate, and highly specialized—is difficult in a world of shrinking research budgets focused on quick results. Nonetheless, Allan Siegel has no regrets. "I consider myself quite lucky that we've been able to carry on for thirty years now," he said. "The cat brain is a beautiful brain to work with."

Emotionally Unavailable

Make two fists with your hands, tucking the thumbs under your fingers. Hold one fist to each side of your head with the heel of the hand pressed to the temple. The fists represent the temporal lobes, the lateral boundary of the limbic system.

Now remove your fists and turn them over so that the fingers face up, as if the brain had been removed from the skull and flipped over. The tips of your thumbs show the approximate size, shape, and position of the amygdala, another key node in the emotional network. The name comes from the Greek word for "almond," and refers to its shape.[19] Below the joint, the base of the thumb represents the hippocampus, curving inward to slot the amygdala into place between the lateral edges of the hypothalamus and the overlying temporal cortex (Figure 4.5).

Finding your way around the amygdala, guided only by the size and shape of its cells, is a bit like sifting through thousands of tiny beads. As Steve Arnold notes, "You have to look carefully. At first, it all looks the same. But if you stare at the region for a while, then you begin to see the differences between cells." In fact, with careful observation, the cells of the amygdala can be sorted into five groups, or nuclei (see the inset for Figure 4.5). Three—the medial, central, and cortical nuclei—crown the upper, or dorsal, surface of the amygdala and contain primarily small to medium-sized oval or spindle-shaped cells.[18,28] This triad, referred to in classical anatomical texts as the "corticomedial" or "centromedial" subdivision, anchors fiber pathways that link the amygdala to the hypothalamus and the olfactory bulbs.[18,28–30] Additional pathways to and from the central nucleus connect this knot of cells to the brain stem cell groups that monitor and orchestrate body functions, such as heart rate, respiration, gastrointestinal motility, and pupil size.[18,28,29] As a result, the amygdala, like the hypothalamus, can issue commands to peripheral response systems, including the autonomic nervous system, the endocrine system, and motor pathways mediating behavioral responses to the environment.

The second group of nuclei, the basolateral subdivision, is so similar to the cerebral cortex that some experts have dubbed this region a "vicarious cortex."[31] Both the cortex and the basolateral nuclear complex increase progressively in size from the bottom to the top of the phylogenetic scale; in humans, this group has expanded to fill nearly the entire "thumb" of the amygdala, reducing the corticomedial group to a fingernail-like cap. The large, pyramidal-shaped cells of the lateral and basal nuclei resemble the pyramidal cells found deep in the cortex, and both connect sensory inputs from the thalamus to areas governing voluntary movement.[18,32] Other basolateral cells resemble the star-shaped neurons found in the outer layers of cortex. In both regions, these cells act as neighborhood gossips, linking adjacent neurons to form local chains of communication.[32]

The architectural similarities reflect an intimate social relationship, for in contrast to the brain stem connections of the corticomedial amygdala, cells in the basolateral division prefer to associate with the overlying cortex. For example, the lateral nucleus, a sensory convention center, entertains inputs from cortical processing areas responsible for all five senses: vision, hearing, taste,

Figure 4.5

The amygdala. The inset is a cross-sectional close-up illustrating the major cell groups.

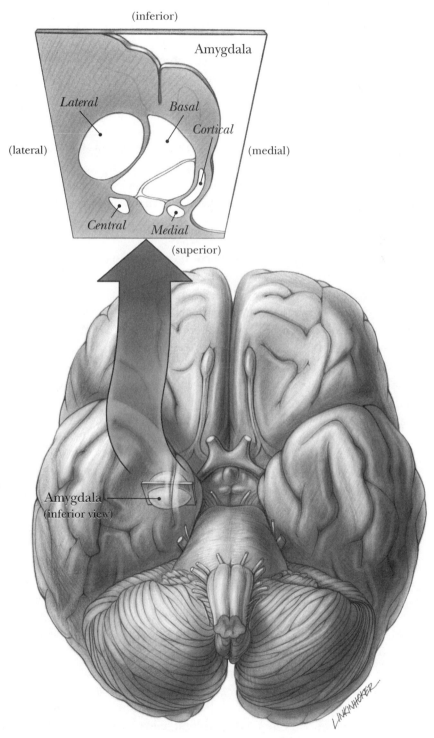

touch, and smell.[33–35] And its partner, the basal nucleus, reports back to the cortex, directing the equivalent of a major highway from the amygdala back to cortical association areas.[35,36]

Within the amygdala, a web of local roads also connects cells within and across the two subdivisions, encouraging information flow between nearby communities. For example, one such pathway leads from the lateral nucleus to the basal nucleus and from there, across divisional boundaries to the central nucleus.[34,36,37] Even shorter connections hop from point to point within the lateral nucleus, shuffling the data from various sensory modalities.[36] These interior connections have a profound functional significance for the amygdala.

Because of its extensive connections with olfactory structures, the amygdala was long assumed to govern the sense of smell. Later studies cast doubt on this theory; in fact, researchers learned that they could destroy the amygdala on both sides of the brain without affecting the ability to discriminate odors.[30] What has become increasingly clear, however, is the key role of the amygdala in emotion, a function well suited to its position at a neural crossroads where brain stem and cortex, recognition and execution converge. As Eric Halgren, of the University of California, Los Angeles, Brain Research Institute, noted in a 1992 review, "The amygdala has all of the right connections with the cognitive neocortex and visceral brain stem to provide the link between them that is central to emotion."[38] The anatomical equivalent of city hall, the amygdala's connections with brain regions that mediate sensation, vital functions, hormone release, memory, meaning, and learning provide a neural forum for chit-chat, expert opinion, debate, and delegation. If, as real estate agents say, "location is everything," the amygdala would be under contract to an emotional buyer the moment a "for sale" sign appeared on its front lawn.

RHESUS MONKEYS, the most common primate in the biological or medical research lab, make Hess's cats appear calm and gentle. Primatologist Frans de Waal notes, "Aggression (including physical violence) is a conspicuous aspect of the social life of rhesus monkeys. . . . The frequency and fierceness of attacks among these animals is amazing."[39] He and coworker Lesleigh Luttrell, observing a closely knit group of approximately fifty monkeys at the Wisconsin Regional Primate Center, documented as many as eighteen separate acts of aggression per monkey over a ten-hour period, a rate of one and a half violent encounters per minute. Equally hostile to humans, rhesus monkeys are also notoriously bad-tempered in the laboratory, readily assaulting a would-be handler.

But in an 1887 report to the Royal Society in London, a young Canadian researcher, Sanger Brown, and his mentor, Edward Albert Schäfer, described a way to control rhesus violence.[40] They reported that surgical removal of most of the temporal lobe transformed wild monkeys from violent offenders into gentle, docile creatures easily handled without repercussions; one animal could "even be teased or slapped, without any attempt at retaliation or endeavoring to escape." However, the procedure could hardly be advocated as a routine strategy for primate domestication, for the lesioned monkeys were not only tranquil, they were nearly insensate. Although the monkey's eyes and ears still functioned perfectly, he could easily pick up and handle objects littering the cage, and he quickly spit out food laced with bitter quinine, Brown and Schäfer wrote that he "clearly no longer understands the meaning of the sounds, sights, and other impressions that reach him."[40] Confronted with a well-known object or cagemate, he behaved as if encountering them for the first time; startled by a sudden loud noise, he placidly went about his business rather than screaming or running away.

Neuroscience, preoccupied with other issues, overlooked the significance of Brown and Schäfer's observations for several decades. It was not until 1937, the same year that Papez published his theory implicating the structures of the limbic system in emotional behavior, that University of Chicago researchers Heinrich Klüver and Paul Bucy repeated the earlier investigators' studies of the behavioral consequences associated with temporal lobe lesions.[41]

Klüver and Bucy confirmed that removal of the temporal lobes did more than simply tame wild monkeys.[41,42] In fact, according to Klüver the surgery precipitated "the most striking behavior changes ever produced by a brain operation in animals,"[43] in particular, "a complete absence of all emotional reactions in the sense that the motor and vocal reactions generally associated with anger and fear are not exhibited." Instead of avoiding unfamiliar objects, a lesioned monkey mouthed or chewed everything he touched, spitting out only items that proved totally inedible. Sexual behavior increased dramatically. Turned loose in a large room containing food, furniture, and toys or other objects, the lesioned monkey explored nothing in detail; instead, it behaved "as if it were ceaselessly 'pulled' from one object to another. . . . It seems . . . that the fact of an object being an object . . . a 'discrete something' . . . is sufficient for eliciting this impulse."[42] But although the animal's attention darted from place to place, responding to even the slightest visual cues, it seemed to have lost the ability to recognize or comprehend what it saw, a deficit Klüver termed "psychic blindness." Struck by this dissociation, he and Bucy wrote, in a 1939 follow-up to their original report, "We may consider the outstanding characteristic of the behavioral changes following bilateral temporal lobectomy to be that they affect the relation between animal and environment so deeply."[42]

It wasn't necessary to remove the entire temporal lobe to create an animal that was emotionally unavailable. Subsequent investigations demonstrated that what came to be known as the Klüver-Bucy syndrome could actually be reproduced by destroying a single building in the temporal neighborhood, the amygdala.[44-46] Complete or even partial lesions of the amygdala, in animals ranging from lizards to lynxes, consistently reduced aggressiveness, fear, and responsiveness and resulted in the "psychic blindness" that preserved color, tone, depth, and texture—but lost the emotional meaning of these features.[45] Animals with a damaged amygdala looked but could not recognize, listened but could not understand.

Psychic blindness wrought especial havoc with the social order. By severing the all-important connection between perception and emotional significance, surgical removal of the amygdala garbled interpersonal relationships beyond recognition, erased critical social cues. A social creature so compromised could lose family, friends, status—even its life.

University of Chicago researcher Arthur Kling wondered about the consequences of psychic blindness outside the artificial environment of the laboratory. To find out, he traveled to the shores of Zambia's Zambezi River, where he trapped seven wild vervet monkeys and surgically removed a section of temporal lobe containing the amygdala.[47] Would these animals with no amygdala be able to survive and socialize in their native environment? Would they give up their place in monkey society, or would they be accepted back in the community?

Kling released the lesioned monkeys—five juveniles and two adults—into one of the three social groups maintaining territories in the area. Tame and passive toward their captors, the "operates" were in trouble with the real world from the moment Kling opened their cage doors. When curious emissaries from the wild group approached and greeted them, they hid in the underbrush or fled, terrified, from tree to tree. One juvenile, accosted by a group of eager playmates who rushed into the open cage, cowered in a corner; another cried piteously as if lost, but ran away at the approach of a concerned adult, finally plunging thirty feet from the top of a tree to avoid the solicitous overtures of a paternal male. Confused and isolated, the lesioned animals were an easy meal for predators. Within seven hours of their release, all but one monkey had vanished.

Had psychosurgeons of the 1960s and 1970s been paying attention, they would have recognized that surgical tampering with the amygdala was also disastrous for humans. In October 1952, a nineteen-year-old Italian boy suffering from debilitating seizures and violent outbursts in which he trampled his younger brother, assaulted doctors and nurses, and attempted to strangle his mother underwent bilateral removal of the temporal lobes.[48] Presiding physicians Hrayr Terzian and Guiseppe Dalle Ore noted that previous treatment measures had no effect on the boy's behavior. The results of the surgery, on the other hand, were profound:

> The most evident and surprising phenomenon . . . was the complete loss of any emotional behavior in the patient. He no longer manifested the slightest rage reactions. . . . The patient, on the contrary, now assumed an extremely childish and meek behavior with everyone and was absolutely resistant to any attempt to

arouse aggressiveness and violent reactions in him. He was completely indifferent toward everyone, including his parents.

Terzian and Ore had inadvertently reproduced, in a human patient, the behavioral syndrome described so vividly by Klüver and Bucy. No longer able to read or write, converse or respond, remember or learn, their unfortunate patient provided chilling confirmation of the emotional importance of the amygdala.

WHY IS THE AMYGDALA so critical to social life? One clue comes from the discovery of neurons in the amygdala that fire in response to faces. Oxford researcher Edmund Rolls has suggested that these neurons, under the direction of the overlying temporal cortex, function as part of the facial identification or "social semaphore" system responsible for coding the emotional significance of faces and expressions.[49] When the amygdala is up and running, its job is to match the sight of an approaching peer with cortical data on past experience, rank, and context; "look up" a corresponding emotional label (e.g., fear, affection, fury); and initiate an appropriate cascade of autonomic, endocrine, and behavioral responses, ranging from the racing heart and surging hormones provoked by the sight of a potential mate to the gut-clenching, stomach-churning, adrenaline-elevating effect of confronting a belligerent superior. Removal of the amygdala deletes the "look-up" process and condemns the individual to a twilight world that confuses friend and foe, real and imagined threat. Without input from the amygdala, social relationships lose meaning, and socially appropriate responses become impossible.

Consciousness Raising

Despite its seemingly mythic potential for warding off monsters, armed assailants, and other forms of evil, not everyone is enamored of the light. At a recent meeting of my local board of supervisors, debate over yet another proposed housing development did not center on the impact of squeezing more children into an already overcrowded school system, the cost of widening an inadequate roadway, or the loss of one of the last remaining slices of green meadow but on streetlights. One supervisor was especially adamant about doubling or even tripling the space between lights. "We don't want it to look like Macy's parking lot," she fumed. "After all, these people are moving here because they want to get out of the city, not be reminded of it."

Environmentally conscious suburbanites reject streetlights, sidewalks, and public parks in the belief that excluding such visible reminders of urban life will also block out the socially unacceptable behaviors they associate with the city. Similarly, Egas Moniz, Walter Freeman, and other advocates of frontal lobotomy believed that by disconnecting the neural suburbs in the cortex from subcortical structures such as the hypothalamus and amygdala, they could prevent emotion from contaminating reasoning and judgment. The lobotomized patient, released from the agitated input of the limbic system, would once again think clearly and act responsibly.

The problem is, you can't see in the dark. On an unlit street, a woman jogging, a teenager biking home from a friend's house, a boy walking his dog have a hard time knowing if the shadowy figure sauntering behind them is a neighbor or a prowler. And without full communication between the frontal cortex and the central core of the limbic network, a brain pressed to distinguish real threat from false alarm can also jump to the wrong conclusion.

THE GEOGRAPHIC BOUNDARY lines of the frontal cortex—that suburban addition to MacLean's limbic system—were not drawn until 1868, when British anatomist Richard Owen partitioned the swell of gray matter at the front of the brain into five subdivisions.[8] The leading arch of cortex that bulges into the forehead he called "prefrontal cortex," a term that subsequently came to mean the entire frontal area. Today the terms *frontal* and *prefrontal cortex* refer to the dorsal and lateral surfaces of the frontal lobes, as well as the band of cortical tissue that runs along the bottom of the brain on either side of the midline, also known as *orbitofrontal cortex.*[50]

Although reports of an association between the frontal cortex and so-called higher mental functions date back centuries, the early localizationists assigned it a role in the administration of intellect and will almost by default. Neither Fritsch and Hitzig, inching their electrodes forward from the motor control regions they had mapped so thoroughly, nor David Ferrier could detect a response to electrical stimulation of this area. Surgical removal of the frontal lobe had no visible effect on movement or perception. But like the emotional derangement caused by removal of the temporal lobes, cortical ablation had profound consequences on social, emotional, and intellectual activity. In the 1880s, Italian neurologist Leonardo Bianchi described a behavioral syndrome after frontal lobe ablation in dogs and monkeys that bore uncanny similarities to the symptoms observed by Brown and Schäfer in monkeys with temporal lesions.[8] Both types of lesions resulted in an animal likely to settle a confrontation—with the experimenter or a cagemate—by withdrawing rather than lashing out. And both reduced normal curiosity and exploratory behavior to a random aimlessness, in which no single object or activity sustained the animal's attention for more than a few moments.

But while some animals with prefrontal lesions became fearless and passive, as though afflicted with a cortical equivalent of the Klüver-Bucy syndrome, others looked more like the socially phobic vervets described by Arthur Kling, terrified of other monkeys and even their human handlers. In addition, loss of the frontal cortex addled intellect and will as well as emotion. While monkeys with temporal lesions were interested in everything, animals that had sustained lesions of the frontal cortex were interested in nothing. Listless and confused, they would dart toward a toy or an open cage door over and over again, only to stop partway, idle, and dash back to their starting point, as if they forgot why they were running in the first place.

AS TERZIAN AND ORE discovered, human beings are not immune to the emotional consequences of cortical damage. Injury to the frontal cortex, in fact, could produce an apathy and mental dullness so profound that David Ferrier is said to have remarked that the relationship of such lesions to "idiocy . . . was a general fact."[8] But in some cases of frontal lobe damage, the deficit, although equally devastating, was more subtle, distorting attention, judgment, and emotion more significantly than cognition. Perhaps the best-known example was that of twenty-five-year-old railroad foreman Phineas Gage, a "trusted and well-liked man," until a stray spark triggered an explosion that blasted an iron tamping rod point-blank into his left cheek and out the top of his head in September 1848.[8,21,51] To the amazement of onlookers, as well as his physicians, Gage—despite the one-and-a-half-inch hole in the front of his brain—did not die on the spot; he ultimately recovered the ability to walk, talk, see, and hear without difficulty. But emotionally, the accident had crippled him: "The equilibrium between his intellectual faculties and animal propensities seems to have been destroyed. He is fitful, irreverent, indulging at times in the grossest profanity (which was not previously his custom), manifesting but little deference for his fellows, impatient of restraint or advice when it conflicts with his desires."[52]

"Gage was no longer Gage," wrote John Harlow, one of his physicians. And the clue to this radical personality change lay in the unique geography of Gage's injury. As documented nearly a century and a half after the accident by Hanna Damasio, Thomas Grabowski, and Albert Galaburda, who photographed and measured Gage's carefully preserved skull, then used these data in conjunction with computer-assisted imaging technology to reconstruct the trajectory of the tamping iron, the rod had spared enough of Phineas Gage's brain to permit life, movement, and speech.[53] But it had fragmented and disconnected most of his frontal cortex, ripping out fibers and cells along with shards of bone. The resulting accidental lobotomy reduced a strong and capable man with a promising future to a disoriented, bad-tempered ne'er-do-well who drifted from job to job, finally dying in penury, forgotten and friendless.

More recently, neurologist Antonio Damasio has described a patient he calls "Elliot," a "new Phineas Gage," who, like his predecessor, lost his personality along with his frontal cortex.[51] In his mid-thirties, Elliot, a happily married and successful businessman, underwent emergency surgery to remove a rapidly growing brain tumor. But the surgical team was forced to remove a section of damaged cortex along with the growth. In doing so, they both saved and ruined Elliot's life.

Formerly a model of organization and self-discipline, Elliot could no longer manage to get to work on time or remain focused once he got there. He lost jobs, squandered his entire life savings, destroyed two marriages. Despite the ever-widening swath of personal destruction spreading behind and around him, he remained bizarrely unperturbed; in fact, he was "far more mellow in his emotional display than he had been before his illness." Impulsive, negligent, and indifferent, the postsurgical Elliot, Damasio notes ironically, "was no longer Elliot."

Leonardo Bianchi believed that prefrontal lesions disrupted what he called "psychical tone," the overall sense of awareness created by the integration of sensory input and "the emotive states which accompany all perceptions," especially perceptions that involved others.[8] As a result, the most important consequence of damage to the prefrontal cortex is the loss of "that sentimentality or feeling for others that we designate as sociality." Rude, tempestuous, and argumentative, patients with frontal injuries—like Gage and Elliot—are always in trouble, drawn irresistibly to every petty altercation, and "psychically blind" to the effect of their inappropriate behavior on others.

Intelligence tests showed that Elliot's language skills, mathematical abilities, and memory were not only normal, but better than those of the average person.[51] He had no problem recognizing faces, proposing morally acceptable solutions to hypothetical social problems, or predicting the probable outcome of those solutions. But in tests of emotional reactivity based on subtle changes in skin conductance—the laboratory equivalent of the familiar lie-detector test—Elliot was an abject failure. And he performed equally miserably in a simple card game that required tracking wins and losses to guide risky decisions, because even repeated penalties failed to elicit the emotional discomfort needed to keep him from making the same foolish mistakes over and over again. Elliot understood the rules of the game, but without the emotional computations of the frontal cortex, they had no meaning for him.

HOW CAN BOTH amygdala and frontal cortex perform the same emotionally charged task? Does the phylogenetically advanced frontal cortex supersede the time-honored computations of the "old" limbic system? Perhaps surprisingly, researchers believe the answer is no, and the reason lies in what Antonio Damasio calls a "tinkerish knack for economy"[51]—the brain's thrifty penchant for recycling proven mechanisms. The techniques for matching perception and emotional significance pioneered earlier in evolution by the amygdala have not been replaced in the prefrontal cortex, argues Damasio, but rather extended and perfected to add depth and flexibility. From a neuroanatomical point of view, survival does not depend on competition but on collaboration.

"REMARKABLY—well, maybe it's not that remarkable—the pathways regulating aggression and those regulating fear are very similar," muses New York University's Joseph LeDoux. As a result, fear conditioning, that model system for studying emotional learning, can serve as a general paradigm for piecing out the relative contributions of the amygdala and cortex—and their connections—to the emotional valuation process.

Like Alan Siegel's maps of the neural circuitry underlying affective and predatory aggression in hypothalamus, LeDoux's efforts to chart the pathways underlying fear conditioning have relied on a combination of experimental manipulations and modern neuroanatomical techniques for tracing the connections between neural neighborhoods. Of course, he's quick to point out, the "advantage of fear conditioning over aggression studies is that you don't have to start in the middle of the circuit, using electrical stimulation to evoke behavior—instead, you can start with the end response and work backward."

The response sequence begins when the rat hears a tone that signifies the possibility of electric shock. Working backward, LeDoux and his collaborators exploited an enzymatic tract tracing technique similar to that used by Siegel to trace the route taken by incoming auditory data. The group added detail to this road map by using discrete brain lesions to interrupt the circuit at various nodes in the network and assaying the consequences for the evolution of an association between stimulus and response.[33,34,37,46,54,55] As expected, they found a well-traveled highway linking auditory relay stations in the brain stem, a critical rest stop in the thalamus known as the medial geniculate body, and cortical regions dedicated to hearing. Synapses later, the cortically digested and reprocessed auditory information rolls down the fibrous highway of the external capsule into the lateral nucleus of the amygdala. A few short jumps, and it has reached the central nucleus, which finally issues directions for mounting the conditioned response.

It all made perfect anatomical sense—except for one small problem: surgical removal of the auditory cortex had no effect on the conditioning response.

LeDoux and his coworkers reasoned that there must be a second pathway ferrying the auditory signal to the output systems originating in the amygdala, one that skirted the cortex entirely. Tract tracing studies demonstrated that this hypothesis was, in fact, correct. A subset of fibers coursing upward from the brain stem hand off their signals to neurons that have established a short-cut out of the midbrain directly into the amygdala. Surgical disconnection of this pathway—unlike cortical ablation—*does* prevent the anxious crouch, the surge in blood pressure triggered by the pairing of tone and shock. Where emotional learning is involved, it would seem that the cortex, with all of its fancy claims of control over "higher mental functions," isn't even needed.

Or is it? When the fear conditioning procedure is expanded to include a second tone (only one of which is paired with the shock), the auditory cortex is not so dispensable.[56] Animals trained under these conditions after cortical lesions, like their counterparts trained in the simple fear conditioning task, still freeze when they hear the shock-related tone. But they also freeze to the irrelevant sound, suggesting that while the circuit that links cortex and amygdala may not be needed to forge the initial connection between threat and response, it is essential to the advanced perceptual skills needed to discriminate more complex situations.

Whether the environment is truly menacing or merely suggestive, the phylogenetically experienced amygdala is at the functional heart of the emotional response network. At one end, its lateral nucleus serves as a central receiving station for information collected by all five senses—both fresh, raw data straight from the thalamus and sensations that have been carefully peeled, seasoned, and processed by the cortex. The tip of a perceptual funnel, the lateral nucleus condenses and directs this stream through the adjoining basolateral nucleus and on to the central nucleus, the source of amygdala-based instructions to the body.

LeDoux refers to these outputs as "the floodgates of emotional reactivity." All requests for action, all decisions to attack, freeze, or retreat, must first pass through this command post. Rat or human, the highway system centered in the amygdala unites a neural collective that ensures survival and preserves the social order.

The frontal cortex has not grown over this system; it has—as demonstrated by the network of connections between cortex and amygdala—grown out of it. And the tragedy of psychosurgery was its failure to recognize this interdependence—that the amygdala needs the frontal cortex to boost its emotional computing power, and that the cortex, conversely, needs the amygdala to express itself.

NEUROSCIENCE INSTRUCTION at the local community college is a pretty basic affair. Information about the brain is sandwiched into courses on anatomy and physiology, introductory psychology, nursing. Educationally, this is a sensible arrangement, for no one attends the community college to become a neuroscientist. These practical students are planning for careers, and they need a concise, professionally relevant overview of brain function, not an in-depth analysis.

Emotional processing conducted over the direct route between thalamus and amygdala is equally stripped down. Designed to construct a survival-oriented, broad-brush picture of the current environment, its discriminative power is low, but its advantage is speed. This is a circuit built for emergencies, that reaches for a weapon at the sight of a stranger skulking in the yard you gave up streetlights to own.

In contrast, students at a large research institution like the University of Pennsylvania don't just want to meet the brain; they want to become intimate.

Fortunately, even as undergraduates, they can lay the groundwork for such a close relationship; in fact they enjoy not only a selection of courses but their own interdisciplinary program, "The Biological Basis of Behavior."[57] And with a graduate program that spans nine departments; includes 145 faculty members, ranging from neuropharmacologists to bioengineers; and a working philosophy that promotes cross-fertilization and collaboration, the dedicated student of the brain can go far beyond the introductory level to probe questions at the cutting edge of neuroscience, adding to, as well as drawing from, the existing knowledge base.

The extended response circuit looping through the frontal cortex also adds perceptual, philosophical, and experiential detail to the simple picture composed in the amygdala (Figure 4.6). In fact, a cortical-based circuit allows the brain to dispense with reality entirely, to begin with a mental image rather than an object. From there, the prefrontal cortex can match image and experience to generate what Antonio Damasio calls a "secondary emotion"—blocks of prerecorded data delivered to the amygdala, where, like a neural CD-ROM, they recreate the emotional pattern of the real stimulus, initiating the same sequence of muscular, hormonal, and endocrine responses.[51]

A cortical circuit also adds flexibility. Experiments conducted by Edmund Rolls and his coworkers at the University of Oxford, in which a whisper-fine electrode is used to record the electrical activity of individual monkey neurons, discovered that cells in the amygdala and orbitofrontal cortex clicked promptly when the animal was presented with a visual cue associated with a food reward.[49] When the task was changed, so that the previously rewarding stimulus

Figure 4.6

Limbic ("low road") and cortical circuits ("high road") mediating emotional information processing.

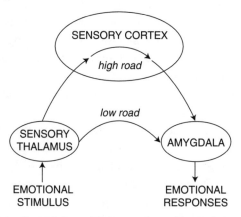

Adapted with permission from J. LeDoux, *The Emotional Brain* (New York: Simon & Schuster, 1996).

now heralded the delivery of salt water rather than the coveted food, amygdala neurons were slow to change their output in line with the new contingencies. Neurons in the orbitofrontal cortex, however, reversed their responses after only one or two exposures to the new paradigm. For real-world encounters that require an alert response to a familiar signal, the rough analysis conducted by the amygdala and its "express route" connection with the thalamus offers distinct advantages. But when the world changes, it's the prefrontal cortex that shifts gears to meet the challenge.

Finally, the cortex layers on a conscious awareness of emotion. Emotional researchers like LeDoux and Damasio emphasize that this perception, or feeling, of fear doesn't originate in the amygdala, but is added later by the cortex, as it combines images and "gut feelings" provoked by fearful responses. LeDoux explains, "The subcortical network that includes the amygdala is basically the way that animals—beginning with invertebrates—detect and respond to danger. But it's not until you become consciously aware that you're in a dangerous situation that you feel afraid." Feelings allow us to draw conclusions about the present on the basis of past experience, further refining the computation of emotional significance and maximizing the probability of an effective, accurate response.

We should take care, however, before overestimating the positive influence of the sophisticated mental processing made possible by the frontal cortex. LeDoux cautions, "Having a larger cortex does allow us to stop and think before we act, but it also allows our thoughts to frighten us. Existential fears are basically a human invention." A large and energetic cortex can be a streetlight, but it can also invent threats where none actually exist.

A Window on the Brain

Tan was only the beginning. For over a century, the case history has been pivotal to our understanding of the relationship between the human brain and human behavior. As Columbia University neuroscientist Eric Kandel notes (with regard to the importance of case histories in extending Broca's work on the localization of language), "Until recently almost everything we knew about the anatomical organization of language came from clinical studies of patients with lesions of the brain."[58] His observation is equally applicable to our understanding of the geography of emotion.

Considered individually, some case studies flash tantalizing glimpses of familiar brain neighborhoods, but skirt dangerously close to the old "one-site, one-function" model. A violent young man rendered docile—and unresponsive—by Terzian and Ore's desperate surgery;[48] a twenty-two-year-old bookkeeper, admitted to New York Hospital with a mysterious disorder that drove her to consume 10,000 calories a day, spike 104 degree fevers, and erupt into violent tantrums—all caused by a marble-sized tumor that had obliterated an entire sector of the hypothalamus;[59] a woman with a rare degenerative disorder literally turning her amygdala to stone, unable to perceive or even recognize fear;[51,60] a teenage arsonist and murderer sporting a nickel-sized hole in the fiber pathway connecting amygdala and cortex.[61] Taken collectively, however, they coalesce into Steve Arnold's network model of multiple nodes and intersecting pathways, an anatomy of neighborhoods working collectively to harness emotion as a tool for making sense of the social environment. This network, like the emotional circuits mapped so carefully in rats, cats, and monkeys, includes not only foundation communities—the hypothalamus, amygdala, and other subcortical nuclei circling the center of the brain—but also outposts in the cerebral cortex, the latest additions to Paul Broca's grand limbic lobe.

IAN MCHARG and his students drew their maps on transparencies, and they added the contributions of social, mental, and physical disease by superimposing maps of the individual pathologies. Since the 1960s, however, mapmaking, like so many other complex tasks, has gone digital. Modern computer mapping software combines cartography and database management to create real-time equivalents of McHarg's meticulously crafted transparencies, overlaying information about climate, population density, and natural resources on electronically generated copies of up-to-the-minute surveyor's maps. The most sophisticated of these geographic information system (GIS) programs dispense with man-made maps altogether, relying instead on digitized satellite images to reconstruct local geography in minute detail.

Thanks to modern brain imaging methods,[62] human neurocartography, like geographical mapmaking, has also gone on-line, resulting in significant benefits for brain researchers. Computed tomography (CT), introduced in the early 1970s, represented the first advance in this digital revolution. In a CT scan, the imaging process is repeated many times in a stepwise fashion across the head; a sophisticated computer program then combines the multiple images into a two-dimensional composite, or "slice," of the brain. Magnetic resonance imaging (MRI) represents an even greater advance in clarity and resolution. In MRI, a powerful magnetic field is used to line up squadrons of hydrogen nuclei in brain tissue. A radiofrequency signal is then switched on briefly, spinning these nuclei out of alignment. When the pulse is switched off, the nuclei flip back into line like subatomic springs, echoing the radiofrequency signal; the characteristic signals from many regions collectively form a palette of energies that paints an exquisitely detailed picture of the living brain. Unlike CT, MRI doesn't require exposure to radiation, doesn't generate a "bone artifact" where the brain meets the skull, and can be used to create three-dimensional images.

CT and MRI are structural techniques, used to locate damage caused by injury, tumors, or disease. But architectural detail and structural integrity aren't the only clues to the state of a neighborhood. Activity—what happens or doesn't happen—represents an equally powerful sign of community well-being. Similarly, changes in brain function can tell a story missed in a structural investigation. Newer imaging techniques—positron emission tomography (PET) and single photon emission computed tomography (SPECT)—have expanded the scope of neuroanatomy to include local brain activity as well as brain structure, giving new life and meaning to the concept of functional localization.

PET, the more sensitive of the two,[63,64] measures the concentration of a physiologically active compound tagged with a positron emitter—a short-lived radioactive isotope that rapidly decays to form particles known as positrons. The positrons collide with electrons to generate a pair of gamma rays that speed out of the tissue in opposite directions, where they are spotted by a ring of gamma-sensitive detectors positioned around the brain. Because active neurons consume copious amounts of glucose, delivered in the blood, positron emitters commonly used in functional brain imaging include radiolabeled glucose, the sugar that serves as the brain's principal energy source, and tracers that measure regional cerebral blood flow. The greater the level of neuronal activity in a given brain region, the greater the level of glucose utilization and cerebral blood flow, and the greater the degree of tracer uptake and positron emission. Conversely, lower levels of activity result in less tracer uptake and a fainter image on the PET scan.

Brain imaging rides a methodological roller coaster, alternately displaying enormous potential—a "window on the living brain"—and posing tremendous challenges. Resolution, while improving all the time, hovers at about 1 millimeter, meaning that even the most expensive MRI systems cannot begin to match the cellular precision achieved by a benchtop microscope at a fraction of the cost.[65] PET and SPECT, while opening the door on brain function, do so slowly, limited to snapping pictures every few seconds at best, a Sunday driver speed that just can't keep up with the millisecond pace of most neurophysiological processes.[66] Yet another limitation lies in what one researcher has called the "time-space problem": the question of whether changes in blood flow and glucose metabolism reflect a long-standing alteration in brain function—perhaps an alteration that has played a major role in the development of violent behavior—or a transient blip, caught on film at an inopportune moment.

Imaging studies of violent perpetrators and their victims are in their infancy. More than one of the dozen or so studies published to date could be fairly criticized as "digital phrenology," glimpses, as one neurologist wryly suggested, "at the bumps on the inside rather than the bumps on the outside." And none has identified the "locus of violence" so keenly sought by psychosurgeons. Yet a closer look at a few of these studies illustrates how, in even a brief glance through this window on the brain, researchers can begin to trace the physical shadow of the neural pathways suggested by case histories.

Those looking to brain imaging for an easy answer to the question of violence will be disappointed. But for those capable of realizing the potential of imaging for charting the functional pathways that make up the living human brain, these techniques offer the power to transform human neuroanatomy forever.

Consider a man identified only as "Case 1," a study in "repetitive, purposeless violent behavior." The child of alcoholic parents, Case 1 himself had a serious drinking problem, exacerbated by a head injury he'd suffered five years earlier. Or consider Case 2, a 45-year-old man given to periodic outbursts of "violent, impulsive behavior." By the time they were admitted to the inpatient unit at the University of Texas Health Science Center in Houston, Case 1 had chalked up several suicide attempts, an impressive record of brutal assaults, and a homicide conviction, while Case 2 had served time for assault, arson, and raping his own sister. Both appeared unfazed by the consequences of this behavior and expressed no guilt or regret over their violent activities.

Cases 1 and 2 represented enough of a threat to merit confinement in the psychiatric unit. But they still made a significant contribution to society—a living map of the human brain. Chronically violent patients described in a 1987 report by psychiatrist Nora Volkow, now of the Brookhaven National Laboratory, and Laurence Tancredi, former director of the Health Law Program at the University of Texas, Cases 1 and 2 participated in one of the first studies to search for localized changes in brain function associated with violent behavior.[67]

Volkow and Tancredi used a combination of CT and PET scanning to detect structural abnormalities and measure regional cerebral blood flow and glucose metabolism in the brains of Cases 1 and 2, two other violent offenders, and four nonviolent age- and sex-matched volunteers. Figure 4.7 shows a sequence of PET images obtained from Case 2. The frontal cortex is dim and silent, especially on the left side. Both blood flow and glucose utilization were also reduced in the left temporal cortex, the gateway to limbic structures, like the amygdala, that facilitate day-to-day survival.

Like Case 2, Case 1 seemed genuinely unaware that he had done anything wrong. PET images from this patient also revealed decreases in glucose metabolism and cerebral blood flow in both temporal cortex and frontal cortex. Neither he nor Patient 1, however, showed any evidence of a structural defect on their CT scans that might account for the changes in function. The disparity warns of the danger in assuming a direct relationship between "normal" behavior and structural integrity, that a well-tended lawn and freshly painted trim are no guarantee that a quiet suburban home doesn't shelter a gruesomely dysfunctional family.

PET scans of the remaining pair of violent offenders, who did regret their hostile behavior afterward, showed a loss of neural activity only in temporal

Figure 4.7

The anatomy of human aggression: A PET study. Images of glucose utilization in the brain of a nonrepentant violent offender. The first nine images represent horizontal "slices" through the brain from top to bottom; the top of each image corresponds to the front of the brain. Active, glucose-consuming regions are white, less active regions are shades of gray, with the most active the lightest shade and the least active the darkest. Note the reduced activity in the frontal and left temporal regions. The last image shows a close-up image taken in a plane at right angles to the others.

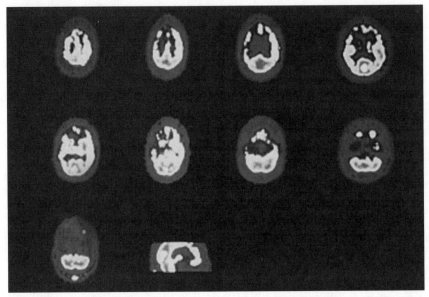

Images courtesy of Nora Volkow, M.D. Reprinted with permission from N. Volkow and L. Tancredi, Neural substrates of violent behavior: A preliminary study with positron emission tomography, *British Journal of Psychiatry 151* (1987): 668–673.

cortex. The frontal cortex of these patients, unlike that of their unrepentant colleagues, remained lit and occupied. But in contrast, their CT scans showed "marked cortical atrophy"; the cortex, while still functional, had simply withered away.

According to Tancredi, about three dozen additional subjects (both offenders and controls) have now been mapped with PET. "Overall," says Tancredi, "we see the same pattern—a disproportionate number of recidivistic violent offenders show a drop in left temporal metabolism. And when these offenders express little guilt or understanding of the moral significance of their actions, we also tend to see less activity in the left frontal cortex." He and Volkow suggest that this derangement in frontal cortical function may "facilitate violent behavior because of [an] inability to understand concepts such as right or wrong or to appraise the consequences of the violent act,"[67] a conclusion in keeping with the analytical and emotional reasoning skills attributed to the frontal cortex by both clinical and basic researchers. Individuals with temporal cortical dysfunction, on the other hand, while subject to "random outbursts of rage and violence, with very poor impulse control,"[67] may retain enough cortical input to appreciate the social significance of their behavior, even if only after the fact.

PET is expensive, even by radiology standards, and so the published report offers a look at only four subjects, only preliminary evidence. But like the circuits painstakingly mapped by animal anatomists, Volkow and Tancredi's functional brain maps suggest that human aggression is also the work of multiple communities, an interactive process that draws on the participation of neighborhoods where the head is on equal terms with the heart, as well as those where learning to stay alive is the primary goal.

Another functional imaging study—the work of Adrian Raine, of the University of Southern California—confirms the importance of the frontal cortex in this network, as well as illustrating how PET and SPECT can be combined with behavioral testing to track regional dysfunction in action. In this 1993 report, Raine used PET to analyze regional brain metabolism in twenty-two murderers and twenty-two nonviolent control subjects while they performed a simple task (pushing a button to register the appearance of the digit 0 in a series of letters and numbers) developed by psychologists to assess frontal lobe function.[68] Glucose utilization was significantly lower in the prefrontal cortex and orbitofrontal cortex of the murderers during this test, a deficit Raine hypothesized "could result in a loss of inhibition normally exerted by the frontal cortex on . . . older subcortical structures which are thought to play a role in facilitating aggression."

One exception surfaced: a serial killer with at least forty-five victims, who, unlike the other murderers, had a near-normal glucose utilization pattern. Raine speculated that in this subject, the relative absence of prefrontal dysfunction might be "consistent with the planned, careful execution of the murders in contrast to the more impulsive acts of the one-time murderer,"[68] suggestive evidence that a robust frontal cortex does not necessarily spell moral integrity.

PERPETRATORS AREN'T the only ones with fascinating changes on brain scans. In fact, MRI provides chilling evidence that violence can leave an indelible mark on the brains of victims as well. A research team headed by J. Douglas Bremner, of the National Center for Post-Traumatic Stress Disorder, Yale University School of Medicine, and the West Haven Veterans Affairs Medical Center, already suspected that this was the case, given reports of memory loss in combat veterans and prisoners of war dating back to World War I. Their own work had demonstrated that traumatized veterans, as well as adult survivors of physical and sexual abuse during childhood, posted significant deficits in short-term memory, especially for the recall of verbal information.[69] But even they were taken aback by the results of MRI scanning in a group of twenty-six veterans with PTSD and twenty-two carefully matched control subjects.[70]

The patients with PTSD had clear evidence of a structural defect that could explain their memory problems; on the right side of the brain, they had an average 8 percent reduction in the volume of a brain region originally assigned to the limbic lobe by Broca—the hippocampus. Even more compelling, scores and scans were correlated: the greater the memory deficit, the greater the reduction in hippocampal volume.[71]

Part of the limbic system for over a century, today the role of the hippocampus in emotion takes a back seat to its importance in the management of short-term memory. Damage to the hippocampus (e.g., during epilepsy surgery) can cause short-term memory deficits similar to those reported by Bremner and his colleagues in trauma victims.[72] And patients with Cushing's disease, an endocrine disorder caused by hyperactivity of the adrenal gland, exhibit deficits in short-term memory that also correspond to MRI-documented reductions in hippocampal volume thought to be caused by prolonged exposure to toxic levels of adrenal stress hormones.[73]

Bremner's results aren't confined to combat veterans. Both his group in West Haven and a second team, headed by Murray Stein, of the University of California at San Diego, have observed similar reductions in hippocampal volume among men and women with a history of chronic physical or sexual abuse.[74-76] Whether trauma is the result of the socially sanctioned violence of combat action, a criminal act, or parental rage, brain imaging demonstrates that one reason some victims cannot simply "get over it" is that the violent act has literally been seared into their brains. John Krystal comments that the structural changes observed by Bremner and others are "just one sign that what the suffering victims have carried around for such a long time is part of something that's very real, as real as breaking a leg." He adds, "Trauma not only leaves physical marks we can observe on the outside—imaging studies suggest that it also leaves physical scars on the inside."

THE CONTINUING REVOLUTION in brain imaging has the potential to tell us even more about the anatomical origins and consequences of violence. For example, by taking advantage of the unique magnetic properties of deoxyhemoglobin, radiologists have developed a functional MRI procedure for charting local cerebral blood flow during neural activity.[65,77] Functional MRI, known to the scientific community simply as fMRI, combines the spatial resolution of traditional MRI with the action-oriented focus of techniques like PET and SPECT. Another new technique, magnetoencephalography (MEG), monitors brain activity by tracking the changes in regional magnetic fields that occur in active neurons—and in milliseconds rather than seconds.[66,78] That's fast enough to keep up with a problem-solving brain. Using fMRI and MEG, neuroscientists can not only differentiate activity levels across brain neighborhoods, they can actually track the sequential activation of a chain of brain regions as they carry out a perceptual or motor task and regional shifts in activity as the brain switches from one task to another—visual evidence of functional networks in action.

Techniques like fMRI dispense with the need for radioactive tracers, meaning they're less invasive, as well as less expensive, than PET, and easily repeated. As a result, they have the potential to add a time dimension to neuroanatomy, allowing researchers to follow neural activity over the long term, in the context of events in the outside world. The impact of a ten-year prison sentence or a novel rehabilitation program, a return to drug abuse or a successful effort to overcome addiction, a gunshot wound, a reconciliation can all be documented to create a biological history of the rise and decay of neural neighborhoods—and a potential answer to the "time-space problem." By looking through the window over and over again, brain researchers may soon be able to see temporal as well as spatial relationships, a neuroanatomical chronicle of the evolving relationship between brain and environment.

ON THE STREETS of Newark, the environment had inspired a progressive devolution in courtesy and motivation. The official street temperature was now 101 degrees. The street wavered and shimmered, baking cars and people to a standstill.

I finished talking to Siegel later than I expected and now had only thirty minutes to get back to Penn Station. My cab, however, was nowhere to be seen. Trapped in the air-conditioned lobby, I paced and fretted while the security guard idly flipped his radio from post to post, scanning the state of affairs across the campus. A heavy-set woman sat down to rest from mopping the floor. She propped her sweaty elbow on the guard's desk.

"Maybe you should think about taking the bus, honey," she suggested.

Dubious, I thought. I watched two buses drift past the entrance gate, followed by a trickle of overheated cars. Still no cab.

"Do you think it's safe for me to take this bus?"

She and the security guard nodded slowly. "This time of day, around here—you'll be okay. It's only about ten blocks. You just ask for the train station when you get on. Just tell him where you're going."

Suddenly the guard's radio crackled with excitement. "We got an update on that car theft up in this lot. There's two people here shot and killed."

OVER THE NEXT two weeks, hundreds of people would die from this heat, many sealed in explosively overheated apartments by their terror of street crime. And in urban medical schools, in Newark and Philadelphia and elsewhere, neuroanatomists, sealed in their guarded and supercooled buildings, would continue to chip away at the biological architecture of that violence, searching for answers. Despite the guards, it is a dangerous profession. The formaldehyde used to preserve tissue sears lungs and eyes. Diaminobenzidene, the chemical road sign that marks fiber pathways, is a potent carcinogen, as are the toluene, alcohol, and benzene in the staining solutions. Radioactivity, carbon steel knives that can slice a finger to the bone, teeth and claws are hazards of daily living. To carry out an MRI scan, you must face a cold, bare room and may confront a mind more fascinated by the clang of the magnet than the anxious beating of your heart. And, at the end of the day, when you finally step out of your laboratory, you may be unlucky enough to find the violence of the real world rolled out for you at the door like a carpet, courtesy of a quirk of geography, a confluence of highways.

Five

BAD CHEMISTRY

In 1995, the U.S. Postal Service distributed over 180 billion letters, postcards, bills, income tax returns, magazines, grocery store flyers, and packages.[1] That's a drop in a bucket, however, compared to the central nervous system, charged with delivering billions of messages to postsynaptic mailboxes every day. The networks of neural highways that link brain regions are the delivery routes of this extraordinary postal system; its medium, a language based on chemistry, rather than words.

Neuroanatomy has disappointed those hoping for an easy answer to the problem of human violence. But perhaps if the answers aren't in the delivery system, they're encoded in the messages themselves. Given that we cannot confine aggression to a single brain region, can we instead measure its impact in units of a single crucial brain chemical?

We want simple answers to complex behaviors, but the message written in the chemical hieroglyphics of transmitters and receptors—like the multilayered architecture of emotional pathways—is far from simple. If neuroanatomy tells us that we must think in terms of circuits, not centers, neurochemistry teaches us that we must learn to think in terms of conversations, not commands.

IN THE EARLY DAYS of neurochemistry, researchers believed that neuronal communication was as straightforward as mailing a letter. The biological equivalent of a pinball machine, the presynaptic cell shot a transmitter message across the synapse, reloaded, and fired again. On the other side of the synaptic cleft, transmitter messages hit or missed their target receptors, bounced off, and fell prey to waiting enzymes. The receiving cell summed the receptor hits, added and subtracted incoming messages, and relayed the conclusion to the next cell. The release-bind-terminate sequence ran again and again as long as the presynaptic cell had something to say, sometimes speeding up or slowing down, but never interrupted by back talk from the postsynaptic partner.

But for people who still answer their mail, a letter is the beginning of a dialogue. And today neurochemists recognize that one of the virtues of neurons is that they are articulate and dedicated correspondents, as eager to reply as they are to receive. In a language crafted from a rich vocabulary of transmitters, clans of related receptors, extended families of second messengers, teams of gene-switching transcription factors, brain cells debate and maneuver, always tailoring their responses to match the changing volume and timbre of the discussion. Far from silent, the postsynaptic cell doesn't just passively soak up transmitted messages; it responds. Neurotransmitters damp down or fine-tune events on the presynaptic as well as postsynaptic side of the synapse, regulating such features as firing rate, transmitter synthesis, and receptor number. These feedback mechanisms balance the relative intensity of intraneuronal communication to maintain a dynamic equilibrium.

Neuropharmacology has been critical to understanding this reciprocity. For example, administering an agonist (a compound that mimics the action of a naturally occurring transmitter) enhances presynaptic activity, initiating compensatory mechanisms on both sides of the synapse. On the receiving end, the postsynaptic cell uncouples and sequesters receptors from second messengers, limiting the impact of the extra input, or even deletes receptors to restore the balance. On the signaling side, the drug activates so-called autoreceptors on the surface of the presynaptic cell that monitor transmitter release and damp down the firing rate accordingly. The net effect is a reduction in activity and responsiveness that prevents the pharmacologic impostor from overloading the synaptic message delivery system.

Behavior is also a dialogue—between past and present, experience and physiology. "The classic way of thinking about the relationship between biology and social behavior, especially aggression, is causal," notes Tufts's Klaus Miczek. "First, you have a biological event, and that leads to a certain kind of behavior. But there's a second side to the coin, which is that behavior itself causes massive changes in the way neurotransmitters are made, how they act on receptors, and ultimately, which nervous system genes are expressed. It's a way of looking at the neurobiology of social confrontation upside down."

Reciprocity, Miczek believes, is as critical to understanding the neurochemistry of behavior as it is to appreciating the interpersonal dynamics of intraneuronal communication: "Instead of only looking at biology as the cause of behavior, we also need to consider the reverse—that being the aggressor or being the victim of aggression is the event that sets neurobiological processes in motion."

He explains that the brain is a painstaking historian, as well as an ardent correspondent, carefully documenting a record of our social successes and failures in the language of neurochemistry. Winning a conflict provokes one pattern of neurochemical changes; losing, another. The cumulative effect of such changes over time molds and shapes each nervous system to meet the specific demands of its own unique environment, to ensure day-to-day survival in that environment, and to preserve and foster key social bonds, while keeping enemies at bay.

Allowing the brain to construct behavior gradually over time can backfire, however. When the demands placed on the individual are unusual or extreme, the drastic accommodations engineered, recorded, and perpetrated by neurochemistry can lead to a progressively less accurate calculation of threat rather than a progressive increase in competency. As the gap between perception and reality grows steadily wider, the risk that adaptive responses will spiral into destructive violence escalates.

The story of the relationship between neurochemistry and aggression is a story of things that can be readily observed and measured: receptor numbers, transmitter levels, attack bites. But it is also a story of factors that are more elusive and processes that have only recently begun to reveal their elegant complexity. A story with two sides, it is ultimately a matter of timing, a question of balance.

Sounding the Alarm

The kitchen table is smothered in newspapers; the window ledge overflows with bills; an avalanche of catalogues has just buried the bedside table. When the mail carrier is dumping hundreds of letters and packages on your doorstep every year, a triage strategy is essential—unless you don't mind losing an electric bill, or tossing out the tax refund check along with the grocery store circulars.

Separating the junk mail from the priority mail is even more important when billions of messages flood the system. The built-in perceptual limits of the nervous system are the initial step in the brain's message triage strategy, the first effort to limit the overwhelming torrent of sensation. Once inside the brain, chemistry takes over. Principal responsibility for the task of culling information is assigned to one family of chemically related neurotransmitters. Known collectively as the *monoamines*—meaning that they sport a single nitrogen-containing appendage known as an "amine group"—these chemical gatekeepers prioritize and categorize incoming sensory data as they hit the neural delivery network. Together, they make sense of an otherwise chaotic environment, and over time they analyze the pattern of our experiences to construct a cognitive framework that characterizes the quality and safety of the surrounding world.

The discovery of these interpretative pathways marked the beginning of a fruitful union between neuroanatomy and neurochemistry. In the 1960s, researchers at the Karolinska Institute in Stockholm discovered that heat and formaldehyde expose an inherent fluorescence in the chemical structure of amine transmitters;[2,3] by baking sections of brain tissue in formaldehyde vapor, the anatomists learned that they could draw amine-containing neurons out of the shadows. Axons as well as cell bodies phosphoresced against the black background of a microscope stage in Day-Glo shades of green and yellow like a neurochemical velvet painting, permitting researchers to locate the origin of monoamine projections and trace their extensive connections.[4]

Histofluorescence—the technical name for the new method of color-coding monoamine neurons—meant that for the first time, neurobiologists could map specific delivery routes served by a particular neurotransmitter.[4] Combined with two other critical advances—*immunocytochemistry,* which locates chemically defined pathways by using fluorescent antibodies to tag transmitters or transmitter synthesizing enzymes, and *receptor autoradiography,* which tags receptors, rather than transmitters, with a radioactively labeled agent that acts as a light source—"chemical neuroanatomy" added a new dimension to functional localization—topographic clues to the physiological significance of particular transmitters.

For example, the distribution of monoamine projections and monoamine receptors suggested that these pathways are intended to reach a wide audience. From base camps deep in the brain stem and midbrain, they dispatch long axons that extend to the boundaries of the cortex in one direction—shooting off side branches to the amygdala, hippocampus, and other subcortical structures along the way—and reach deep into the cerebellum and spinal cord in the other. Nearly all regions of the brain receive at least one type of monoaminergic input; some, such as the cortex, accept inputs from all. The result is a dense web of overlapping monoamine fibers, a physical and neurochemical filter between the brain and the outer world.[4-6]

The members of the monoamine transmitter family have more in common than their collective good looks. Each corresponds with a cohort of structurally similar receptors; each receptor, in turn, activates a preferred "second messenger" protein. The second messenger translates receptor binding into action: changes in the excitability of the postsynaptic cell, activation of additional intracellular proteins, initiation of the cascade of reactions that ultimately triggers gene-switching transcription factors.[4,7] For example, the monoamine norepinephrine—the transmitter of record in the sympathetic nervous system that helps to coordinate the fight-or-flight response—delivers messages to one of four types of receptors, grouped into two "superfamilies": the so-called alpha receptors and beta receptors.[4] Receptor diversity increases flexibility, allowing norepinephrine access to multiple second messengers. In addition, it fosters reciprocity. Released norepinephrine that latches onto presynaptic alpha autoreceptors advises the presynaptic cell when it's time to turn down the volume. This autoreceptor-driven feedback mechanism, coupled with compensatory actions on the other side of the synapse, works to keep the overall level of activity at an even keel. In contrast, events that jam the synaptic airwaves—such as surgical destruction of the norepinephrine neurons—push up the number of receptors and encourage them to hook up with all available second messengers to make maximal use of any remaining dribbles of transmitter. Thanks to these tuning mechanisms, monoamine pathways operate with a flexibility and responsiveness that can shape perception and behavior to match the vagaries of an ever-changing environment.

SOME LETTERS spell trouble before you've even peeled up a thumbnail-sized corner of the envelope—a return address from a law firm, January credit card bills, anything from the IRS. By the time you start to read, your heart is pounding, your face is flushed, and your stomach is churning.

An observer might say you'd "gotten your adrenaline up." A monoamine secreted by the adrenal gland, adrenaline, or *epinephrine,* is well known for its ability to crank up heart rate, respiration, and metabolism in the face of an emergency. But epinephrine is only half of the story. It's the sympathetic nervous system, manned by the chemical precursor of adrenaline, norepinephrine, that alerts the adrenal gland in the first place—and the heart, lungs, vasculature, stomach, and muscles as well—triggering the familiar symptoms of emotional arousal.[8]

The perception of threat, however, is a brain process. Alarming events, disturbing thoughts, even the apprehension that begins with the arrival of an ominous envelope, precipitate emotional reactions that wake up the sympathetic nervous system and trigger defensive responses. This neural aspect of fight or flight, like its peripheral counterpart engineered by the sympathetic nervous system, also relies on norepinephrine.

The prototypic monoamine transmitter, norepinephrine takes a little and goes a long way. From two elite cadres of neurons based deep in the brain stem, each comprising no more than twelve thousand cells (the *locus ceruleus,* or "blue nucleus"—so called because of its bluish-gray cast in human brain tissue),[4] forebrain-bound noradrenergic axons carpet target sites in the hypothalamus, hippocampus, amygdala, and especially, the cerebral cortex (Figure 5.1). Here they shoot up through the six layers of cortex like seedlings grasping for sunlight. Near the outer surface, the fibers split at right angles to the stems, each branch chaining across the cortex to form a neurochemical Internet linking

Figure 5.1

Noradrenergic pathways in the human brain.

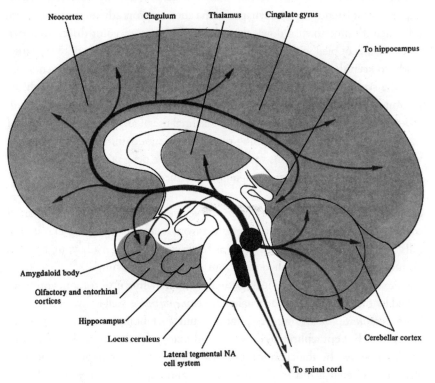

Reprinted with permission from L. Heimer, *The Human Brain and Spinal Cord* (New York: Springer-Verlag, 1983). NA = norepinephrine.

neuron to neuron.[6] Thanks to this elaborate fiber network, an event that activates the locus ceruleus is capable of simultaneously ringing up the entire cortex—just what you'd want in an alarm system.

THE INCH-THICK packet of paper had to be posted to a West Coast colleague that day, and the mail carrier had just rolled around the corner. But I didn't panic; I figured I could easily sprint out to the main road and the local mailbox before his final pickup.

I had no way of knowing that I'd be thwarted an hour later, not by my own dawdling, but by postal terrorism. The local mailbox, I learned, no longer accepts stamped mail over sixteen ounces. The new regulation is one of several designed to help postal inspectors in the delicate task of monitoring incoming mail for letter and package bombs or other potentially hazardous items.[9]

The noradrenergic pathways based in the locus ceruleus form a neuronal postal inspection system, assigned the defensive task of scanning the torrent of messages coursing into the nervous system and red-flagging those critical to survival. During sleep, or while you're going about the humdrum activities of daily living, locus ceruleus neurons mark the uneventful passage of time with a steady background rhythm.[10,11] An unexpected flicker, a sudden noise, triggers a warning burst of impulses; and events that cause pain or stress—cold, electrical shock, restraint, physical injury—sound a full-fledged alarm. The previously sedate neurons fire a barrage of action potentials, blazing away at rates two-, three-, even tenfold greater than baseline.[4,10,12,13] Norepinephrine liberated by all this activity floods synapses in the cortical and limbic target regions, transmitter synthesis escalates to keep up with demand,[11] and breakdown enzymes work overtime to keep the synapse clear of excess norepinephrine.

The alarm response also trickles backward to the neuronal genome, where it switches on c-*fos* and other immediate early genes that code for DNA-manipulating transcription factors.[14] These alarm-activated transcription factors regulate future expression of receptors, transmitter-synthesizing enzymes, and other proteins crucial to synaptic structure and function. As a result, untoward events do not just upset neuronal communication in the present, but can remodel the brain's early warning system, fine-tuning sensitivity to match the relative safety of the surrounding environment.

The induction of transcription factors is only the beginning, the first retort in the give and take between the brain and the outside world. In contrast to the noisy warning triggered by acute stress, repeated exposure to danger, like the ongoing threat of a Unabomber at large, prompts a clandestine overhaul of the noradrenergic alarm system. On the surface, the circuit appears to be winding down and losing interest. Acute responses—firing rates, transmitter release—gradually slow and stabilize, and levels of norepinephrine drift back to baseline as stores are replenished.[11] But underneath the calm demeanor, noradrenergic neurons have actually shifted into overdrive. Manufacturing capacity is quietly enhanced, as levels of norepinephrine-synthesizing enzymes are ratcheted upward. Baseline firing rates are reset to a new, higher value, while the number of postsynaptic receptors—perhaps in an attempt to keep a lid on the system—declines.[11]

The result is a security system that tolerates familiar hazards but responds more vigorously than ever to the unknown. Should a new and unfamiliar danger pop up, the edgy noradrenergic neurons explode in a burst of activity, as if the initial stress response had taken a martial arts course. Persistent threat has kindled, or sensitized, norepinephrine-containing neurons until they overreact to every provocation. A recent study conducted by Michael Zigmond and his colleagues at the University of Pittsburgh illustrates the consequences of this stress-induced sensitization of the noradrenergic alarm system.[15] The team confined one group of rats to a chilly room for three weeks, while a second control group was housed in a low-stress, ambient temperature environment.

When both groups were tested for their subsequent reaction to a mild electric shock, the postshock surge in norepinephrine release within the cerebral cortex was more than two-and-a-half times greater in the rats previously stressed by cold exposure.

Confronted with a clear and present danger, the central alarm system based in the locus ceruleus—and its peripheral counterpart, the sympathetic nervous system—act in concert to mobilize the internal resources needed to mount a rapid and effective defense. Hahnemann University researcher Gary Aston-Jones, one of the first to link the noradrenergic pathways to arousal and vigilance, has proposed that the sympathetic component of this brain-body network "prepares the animal physically for adaptive responses . . . to urgent stimuli," while centrally, parallel activation of the noradrenergic neurons in the locus ceruleus "prepares the animal cognitively" for these responses.[12] The locus ceruleus, in other words, "serves as the cognitive limb of the global sympathetic nervous system."[12] It is the thinking element of an integrated response network dedicated to personal safety.

In the short term, alarm responses masterminded by norepinephrine neurons keep rats safe from predators and electric shocks, humans safe from potential assailants. In a chronically chaotic environment, long-term changes in noradrenergic function continue to mold responses to ensure survival. But as anyone who's ever been rudely awakened by a trigger-happy car alarm or a watchdog that takes its job too seriously will agree, overzealous reactions also have a downside. Under the right circumstances, Michael Zigmond points out, the behavioral sensitization of the global sympathetic nervous system that follows in the wake of repeated threat may well represent "too much of a good thing," harming, rather than protecting, the endangered individual.[11]

TRAUMA REPRESENTS the far side of the stress spectrum. As a result, it's not surprising that a few moments on the receiving end of violence can have profound and long-lasting effects on norepinephrine, the brain's survival transmitter, or that PTSD, the ultimate survival response, should be associated with abnormalities in noradrenergic function. Patients with PTSD, for example, post significantly higher urinary levels of norepinephrine than normal volunteers or even patients suffering from other types of psychiatric disorders, such as schizophrenia.[16,17] And changes in noradrenergic function aren't limited to transmitter levels. Stress researchers have also charted decreases in noradrenergic alpha receptors, decreases in the productivity of postsynaptic second messengers, and a decline in the activity of the breakdown enzyme monoamine oxidase.[18] Thanks to reciprocity and feedback, trauma does more than simply turn up or turn down the activity of noradrenergic neurons; it disrupts function throughout the brain's alarm system.

A cardiologist tracking down the cause of chest pain in an overweight, middle-aged man typically begins the search with a series of baseline measurements: blood pressure, pulse rate, heart sounds, perhaps even an electrocardiogram. These values provide the doctor with a snapshot of how well this cardiovascular system is carrying out the normal activities of daily living and how it differs from the systems of healthy adults who aren't in pain. But that's only one way of looking at the problem. By monitoring the same signs during and after exercise, a diagnostic enterprise commonly known as a stress test, the cardiologist can evaluate how well the heart responds to a challenge. Together, baseline observations and challenge responses offer the physician a more complete picture of the state of the heart than either diagnostic strategy could ever do alone.

Similarly, measuring transmitter levels and receptor densities is only one way of defining the relationship between behavior and neurochemistry, a snapshot approach not unlike that provided by structural brain imaging. To examine how the nervous system responds under pressure, clinical investigators need a technique that can also assay function. They have turned to pharmacology, using drugs keyed to particular neurotransmitter systems or specific receptor subtypes as tools to challenge that system. The behavioral response to drug provocation adds a functional dimension to neurochemistry, unmasking covert alterations in neuronal function, backing up the associations suggested by observational studies, and further clarifying the behavioral impact of specific changes in each step of the transmission process.

ONLY FIVE MINUTES after the experiment began, the subject, a Vietnam veteran participating in a study of stress-induced sensitization at the National Center for Post-Traumatic Stress Disorder's Veterans Affairs Medical Center in West Haven Connecticut, was so distraught that he warned an attending nurse to get out of the way in case he needed to run for cover: "I feel like I'm picking up dead bodies the centrifuge sounds like a helicopter . . . A chopper is shooting at us, we're trying to shoot back at it! One of the guys' head is shot off! Brains are coming at me! I smell burnt flesh . . . I feel scared, I can't hear what's going on . . ."[19]

The veteran's vivid flashback is an example of how neuropharmacology can expose the functional status of neurotransmitter systems underlying emotional behavior. The drug used in this neurochemical stress test was yohimbine, a tropical plant derivative that activates noradrenergic neurons by jamming the presynaptic autoreceptors that moderate firing rates.[20] More norepinephrine reaches postsynaptic receptors mimicking the neural alarm tripped by a real threat. In this study, twenty veterans undergoing treatment for PTSD, as well as eighteen healthy men recruited from the local community, volunteered to challenge their noradrenergic response systems and were randomly assigned to receive yohimbine or an inert placebo solution.

The drug pushed up blood pressure and activated the norepinephrine system even in the healthy subjects. But these reactions were mere pharmacologic inconveniences compared to the physiological and behavioral firestorm that yohimbine provoked in the PTSD patients. Fourteen of the nineteen patients experienced severe panic attacks; eight also reported vivid flashbacks. Other symptoms—intrusive thoughts, emotional numbness, guilt, anger, grief—rose to a crescendo. Blood pressure and heart rate soared, particularly in the patients who developed panic attacks.

Transmitter and receptor measurements linked PTSD to systemic changes in norepinephrine pathways. Challenge studies, like the West Haven yohimbine study, confirmed that this link has functional significance. By disconnecting the feedback mechanisms working to keep the sensitized alarm system in check, the drug uncovered an underlying instability, a "dysregulation" of the sympathetic nervous system that precipitates "exaggerated behavioral responses more appropriate to emergency situations."[20]

PAINFUL TO REMEMBER, trauma cannot be forgotten. Symptoms like flashbacks and intrusive memories are a cruel perversion of emotional memory, the dark side of a neural process that can also preserve every beautiful detail of the birth of a child. So are the intense feelings of fear, rage, and panic that trauma survivors experience when they're confronted with the most trivial similarities between events in the present and the original act of violence. Such unwarranted generalization to what researchers call "trauma reminders," like my terrified response to a lost and befuddled Londoner, trap victims of violence in a twilight world where their bodies "continue to react to certain physical and emotional stimuli as if there were a continuing sense of annihilation."[21]

"Emotionally charged events," says Larry Cahill, an expert on learning and memory at the University of California, Irvine, "alter the storage of emotionally relevant experiences." But exactly how does violence distort memory? Why does fear lock in details and how does it suspend them so close to the surface of consciousness? The answer, neurobiologists believe, lies in the intersection between chemistry and anatomy, in the effects of a noradrenergically mediated state of emergency on one of the key nodes in the limbic network dedicated to emotion, the amygdala.

The startle response that makes you jump and blink at a sudden sound or flash of light—the fastest known reflex in the human nervous system—has been our window on the mystery of how the brain links memory and emotion. Like blood pressure, heart rate, and activity level, the startle reflex can be conditioned to a benign stimulus.[22] With the same training technique Joseph LeDoux employed in mapping the neural circuitry of fear, a rat is trained to associate a light and an electric foot shock. Then the animal is transferred to a specially floored cage that measures the force of his landing when he jumps, startled, at a sudden blast of noise piped into the cage. The test cage is dark at first, and the magnitude of the startle response no greater than that of an untrained rat. But when a light is turned on in the cage, the same unexpected blare nearly sends the conditioned rat through the roof.

Fear conditioning also potentiates the startle reflex in humans. For example, researchers at the National Center for Post-Traumatic Stress Disorder reported that eye blink responses to a blast of white noise were significantly higher in Desert Storm veterans suffering from PTSD and in women who were survivors of childhood abuse, compared to nontraumatized control subjects.[23,24]

In rats, there's an easy way to disconnect this link between trauma and startle, and it doesn't require years of therapy; simply remove the amygdala. Remember that this structure, as shown so elegantly by Joseph LeDoux and suggested decades earlier by the results of Klüver and Bucy, integrates sensory input and cortical opinion to guide appropriate response strategies; in the case of fear-potentiated startle, it alerts brain stem centers that direct the rat to jump higher. In studies of the anatomical basis of fear-potentiated startle, Michael Davis at Yale University has shown that surgical removal of the amygdala normalizes the conditioned startle response, further evidence that eliminating the executive function of the amygdala can neutralize the power of fearful objects or events.[22]

A tributary of the norepinephrine pathway originating in the locus ceruleus feeds into the amygdala. But the amygdala is also the end of the line for a diffuse community of norepinephrine-containing neurons that lie outside the locus ceruleus, in surrounding brain stem centers that coordinate body functions like breathing and blood pressure and that receive information about the state of these functions from the peripheral nerves.[4,5] Thanks to this second line of communication, Cahill's colleague James McGaugh suggests that norepinephrine is perfectly placed to exert an influential role on the amygdala's emotional valuation of sensory information.[25]

When danger strikes, news of the physical consequences that follow environmental activation of the sympathetic nervous system travels quickly to the brain stem norepinephrine neurons, as well as the locus ceruleus—and they immediately alert the amygdala. It's this signal—that something in the environment has made the heart beat faster—that prompts the amygdala to slap a warning label on data from the incoming sensory stream. When the entire experience is transferred to memory, the labels stay put, branding details of the traumatic event into the memories. Like loose threads on an old sweater, these amygdala-enhanced details catch on every outstanding resemblance to the original trauma, pulling fragments of the painful memory up to conscious awareness over and over again.

Whether we're neurochemists or judges, therapists, policymakers, or just concerned citizens, we'd better not forget the lessons of personal history. The consequences of violence, the terror of seconds suspended between death and life, do not vanish—they have been chemically burned into our memories. Impossible to erase, they are all too easy to reawaken.

Going Postal

Homemade pipe bombs aren't the only hazards postal workers face. In fact, sorting and delivering the mail can be a very risky business. About twenty-seven hundred mail carriers are attacked by unruly dogs every year.[26] Thousands more are assaulted by unruly customers. And over the last decade, shooting incidents at post offices around the country, many carried out by disgruntled employees, have claimed the lives of thirty-five people, meaning that murder is second only to traffic accidents as a cause of on-the-job death for post office employees.[27]

Chemical messengers charged with the delivery of life-or-death information are also at risk for occupational aggression. Anger, after all, like fear, is a reaction to threat, a physiological call to arms that revs up the heart, coils muscles, and sends blood pressure soaring. Given their role in driving these responses and gauging the safety of the outside world, it's not surprising that the monoamine transmitters labeled by Burr Eichelman, former chair of psychiatry at Philadelphia's Temple University School of Medicine, the "aggressive monoamines"[28] have long been implicated on the fight, as well as the flight, side of the response equation.

Research on the neurochemical origins of aggression, in fact, began with norepinephrine, the monoamine system that sounds the alarm in response to perceived danger. Studies reaching back to the early days of aggression research, to the old animal models that used foot shock to provoke fighting, first showed that stress and pain can precipitate aggressive responses as well as potentiate the startle reflex, and that stress-induced fighting, like fear potentiation, is associated with significant changes in brain norepinephrine systems. For example, in a study conducted more than two decades ago, Eichelman and his associates, then at NIMH, stressed rats by physically restraining them for two hours every day over a month-long interval.[29] At the end of this ordeal, pairs of animals faced off on an electrified grid. Compared to similar pairs that had remained undisturbed in their home cages, the immobilized rats clashed nearly three times more frequently, and their brains produced twice as much norepinephrine. Even after a four-week recovery period, fighting and norepinephrine synthesis remained higher in the rats that had been restrained. These results, subsequently replicated in mice and monkeys, were the first evidence that threat could alter noradrenergic function or that stress-induced sensitization—the persistent neurochemical and behavioral activation that follows trauma—could be expressed as defensive aggression, as well as terror.

Alan Siegel was one of the investigators who followed up on Burr Eichelman's groundbreaking observations that aggression—particularly defensive, or protective, aggression—goes hand in hand with activation of the brain's norad-renergic alarm system. Working from his painstakingly detailed maps of the neural circuitry mediating defensive aggression, Siegel and coworker Jeannette Barrett implanted guide tubes in a cat's medial and anterior hypothalamus, injected a minute amount of norepinephrine into the anterior site, and then measured the amount of current needed at the medial site to trigger the hissing, spitting, and fur-raising signs of affective defense.[30,31] Norepinephrine lowered the attack threshold; the higher the amount of catecholamine injected, the less current needed to elicit the defensive behavior.

However, facile attempts to "explain" aggression as the result of "too much" norepinephrine do not begin to reflect the true complexity of the relationship between this transmitter and the nuances of aggressive behavior, any more than norepinephrine levels alone can fully account for the exaggerated startle responses, flashbacks, and panic attacks of PTSD. Reciprocity is one reason. Neurobiology examined "right side up"—the classical, linear way—does indeed show that events that enhance norepinephrine function, such as Siegel's local microinjections, exacerbate aggressive behavior. But conversely, neurochemistry examined "upside down"—Klaus Miczek's way—reveals that a hostile confrontation itself can increase both the overall level of brain norepinephrine and the rate at which the transmitter is made and utilized.[32] On a neuronal level, the stress of conflict reprograms the entire noradrenergic synapse as effectively as foot shock, restraint, cold exposure, or the horror of combat. Production capacity goes up, but so does norepinephrine release; beta receptor density, on the other hand, goes down.[33]

When drug challenge studies are carried out in an ethologically relevant social setting, investigators can analyze the complex changes in each step of the aggressive encounter, using a combination of pharmacology and behavioral biology to dissect the chemical mechanisms of attack and defense. For example, such studies have demonstrated that the administration of propranolol, an antihypertensive drug that blocks norepinephrine's action at beta receptors, specifically reduces attack behavior, leaving defensive and escape behaviors untouched.[34]

Real-time analysis of the aggressive encounter suggests that the secret of propranolol's antiaggressive effect lies in its power to alter the characteristic sequence of conflict behaviors. Using a computer-aided video monitoring system to follow and record the moment-to-moment behavior of both combatants, Miczek and long-time associate Tom Sopko have charted the effect of propranolol on the attack bursts they call epochs, the rapid series of threatening gestures that include bites, body blocks, or pursuits and that account for more than 90 percent of all aggressive activity in a typical mouse fight.[35] The videotapes show that a resident mouse on beta blockers doesn't just attack less often; he also attacks differently. Instead of pounding his opponent, he seems content with a single half-hearted thrust or snap. The higher the dose, the longer the interval between aggressive epochs and the fewer aggressive moves per epoch.[35]

Norepinephrine, the results imply, is more than a switch; it regulates a sophisticated neural mechanism that determines not simply whether behavior will occur but what it will look like.

TRACKING THE NORADRENERGIC correlates of violence in humans confirms that the relationship between monoamines and behavior is more complex than a one-to-one correspondence between transmitter levels and criminal records. A study carried out in 1979 by a team at NIMH that included Fred Goodwin measured norepinephrine breakdown products in cerebrospinal fluid (CSF) drawn from military personnel remanded for psychiatric treatment and normal volunteers; then the researchers correlated these values with the subjects' lifetime history of aggressive behavior. Among the patients, higher norepinephrine levels were associated with higher aggression scores.[36] Another series of studies found a similar correlation between urinary norepinephrine levels and aggression in violent offenders confined to a maximum security psychiatric hospital.[37-39] Conversely, over a dozen published studies, including two placebo-controlled clinical trials, have demonstrated that drugs like propranolol that block norepinephrine action can reduce aggressive behavior in humans who have suffered a stroke, head injury, or other neurologic insult, as well as in rats.[40-43]

Other reports, however, have observed *decreases* in CSF and urinary norepinephrine levels among violent offenders.[44] Drug studies have hinted that propranolol is not always effective in curbing aggressive behavior. The dissension highlights the critical importance of timing and personal history, for the differences may well represent study-to-study variation in the early life experiences, frequency of violent encounters, or current environment of the participants. More important, the discrepancy between human neurochemical studies may reflect our relative ignorance of the typology of human violence; the differences, signs of a differential responsiveness that distinguishes two important types of violent behavior—one characteristic of people who feel unnecessarily threatened and the other of people who don't seem to feel anything at all.

SHIFTING INTO HYPERDRIVE isn't the only way to miscalculate threat. As my father observed, a machine that isn't switched on isn't going to get the job done at all. In contrast to violent individuals who overreact, others seem to be asleep when it comes to recognizing society's limits on aggressive behavior. Oblivious to parental intervention, they rapidly graduate to flaunting school authorities, neighbors, the local police. By adulthood, they've developed a revolving door relationship with the criminal justice system. Released after serving time for a bar brawl, they're back in prison a few months later for shooting a clerk during a convenience store robbery; warned to stop harassing a neighbor, they pick another fight. Their girlfriend gets a protection order, and they're back on her doorstep as if nothing happened.

An expert on criminal behavior, like forensic psychiatrist Robert Simon, would call the callous belligerence of such offenders antisocial. When their behavior degenerates from bad to outrageous, we in the community often call them psychopaths. The first type of excessive aggression to be formally defined by the psychiatric community, *psychopathy*, later called *sociopathy*, and known in the current guidebook, the fourth edition of the *Diagnostic and Statistical Manual of Mental Disorders* (DSM-IV), by the less lurid term *antisocial personality disorder*,[45] is, in Simon's words, "narcissism . . . malignantly transformed into living, breathing evil."[46] Its essential feature is a psychological dead space where human compassion ought to be found. Incapable of remorse and slow to learn from their mistakes, the antisocial flaunt an open disregard for the rights of others, viewing the world as a "giant dispensing machine from which they obtain goodies without giving up any coins."[46]

Simon sees the lawlessness of the antisocial personality as a face-off between this insensitivity and "stimulus hunger"—a burning need for constant stimulation, the struggle of a person who feels nothing to feel something. At the neural level, antisocial detachment may be the consequence of a sluggish alarm system, a sympathetic nervous system that seems to have forgotten its responsibility for responding to urgent events and a brain that is far too slow in detecting them.

University of Southern California researcher Adrian Raine has been studying the relationship between sympathetic nervous system activity and antisocial behavior for more than a decade and has found that antisocial and psychopathic individuals, for all their hostility and acting out, are actually slower to respond emotionally than the rest of us.[47–49] Their skin conductance ("lie detector") responses to emotionally significant visual images (such as an accident) or unpleasant stimuli (such as uncomfortably loud sounds) are lower than those of socially competent individuals. In contrast to the exaggerated startle responses of PTSD victims, antisocial individuals are unusually slow to connect stimulus and response. Raine calls them "poor conditioners."

Even when they're just sitting around, antisocial individuals are more low-key than the average person—long before they're old enough to be tried as adults.[49–51] Thirteen separate studies have found consistent and significant correlations between resting heart rate and antisocial behavior among children and adolescents.[47] In fact, the magnitude of the correlation (statisticians call it the "effect size"), as Raine points out in a recent review, between heart rate and antisocial behavior in adolescent troublemakers is stronger than the relationship between age and height in adolescent girls.[47]

Antisocial behavior is defined broadly in these studies; it includes verbally abusing parents or smashing school windows as well as assault. But when subjects are broken down into violent and nonviolent offenders, the difference in resting heart rate, compared to nonviolent control subjects, is even greater. In a recent reanalysis of data from a 1990 study linking decreases in heart rate and the development of antisocial behavior among a group of 101 English schoolboys, Raine found the lowest heart rates in the boys ultimately convicted of violent offenses.[50] A much larger study, of more than eleven hundred children on the Pacific island of Mauritius, reported that a low resting heart rate in children as young as age three could be correlated with the incidence of aggressive behavior at age eleven.[51]

Over twenty years ago, J. A. Gray suggested that noradrenergic pathways were critical to what he called the *behavioral inhibition system:* the neural circuitry that suppresses behavior followed by an unpleasant outcome.[52,53] Physiological arousal cues this system to hold bad behavior in check. Raine believes that the reason antisocial aggressors "just don't get it"—regardless of social disapproval, personal failure, even incarceration—may be related to the lower responsiveness of this noradrenergic inhibitory system, an insensitivity suggested by their damped-down heart rates and confirmed by their half-hearted responses to stress. Without the emotional arousal that normally locks environmental cues to outcomes, the antisocial individual never learns to anticipate the consequences of his aggression or that his actions have an impact on the welfare of others. Instead, his complacent nervous system betrays him over and over again to the same mistakes, while his inability to comprehend the feelings of others progressively erodes social relationships. His guilt-free violence represents one distinct type of human aggression, a form distinguished by an underreactive autonomic nervous system.

THE COMMON GROUND between the touchiness of PTSD sufferers and the cold-bloodedness of the antisocial is a neurochemical survival strategy gone haywire. When the noradrenergic system charged with spotting threat, warning, and danger is compromised or unbalanced, our perceptual safety net is breached. The result is a nervous system that either overreacts or, worse, ignores real peril and substitutes contempt for empathy.

Rewarding Good—and Bad—Behavior

The same postal service that can inadvertently deliver a pipe bomb can also render the dream of a lifetime. Perhaps that's why so many Americans are so willing to respond to the latest entreaty from Publisher's Clearinghouse and why hundreds of thousands fall for postal scams that promise cars, televisions, and jewelry to a few lucky winners. Lured by dreams of a multimillion dollar payoff or a dream vacation for two, who wouldn't just drop the reply card in the mail, pay a "small processing fee," or dial the toll-free number?

Our vulnerability to reward is built into our brains, courtesy of a second monoamine transmitter, d̲ihydr̲o̲xyp̲henylethyl̲a̲mine, better known by its short name, *dopamine*. If the motto of the brain's norepinephrine system is "Better safe than sorry," the corresponding principle on some dopamine highways might be summed up as, "If it feels good, do it."

Like norepinephrine, dopamine begins life as tyrosine, one of the amino acid building blocks of dietary protein, but it exits the biochemical assembly line one step earlier. And like their noradrenergic counterparts, the long-range neurons that use dopamine to communicate have their origins in the brain stem, in a parallelogram of dark-colored cells known as the *substantia nigra* and the surrounding *ventral tegmental area* that collectively give rise to three distinct forebrain-bound pathways (Figure 5.2).[4,5] The most intensively studied of these projections infiltrates a community of subcortical nuclei critical to voluntary movement (the so-called *basal ganglia*); their inexplicable death results

Figure 5.2

Dopaminergic pathways in the human brain.

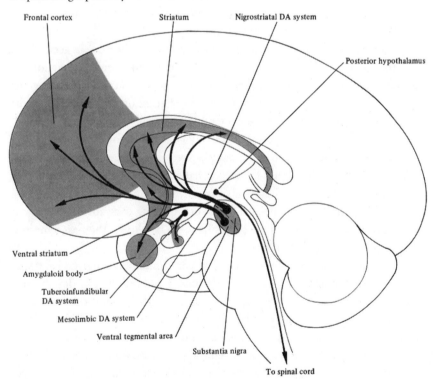

Reprinted with permission from L. Heimer, *The Human Brain and Spinal Cord* (New York: Springer-Verlag, 1983). DA = dopamine.

in the rigidity, tremor, and hesitancy that characterize Parkinson's disease. The other two pathways are based largely in the ventral tegmental camp and convey dopaminergic messages to emotionally familiar territories: the amygdala, the prefrontal cortex. Motor, limbic, and cortical projections correspond with five distinct receptors, at least three of which are known to activate or inhibit second-messenger proteins.[4] And like noradrenergic transmission, dopaminergic communication does not take place in a vacuum, but initiates reciprocal exchanges between pre- and postsynaptic cells, designed to balance activity on both sides of the synapse.

It's the mesolimbic ("from middle brain to limbic system") pathway that forms the dopaminergic foundation of the brain's internal reward network. Psychologist James Olds pioneered the discovery of brain neighborhoods devoted to reward, reporting in 1954 that electrical stimulation of certain sites in the hypothalamus could entice a rat to press a bar as effectively as—in fact, far more effectively than—a food reward.[54-56] Subsequent experiments demonstrated that stimulation outside the hypothalamus, at stopovers all along the mesolimbic dopamine pathway—the amygdala, the anterior cingulate cortex, and especially, the *nucleus accumbens,* another newcomer to the limbic system that straddles the boundary between the motor nuclei of the basal ganglia and the frontal cortex—was also highly reinforcing.[57]

More than a hedonistic "pleasure center," however, the nucleus accumbens and other structures comprising the reward system link events and actions to positive outcomes to compute a "reinforcement value," much as the amygdala matches sensory stimuli and experience to calculate emotional value. Food, for example, has significant reinforcement value (especially when you're hungry), and Maryland crab cakes or freshly baked chocolate chip cookies have a greater value than a sandwich from the company cafeteria. Self-preservation is also immensely rewarding. As a result, behaviors that pay off by promoting survival— fending off an intruder, appeasing an attacker, running away—are likely to be assigned a high reinforcement value by the dopaminergic reward computer, ensuring that these responses are the ones likely to be chosen the next time you find yourself in similar circumstances.

Pharmacology demonstrates that chemical stimulation can overload dopamine pathways much as activating drugs such as yohimbine can destabilize a trauma-sensitized noradrenergic alarm system. The behavioral result, in both cases, is self-defensive overreaction. Early animal research on the role of dopamine in aggression, for example, found that flooding this transmitter system with large doses of dopamine's chemical precursor, *l*-dopa, incited laboratory rats and mice to fight fiercely.[42,58] However, a closer look at "*l*-dopa rage" revealed that the uncoordinated and fragmented behavior provoked by the drug was hardly the typical picture of rodent hostility; the animals displayed none of the well-choreographed sequences usually seen in resident-intruder encounters, but lashed out indiscriminately at their opponents, then cowered defensively. Subsequent studies with the drug apomorphine, which mimics the action of dopamine, described rats who reared, screamed, and boxed at each other with their forepaws, the familiar defensive pattern of animals provoked into "fighting" by electric shock paradigms.[42]

Apparently, overwhelming the "pleasure center" is anything but pleasurable. Pushing the dopamine—as well as the norepinephrine—system to its limits elevates daily interaction with the world from a struggle for existence to a pitched battle for survival, a perpetual conflict with an environment that seems not only unsafe but unpleasant. When every encounter represents a threat and every experience is disagreeable, can it be any wonder that hostility is the end result?

EXAMINED WITH MORE subtlety than is possible with powerful activating drugs like apomorphine, dopamine function turns out to illustrate the importance of social history to neurochemistry rather convincingly. Twenty years ago, Ken Modigh, the same investigator who reported that fighting elevates brain norepinephrine, noted that dopamine and the rate of dopamine synthesis and release also rise after a hostile encounter.[32] More recently, Klaus Miczek's group found that when an intruder mouse confronts an angry resident for the first time, the rate of dopamine utilization in the mesolimbic pathway surges.[59] But by their tenth encounter, they utilize dopamine at the same rate as mice that have never fought at all (Figure 5.3). Over time, repeated encounters are characterized not by a static neurochemistry that "causes" behavior, but by a dynamic readjustment in dopaminergic activity that represents a progressive reshaping of the dopamine system by experience.

That's not the whole story. Defeat is a powerful stressor. And stress insidiously sensitizes dopaminergic, as well as noradrenergic, pathways supplying the limbic system and cortex. (See, for example, Gresch et al.)[15] Miczek notes that under a seemingly placid neurochemical surface, dopamine function in mouse victims remains off-center, sensitizing the losing animal "at some time point in the future to various types of dopamine challenges, either environmental pressures that act via dopamine systems or dopaminergic drugs." How does victimization change neurochemistry in such a long-lasting way? One possibility is that the molecular consequences of defeat stress do not end at the synapse, but reach all the way to the genome. Like the stress caused by restraint, extreme cold, foot shock, or trauma, social stress has the power to turn genes off and on. Within an hour of losing a fight, the experience has switched on the immediate early gene c-*fos*, presumably in preparation for activating additional genes.[60] On the surface or deep within the cell, the complex and elegant relationship between neurochemistry and aggressive behavior illustrates Miczek's contention that timing and history are critical to understanding the neural foundations of behavior.

Figure 5.3

Mesolimbic dopamine neurons use more dopamine (shown as the ratio between levels of the major breakdown product of dopamine, and levels of the parent transmitter) after the first confrontation between a resident mouse and an unfamiliar intruder. After ten encounters, dopamine utilization normalizes.

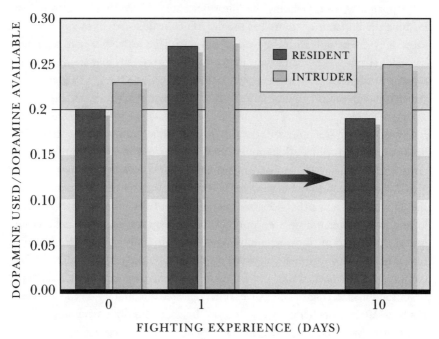

FIGHTING EXPERIENCE (DAYS)

Adapted with permission from M. Haney, K. Noda, R. Kream, et al. Regional 5-HT and dopamine activity: Sensitivity to amphetamine and aggressive behavior in mice, *Aggressive Behavior 16* (1990):259–270. Copyright © Wiley-Liss, Inc., a subsidiary of John Wiley & Sons, Inc.

Acting on Impulse

Eighteen-year-old Jonathan Dieck didn't wait around for cheap thrills to land in his mailbox. He and his bowling buddies, Keith McMahon and Brian Cerebe, cruised the North Philadelphia neighborhood of Port Richmond after dark with a pair of young admirers, creating their own good times.[61–63]

Equipped with walkie-talkies, a radio scanner for monitoring police and fire reports, and a box of toy plastic eggs filled with lighter fluid, the group piled into McMahon's car and divided into two teams: the "stars" and the "half-stars." Stars drenched that night's selected target with gasoline or pelted it with the loaded eggs; the half-stars back in the car kept watch. A touch from a pocket lighter ignited rivers of flame. The stars raced back to the car, and the group circled the block to enjoy the unalloyed thrill of the fire sweeping upward, the explosive rush as the fire streams turned a wall into flame, and the swarm of yellow-slickered firemen, pumping useless jets of water into the skeleton of the ravaged building.

The fun escalated as the targets grew bigger. And in the hours just before dawn on the morning of March 13, 1996, the cocky young arsonists, no longer content with low-budget productions, hit on the recipe for a blockbuster. Dieck and McMahon looked on as one of the younger boys doused an illegal tire dump moldering under Interstate 95 with a backyard barbecue fuel known as "blue block," tossed in a wad of smoldering paper towels, and ignited an eight-alarm inferno.

The intense heat of the burning rubber melted the steel rods that supported the overpass, fogged the surrounding neighborhood with an acrid, oily black smoke, exasperated commuters for nearly five months, and cost Philadelphia taxpayers about $6 million. City police and fire officials, backed by federal agents from the Bureau of Alcohol, Tobacco, and Firearms, were soon hot on Jonathan Dieck's trail. Even his mother suspected the truth; after investigators dragged Jonathan in for questioning, she pleaded with him to confess. Dieck's answer was one last wild ride. On July 31, Jonathan Dieck, the boy who was mesmerized by fires, rammed a stolen Jeep Cherokee into a tree, dying in a shower of glass and a pool of dripping gasoline.

Arson fascinates aggression researchers as well as firesetters. Angry young men like Jonathan Dieck, driven to torch warehouses and tire dumps by impulse rather than insurance money, led these investigators to a third monoamine neurotransmitter, serotonin. Complex and sometimes controversial, the resulting alliance between serotonin and violence has had a profound influence on neurochemical research, research that has helped to identify a second type of human aggression.

Serotonin is a slightly larger and more complex molecule than norepinephrine and dopamine and belongs to the chemical class known as indoleamines, synthesized in the body from the dietary amino acid tryptophan. The object of a hundred-year quest by physiologists, serotonin began its biological ascent with an identity crisis, when American and Italian researchers simultaneously isolated a new chemical that caused powerful contractions of the smooth muscle that forms blood vessels and lines the wall of the gut.[4] Both chemicals ultimately proved to be the same substance: the indoleamine 5-hydroxytryptamine. The American name stuck, and 5-hydroxytryptamine became serotonin, or "5-HT," for short.

Serotonin is a busy and ubiquitous substance that occurs in plants as well as in animals and humans. In mammals, it can be isolated from blood platelets, where it controls blood vessel diameter and causes the platelets to clump, initiating blood clot formation; from mast cells, where it plays a role in pain perception and inflammatory responses; and from the enterochromaffin cells of the gut, where it controls intestinal tone and motility. Less than 2 percent resides in the brain,[4] but that 2 percent is distributed in a way that maximizes serotonin's potential influence on neural function.

This transmitter could easily be the work of a neuroarsonist; it glows in the dark with a golden luminescence, kindled by the same formaldehyde-based reaction that lights up catecholamine pathways. The sparkling lights that are serotonin-rich cell bodies converge in nine distinct clusters known as the *raphe* (meaning "seam") *nuclei,* which straddle the midline across the rear of the brain.[4,5,64] Neurons in the most anterior groups cast out overlapping fiber cables to limbic structures, such as the amygdala, hypothalamus, and hippocampus; the basal ganglia; and the cerebral cortex, where the projections fan out in a diffuse meshwork that envelops regions associated with reasoning, feeling, remembering, and responding (Figure 5.4). Anatomically, therefore, serotonin, like its catecholaminergic cousins, norepinephrine and dopamine, is perfectly

Figure 5.4

Serotonin-containing pathways in the human brain.

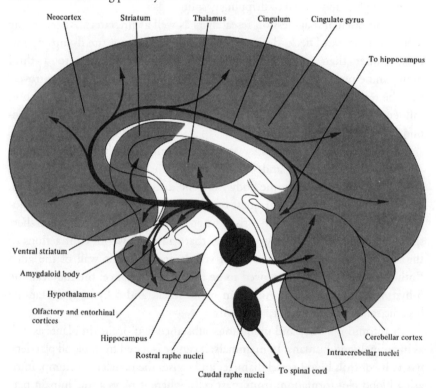

Reprinted with permission from L. Heimer, *The Human Brain and Spinal Cord* (New York: Springer-Verlag, 1983).

placed to coordinate logical, emotional, and sensory input with the complex behavioral responses needed for social interaction.

The introduction of the antidepressant Prozac in 1987 elevated serotonin to celebrity status in the neuropharmacology world. Unlike older antidepressant medications—sloppy, indiscriminate compounds with a taste for several transmitters and their receptors—Prozac and related compounds were designed to specialize in blocking transmitter removal at serotonergic synapses. As a result, they were better tolerated by many patients, revolutionizing the treatment of depression, some anxiety disorders, obsessions and compulsions, and even eating disorders like anorexia and bulimia. The resulting commercial success of the SSRIs (selective serotonin reuptake inhibitors) spurred an enormous increase in research on the biochemical, physiological, and behavioral actions of serotonin.

One outcome of this intense scrutiny has been a recognition of the importance of timing and reciprocal interactions to understanding serotonin transmission. From the first dose, SSRIs jam the uptake mechanism that normally recycles serotonin back into the presynaptic neuron.[65] As a result, levels of serotonin in the synapse are increased. Continued use enhances transmission even further, as the drugs damp down autoreceptors. Ultimately, both actions are overshadowed by long-term changes on the postsynaptic side of the synapse, as the number of serotonin receptors declines.[66,67]

The far-reaching monoamine pathways work to construct a global picture of the world, with norepinephrine pathways assigned to safety, dopamine pathways to reward. Serotonin pathways mediate ambiance. Midbrain-to-forebrain projections using this transmitter appear to have a "pacemaker" function, firing in a steady tonic pattern not unlike the ubiquitous Muzak piped into grocery stores and doctors' offices.[4] The continuous serotonergic background suppresses activity in target cells of the cortex and limbic system, superimposing an interpretative overlay that prevents excessive reactions to sensory information and coordinates an even-handed response just intense enough to meet the demands of the situation.

"No other transmitter," wrote Klaus Miczek in a 1992 National Research Council report, "has been more intimately implicated in the neurobiologic mechanisms of aggressive and violent behavior than serotonin."[41] Italian researchers may have lost out in the naming race, but they were among the first to link their newly discovered transmitter to aggression. Starting from observations that male mice housed individually often became aggressive, Luigi Valzelli and his colleagues at the Istituto di Recerche Farmacologiche Mario Negri in Milan began to study the neurochemical consequences of lifestyle in the mid-1960s and found decreases in serotonin or serotonergic breakdown products in mice that became aggressive after prolonged isolation.[68,69] Subsequent animal studies described an inverse correlation between serotonin levels in the brain and the defensive aggression provoked by foot shock (the lower the serotonin concentration, the more vigorous the aggressive response).[70,71] Experimental manipulations that depressed serotonin, such as eliminating tryptophan from the diet, blocking synthesizing enzymes, or destroying serotonergic neurons, were reported to facilitate aggressive behavior, while measures that pumped up the amount of serotonin in the synapse—by promoting synthesis and release or by blocking enzymatic breakdown—reduced aggressive responses.[72]

But a closer look suggests another dimension to the serotonin story, one that implies a special relationship with a particular type of aggression and a mechanism focused on perception rather than response. The aggression of serotonin deficits is not the cool, often premeditated, "use 'em up and spit 'em out" cruelty of the antisocial psychopath, but the knee-jerk overreaction of the hothead. It is a violence characterized by too much emotion rather than too little, and painfully aware of the surrounding environment.

Valzelli's observations already hinted that destabilizing serotonin function changed behavior by changing an animal's outlook. His serotonin-depleted mice were jumpy, tense, and difficult to handle even before they confronted an opponent. More sensitive to pain and touch, they squeaked louder with less provocation, darted hyperactively around their cages, and even displayed brisker reflexes than group-housed mice, who had more typical serotonin levels.[73] Lowering the serotonin volume, in other words, seemed to have the same effect as getting trapped behind a city bus in rush hour traffic—it frayed an animal's nerves until it was edgy and bad tempered enough to strike out at the slightest provocation.

DEPRESSION HAS an unhappy way of sliding over into violence. About one in every six depressed patients ultimately commits suicide,[74] and about half of the thirty thousand suicides that occur in the United States every year can be attributed to depression.[75] Conversely, violence has a tendency to end in desperation and suicide. Murderers, for example, select themselves as their final victims at a rate almost seven hundred times greater than law-abiding citizens; one survey of the suicides in a Detroit prison revealed that almost 60 percent occurred among the small proportion of men serving time for homicide.[76]

The death pact between suicide and violence is written in the language of serotonin. In the late 1960s, researchers probing the neurochemistry of depression and suicide homed in on the role of monoamines, already implicated in depression by virtue of the known pharmacological effects of the first antidepressants. Autopsy studies that compared postmortem levels of neurotransmitters in the brains of suicide victims and controls who had died of other causes were the first to point a finger specifically at serotonin. But postmortem data did not convince everyone; the rapid deterioration in brain integrity that begins at the moment of death can wreak neurochemical havoc before it is halted by a researcher and some dry ice. It took a study of serotonin function in living patients to verify this monoamine's culpability in self-directed acts of violence. A research team at the Karolinska Institute in Stockholm, headed by Marie Asberg, measured amounts of the serotonin breakdown product 5-hydroxyindoleacetic acid (5-HIAA) in samples of CSF taken from dozens of depressed patients, including some who had attempted suicide.[77] In what would ultimately prove to be one of the most consistently replicated findings in experimental psychiatry,[78] these investigators found that patients with a history of suicide attempts had lower 5-HIAA levels than those who had never gotten this desperate. And the lowest levels of all were found in those who had selected a violent method, such as shooting themselves—patients who, Columbia University's John Mann observes, are "alive by accident"—to try and end it all.[79]

But violent suicide wasn't just the tragic result of a serotonin shortage. Subsequent work by Mann demonstrated more extensive derangement of serotonin function in violent victims. Together with associate Michael Stanley, Mann analyzed serotonin interactions with postsynaptic 5-HT_2 receptors and presynaptic reuptake sites in the frontal cortex of people who had died after a violent suicide attempt and found changes on both sides of the synapse—but in opposite directions. Reuptake sites on serotonergic neurons themselves were reduced,[80] but the number of postsynaptic receptors was significantly increased.[81,82]

No neuronal event occurs in isolation. Crises that disrupt function on one side of the synapse are perpetuated on the other. A process captured in units of transmitter is unlikely to stop there; closer inspection may reveal that it has spread to engulf an entire pathway.

A team headed by Gregory Brown of NIMH went on to show that serotonin's influence wasn't limited to self-inflicted violence. They demonstrated that low CSF levels of serotonin and its metabolite 5-HIAA were correlated not only with previous suicide attempts, but also with a life history marked by aggressive outbursts toward others.[36,83] Additional studies conducted in the decade following Asberg's landmark observation—in depressed psychiatric patients, criminals, and even healthy volunteers—also reported lower CSF 5-HIAA levels among those who were hostile, suicidal, or violent.[78,84]

Low serotonin and human aggression seemed to go hand in hand. But Asberg's and Brown's results still left an important question unanswered: whether changes in serotonin activity correlated better with the aggressive act itself or (as the animal research suggested) with an overly sensitive perceptual system, primed to overreact to the slightest threat, the mildest insult.

Finnish researchers Markku Linnoila (who would ultimately become scientific director of the Division of Intramural Clinical and Biological Research at the National Institute on Alcohol Abuse and Alcoholism) and Matti Virkkunen attempted to clarify the relationship of serotonin, aggression, and impulsivity in a series of landmark studies that began with a survey of thirty-six men serving time for murder or attempted murder. From case reports and prison records, the two investigators classified twenty-seven of these violent offenders as either "impulsive"—meaning that their crimes were neither premeditated nor part of a robbery or burglary—or "nonimpulsive." Those assigned to the impulsive group had significantly lower concentrations of the serotonin metabolite 5-HIAA in CSF samples than the nine "nonimpulsive" killers. And the lowest levels of all were posted by offenders who had not only killed someone else but had also, at one time or another, tried to kill themselves.[85]

The result supported the idea that the aggression linked to alterations in serotonin was an impulsive overreaction to provocation, not a cold-blooded act of revenge, spite, or expediency. To verify this interpretation, Linnoila and Virkkunen needed a better way of separating impulsiveness from the actual infliction of physical injury. They turned to a group of Finnish arsonists. The scientists reasoned that arson, an impulsive crime usually free of murderous intent, might hold the answer. Comparing CSF samples drawn from the arsonists, a similar group of violent offenders, and healthy controls, they found that the levels of 5-HIAA were lowest in the fire setters.[86] Serotonin levels, they concluded, primarily reflected the ability to stop and think, and reductions in serotonin warned of a disastrous misconnect between the brain and the environment.[86,87]

SCOTT HAD NEVER appeared in court, never shot a gun or pulled a knife, but his aggressive behavior was ruining his life. Trim, with the keen intensity of middle-aged men who run five miles at dawn and don't put down the phone until after midnight, some of Scott's competitors saw his "take-no-prisoners" attitude as the key to his success as an environmental engineer. But friends, family members, and customers who had been on the receiving end of his periodic rages held a less sanguine opinion. Ultimately, Scott's wife decided that she'd endured enough screaming matches. Exhausted and frightened that he'd finally "lose it" completely one day, she told him she wanted to end their marriage.

Emil Coccaro, who directs the Clinical Neuroscience Research Unit at Philadelphia's Medical College of Pennsylvania, sees the impulsive outbursts of people like Scott as the defining behavioral feature of a second subtype of aggression—and changes in serotonin function as a crucial element in its underlying neurochemical mechanism. Coccaro has been studying and categorizing human aggression for more than a decade, and he has concluded that this combination of behavior and biochemistry distinguishes impulsive aggression from the calculated violence he calls "premeditated."

"The difference between people who are impulsively aggressive and those who aren't impulsive is that impulsive aggression doesn't get you anywhere," Coccaro explains. "Take the guy who goes into a store with a gun. If he shoots someone in the process, it's not socially appropriate behavior, but it is consistent with his goal—stealing money. But people who are impulsively aggressive just blow up. It gets them into trouble, but it doesn't get them much of anything else."

The impulsively aggressive strike first and ask questions later; their motto might be, "The best defense is a good offense." For them, life is a minefield riddled with threats and insults that demand a forceful response. And the individuals most likely to ignite their hair-trigger rages are not strangers but family members, friends, and coworkers.

"We see lots of people in the community who have problems with impulsive aggression and they're not all criminals," Coccaro notes. "Society may worry more about the guy who goes into McDonald's with an assault weapon. But the people who are impulsively violent—they're the ones we're more likely to confront."

Like "territorial" and "intermale" aggression, impulsive and antisocial aggression are not mutually exclusive categories. Antisocial individuals can also display episodes of impulsive behavior. They lose control because the rules that govern human social behavior are meaningless to them and cannot constrain their behavior. But many impulsive aggressive individuals—the people Coccaro says "just blow up"—do believe in the rules—until their hypersensitive nervous systems cross paths with a perceived threat.

Coccaro explains, "They're hotheads—the people who scream and shout, who throw things, who hit their wives or their kids." In contrast to the antisocial, hostile impulsive individuals are too easily aroused, rather than too slowly. They show it, in their flushed faces and tense muscles. And they feel it, in an overwhelming rush of uncontrollable anger. Matthew Stanford, a psychologist at the University of New Orleans who studies impulsive aggression in college students and prison inmates, says, "They'll tell you that just prior to losing control, they feel stressed and very agitated. They show clear physiological responses, and they perceive these responses as anger."

Hostile impulsive aggressors differ from antisocial aggressors in one other important way: they're sorry. They often admit to feeling guilty and ashamed after they've hurt someone; many, in fact, insist they never wanted to be violent in the first place. But by the time they realize their behavior is wrong, it has already happened. "I know it's coming," Stanford says his impulsive aggressive subjects tell him, "and I just can't stop it."

Coccaro's definition of impulsive aggression has its roots in the diagnostic category that the DSM-IV calls "intermittent explosive disorder (IED),"[45] a descendant of Mark and Ervin's dyscontrol syndrome. As a guideline for classifying human aggression, IED leaves a lot to be desired. Loosely defined as a periodic eruption of a compulsion to hurt people, places, or things, the IED label, as it currently stands, is more like a wastebasket than a benchmark, an orphan category relegated to the back of the diagnostic manual as an option of last resort for hostile behavior "not elsewhere classified." Researchers and clinicians alike agree that the IED criteria do not really describe the typical "explosive" individual they study and treat; in fact, some experts estimate that the current definition misses as many as 90 percent of individuals who fly off the handle on a regular basis.

Coccaro, fellow psychiatrist Ernest Barratt of the University of Texas, and a core group of dedicated colleagues want to change that. They've been working to revise, clarify, and extend the definition of intermittent, explosive aggression to correspond more closely with the key features of impulsive aggression they've observed in hundreds of subjects, ranging from violent offenders to bad-tempered attorneys. These features are summarized in Table 5.1. Their goal is twofold: to begin to construct a sensible typology for human aggression and to fashion a precise, easy-to-use, reliable descriptive tool that ensures that epidemiologists, neurochemists, and clinical researchers studying impulsive aggression are all on the same wavelength.

In contrast to IED, impulsive aggression, as defined by these researchers, corresponds more closely to the edgy irritability they see in the real world. And it can be tied to the changes in the serotonin system they see in the laboratory. More than a "serotonin deficiency," impulsive aggression—like PTSD—is likely to involve a global disruption of monoamine function, a progressive and ultimately maladaptive realignment of receptors, responses, and feedback mechanisms.

Coccaro and his coworkers have found evidence of such wide-ranging changes in serotonin function among their impulsive aggressive patients. To do this, they took advantage of a peripheral measure of serotonin activity in the brain: blood levels of the hormone prolactin. Serotonin projections to the hypothalamus deliver a chemical messenger known as a releasing factor to the pituitary gland; pituitary cells, in response, discharge a pulse of prolactin into the bloodstream. To measure the functional integrity of serotonin responses, Coc-

Table 5.1. Criteria for Impulsive Aggression

- Recurrent episodes of verbal or physical aggression toward other people, animals, or property
- At least two outbursts per week for at least one month
- Not premeditated
- Intensity of aggressive behavior is out of proportion to the provocation
- Causes aggressor marked distress or interferes with job and relationships

Adapted from E. F. Coccaro, R. J. Kavoussi, M. E. Berman, et al., Intermittent explosive disorder: Revised criteria for aggression research, *APA New Research Abstracts NR564* (1997), 220.

caro's team prodded the serotonin-containing neurons with an activating drug, fenfluramine (which simultaneously stimulates serotonin release and blocks reuptake), then measured the magnitude of subsequent prolactin release.[78]

Aggressive primates secrete significantly less prolactin in response to a fenfluramine challenge than do nonviolent members of the social group.[78] And in humans, Coccaro found a similar relationship between blood levels of prolactin after fenfluramine and impulsive aggressive behavior, including suicide attempts, direct assaults, self-reports of hostile behavior, and verbal tirades.[84,88,89] The decrease could not be blamed on a pituitary malfunction, for direct stimulation of pituitary cells resulted in prolactin levels no different from those of nonaggressive control subjects. Nor was it a flash in the pan: eleven of thirteen studies conducted to date by Coccaro and others have reported disruptions in the serotonin-mediated control of hormone release.[78]

Together, the observational studies of Linnoila and Brown, postmortem data, and drug challenge studies suggest a complex relationship between serotonin and aggressive behavior that spans the synapse. Equally important, serotonin data suggest that there's something special about impulsive aggression, not only on the behavioral level, but below the surface, at the level of the neurochemical mechanisms that are the functional grammar of the nervous system.

The relationship between serotonin and aggression, like that between norepinephrine or dopamine and emotional behavior, is the result of an ongoing interaction between brain and environment. The amygdala of a mouse that has just enjoyed his first victory over an intruder, for example, releases and processes more than twice as much serotonin as the amygdala of a mouse with no fighting experience. But with additional confrontations, the thrill of victory fades, and so does the robust serotonin response.

Similarly, subordinate vervet monkeys have lower plasma serotonin levels than their superiors that achieve an exalted social status after a history of winning critical battles over food, resting places, and access to females.[90,91] But if the subordinate begins to rise through the social ranks, his serotonin levels go up as well. Can the social climber credit his success to a surge in serotonin, or has the change in his social standing pumped up transmitter levels? The answer may well be both: every successful challenge has a positive effect on serotonin function, and the gain in transmitter levels tips the balance of power in favor of continued social striving. Each cycle updates the neurochemical profile to correspond to the animal's new position in the social hierarchy, a chemical record of his social successes and failures.

We can track this dynamic process in the shifting pattern of correlations between behavior and neurochemistry, but until recently, we could not observe what happened at the most critical point: during the actual aggressive encounter. To do that, researchers needed the neurochemical equivalent of a videocamera, a way of tracking changes in transmitters and receptors in real time. Klaus Miczek believes he has just the ticket.

A delicate plastic tube no thicker than postal twine and a tiny filter smaller than the head of a pin are Miczek's solution to escaping what he calls "the correlational box"—the problem of reconstructing the past from post hoc measures. The filter perches at the tip of a whisper-fine probe threaded through a guide tube, or cannula, similar to that Alan Siegel used in his mapping studies of predatory and affective aggression. Postdoctoral fellow Annemoon van Erp positions the cannula within the prefrontal cortex of rats, who are then housed with a breeding female or in bachelor groups of four or five animals. The rats adapt quickly to the probe; soon after the surgery, they're going about their normal daily activities of eating, grooming, and even fighting.

The probe is actually a hollow shaft housing two side-by-side plastic tubes. A physiologically correct, neurotransmitter-free fluid flushed gently through the tubes bathes the membrane at the end of the probe. On the brain side of the probe, the extracellular fluid that surrounds neurons is loaded with norepinephrine, dopamine, and serotonin, creating a concentration gradient. The monoamines flow from high to low concentration across the membrane and are gently pumped back up through the probe, through the loops and coils of tubing, and into a test tube mounted on a slowly turning wheel. The wheel advances to the next tube every ten minutes; collectively, the drops of fluid collected over the course of many tubes comprise a neurochemical record of an entire confrontation.

An experienced resident rat isn't deterred by his plastic tether; he doesn't hesitate to attack a hapless intruder. But the chemical record shows that the assault isn't preceded by a drop in cortical serotonin levels. In fact, serotonin concentrations in the first tubes, filled with fluid collected just before the attack—as well as in the tube collected during the brief moments of actual physical contact—remain surprisingly quiescent, hovering peacefully around baseline levels. It's only after the cage bedding has settled, and the resident is busy rearranging his fur, that serotonin levels begin to decay, reaching a low point about an hour *after* the confrontation.[92]

When the intruder is the one with the cannula, a different picture emerges. His cortical serotonin levels rise from the first moment he's threatened. Even if he's safely ensconced behind a screen protecting him from direct physical contact with the resident, serotonin concentrations shoot up—and stay there—during the entire confrontation.[93]

Timing is everything. If Miczek's group had sampled transmitter levels after the encounter, they would have observed elevated serotonin levels in the intruder rat and unusually low levels in the resident, a result they could well have interpreted as evidence that aggression is "caused" by a serotonin deficit. Instead, they had a real-time view of the neurochemistry and learned that the serotonin decline takes place during the recovery phase following the attack, rather in the preceding moments. This important study warns that we ignore the circular relationship between brain and behavior at our peril, that correlations are clues but not necessarily causes. Alterations in brain chemistry show us where the brain has been as well as where it is going next, as it collaborates with the environment to shape responses that ensure survival.

Feeling No Pain

In 1974, if you wanted to learn more about pain—mental or physical—Baltimore was the place to go. Here, in a warren of overcrowded laboratories at the Johns Hopkins University Medical School, psychiatrist Solomon Snyder had just made an extraordinary discovery that was revolutionizing not only the study of pain-killing drugs, but the study of neurochemistry in general.

The problem of how narcotic analgesics block pain was one of the top ten questions in neuropharmacology. Research on hormones, on drugs that blocked communication between nerve and muscle, and on heart and blood pressure medications that altered activity in the sympathetic nervous system had already demonstrated that the first step in the action of all of these drugs was their interaction with specific receptors. Many researchers believed that receptor binding was the essential feature governing the effects of all drugs and that the physical and chemical properties of specific receptor populations defined drug specificity and potency.

But if morphine and related narcotic drugs, like codeine and heroin (collectively known as opiates), acted via a receptor, it wasn't one already known to neuroscience. Snyder (and a core group of like-minded competitors) believed that the opiates interacted with their own unique receptors. The idea was compelling on paper but proved technically elusive to verify.

Then Snyder and graduate student Candace Pert hit on the idea of fishing for the receptor with a chemical hook fashioned by adding a radioactive tag to naloxone, a drug that blocks the action of morphine. It was an inspired decision. The labeled antagonist homed in on the novel binding site with a single-minded tenacity, and the Hopkins group finally confirmed the existence of an opiate receptor.[94]

Where there's a receptor, there ought to be a transmitter. The success of Snyder's quest to find opiate receptors naturally suggested that the brain itself made an endogenous morphine—a substance keyed to the newly discovered receptor. However, it took three additional years after the publication of the discovery of the opiate receptor in 1973 for John Hughes and his supervisor, Hans Kosterlitz of the University of Aberdeen, to track down the missing transmitter. And it wasn't another small molecule amino acid derivative, like norepinephrine or serotonin. Instead, the "brain's own morphine"—Hughes and Kosterlitz dubbed it *enkephalin*—was a protein, or to be more specific, a small snippet of a protein, a fragment known as a peptide.[95]

Enkephalin, neurobiologists would discover, was only the tip of a very large iceberg. To begin with, enkephalin itself turned out to be not one peptide but two, a pair of near-look-alike fragments that differed by only a single amino acid. And the enkephalins had relatives. All told, the brain manufactures more than a half-dozen other morphine-like peptides, or "endorphins" (from <u>end</u>ogenous <u>morphine</u>), each custom designed to recognize a distinct subset of the burgeoning opiate receptor family. The peptide story didn't stop with the opioid peptides either. During the "decade of neuropeptides" following the discovery of the enkephalins, more than fifty biochemically active peptides were found in the brain, a list that is still growing.[96]

THE MODERN OFFICE functions with a fax machine to back up the mail. A client hasn't received your latest proposal, and he's making a decision today. Another copy can be on his desk in a few minutes. You forgot to include an invoice? Your secretary can fax one immediately to accounts payable, then follow up with a hard copy.

In the brain, neurons rely on peptides to be their fax machines, to expedite, clarify, and override the messages passed from cell to cell by the monoamines and other small-molecule transmitters. More parsimonious than the corporate world, however, peptide communication doesn't require extra phone lines; it simply uses the neural delivery routes already in place, often sharing the same neuron as the nonpeptide transmitter it's assisting. One of the two enkephalins, for example, "colocalizes" with both serotonin and norepinephrine. Released along with their transmitter partners, peptide follow-up messages back up the directive issued by the primary transmitter; in the words of one classic neuropharmacology primer, the "peptide usually embellishes what the primary transmitter . . . seeks to accomplish."[4]

Peptide "contransmitters" add a new layer of complexity to neurochemistry. To understand the chemical origins and consequences of behavior, neurobiologists must now consider not only traditional neurotransmitter systems but also the role of their peptide sidekicks. When it comes to understanding the neural consequences of violence, that means looking beyond norepinephrine and serotonin, to the opioid peptides that incited the peptide revolution.

Pain is on the offensive front line in the struggle for survival. Touch a hot stove or step on a pin, and a spinal reflex whips your hand or foot away before your cerebral cortex can even label the sensation, much less identify its source. But the rapid response to pain—like all other knee-jerk reactions—isn't always helpful. When the source of the pain is a life-threatening attack, time and energy focused on attending to the agony of a shattered bone, a gaping wound, are precious seconds not devoted to the critical business of fight or flight.

That's where the brain's homemade analgesics can be a real life-saver. Accident and crime victims often describe a pain-free window immediately following the traumatic event. Because this response can be readily blocked by naloxone and other drugs inhibiting the interaction between opioid peptides and their receptors, neurobiologists believe that we can thank endogenous opiates for "stress-induced analgesia."[97] Released in response to the autonomic alarm triggered by an attack, enkephalins and endorphins damp down pain sensations long enough to allow the victim a chance of escape.

Stress-induced analgesia can also ease the agony of defeat when escape proves impossible. An intruder mouse, facing the heated attacks of an outraged resident, wisely attempts to run away, for the attacker will continue to press his point if given the chance. Cornered, the intruder's best chance of avoiding an onslaught is simply to give up, crouching on his hind legs, forepaws upstretched and ears flattened.[35] Residents lose interest when presented with such an obvious admission of defeat, wandering off to concentrate on grooming or riffle through the cage bedding in search of forgotten tidbits.

The unlucky intruder certainly appears discomfited by his close brush with serious injury. But physically, he's feeling no pain.[35,98,99] After a single confrontation, defeated intruder mice are so much less reactive to pain than their victorious opponents that they take three times longer to respond to a painful stimulus. Defeat—not the severity of the attack—is the key to this pain insensitivity. Regardless of whether the intruder has been bitten two times or thirty times, the analgesic response kicks in.

The induction of defeat-induced analgesia corresponds to a surge in endorphin and enkephalin peptides, suggesting that the analgesic response is the result of the mobilization of these natural painkillers.[60,72,100] Within two hours, news of the defeat has trickled down to the gene level, switching on genes that code for the production of opioid peptides. An enkephalin-mediated response ought to be stopped in its tracks by an opiate receptor blocker—and the antagonist naloxone blocks defeat analgesia. Intruder mice treated with the opiate blocker before their encounter with a resident still surrender, but they no longer develop defeat analgesia.[35,98,99]

IN HUMANS, the neurochemical remodeling quietly initiated by trauma can resurface in the form of altered pain sensitivity months or even years after the catastrophic event. Roger Pitman, of the Manchester, New Hampshire, Veterans Affairs Medical Center, and researchers from the Department of Psychiatry at Harvard compared analgesic responses in eight Vietnam veterans diagnosed with PTSD and eight symptom-free veterans after they watched a particularly intense fifteen-minute segment of the Vietnam war epic *Platoon*.[101,102] The PTSD sufferers reported that they felt little or no pain when touched briefly with a hot probe; in fact, pain intensity ratings were even more effective at distinguishing

the men with posttrauma symptoms from their healthy counterparts than physical signs of fear and arousal, such as heart rate or skin conductance. Naloxone
blocked the analgesic response, clinching the involvement of brain opiate mechanisms. Pitman and his colleagues concluded that an "opiate response to traumatic re-exposure"—defeat analgesia—may be the underlying basis of the
psychic numbing that is a hallmark symptom of PTSD.[102]

The brain's natural response to the experience of violence is to shield the
body from additional pain. But just as the memory of trauma can push the
fight-or-flight reaction of the sympathetic nervous system from adaptive to
overreactive, it can push pain relief to the point of immobilization, freeze terror at the expense of turning emotional sensitivity entirely to stone.

Bad Chemistry?

What are we to conclude about the relationship between neurochemistry and
aggression? By charting the course of the intricate exchanges between transmitters and receptors, enzymes and breakdown products, monoamines and peptides, can we begin to translate the narrative of emotional behavior—or are our
analyses of serotonin metabolites, drug-induced flashbacks, pain-killing responses to defeat mere chemical phrenology, an attempt to blame aggression
on "bad chemistry"?

Biology must include a recognition of history, timing, and environment.
When the study of neurochemical mechanisms is expanded to include an up-
side-down approach to the ways in which behavior can alter neurotransmitter
systems, as well as the impact of chemical variations on behavior, we can
achieve a balanced, more realistic understanding of the relationship between
neurochemistry and aggression. If we admit to complexity, we do not have to
surrender to what one investigator has tartly labeled "intellectual nihilism." We
can be humbled by the potential inherent in a brain capable of communicating
in dozens of chemical dialects, hundreds of receptor variations, without concluding that this sophistication closes the door to understanding.

Chemistry is a language, not a "cause." Social history and feedback mean
that the relationship between behavior and chemistry is circular rather than
linear, and that experience is not only relevant to understanding neurochemical mechanisms of aggression; it is critical. At the synaptic level, a similar reciprocity between activity and response, transmitter levels and synthetic
enzymes, release and receptor density suggests that if the basic structural unit
of the brain is a network, not a "center," the basic chemical unit is the synapse,
not a transmitter. Measuring isolated elements—serotonin metabolites, for example—will highlight the contribution of these "key nodes" to neurochemical
networks. But to understand the intersection of behavior and chemistry com-

pletely, we must know the state of all the nodes, how the entire system has changed over time.

IF THE CHANGES in pathways using the monoamine transmitters norepinephrine, dopamine, and serotonin seem especially important to understanding aggression, that's not surprising, given the role of these transmitter systems in monitoring safety, reward value, and mood. But the origins of violence are more than a serotonin deficiency, and the consequences more profound than a short-lived spike in norepinephrine release. Both reflect more subtle and widespread changes in monoamine transmitter systems, changes that affect the entire transmission process, and peptide modulators as well as the primary transmitters. The progressive reconfiguration of these critical sensory-filtering and mood-setting pathways—the cumulative effect of months or even years of social encounters, of conflicts won and lost—shapes the way we view the world. If skirmishes have been rare, and resolved in our favor, we are likely to feel secure and confident. A less salutary history, and we may come to see the environment as dangerous, our neighbors as hostile and undependable, the appropriate response as running away, giving up, or shooting first. In the worst of cases, we may conclude that the outside world is irrelevant and unfulfilling, and that aggression is an engaging option in an otherwise emotionally flat existence.

The sensitivity of brain monoamine systems is the foundation of how we perceive and react to threat. In the nuances between the chemistry of these pathways and behavior, investigators are beginning to discriminate overreaction from underreaction biochemically as well as behaviorally, and to identify neurobiological features that differentiate antisocial from hostile, impulsive violence. These differences have profound implications not only for the classification of human aggression, but also for the development of intervention strategies that address the root of the problem, rather than fulfilling a desire for retribution.

Six

RAGING HORMONES

Before the telegraph crossed plains and mountains, message delivery in the American West relied on brave men and speedy horses. The Pony Express, a nineteenth-century version of a strategy that originated in the Persian Empire, transformed the relay race into the country's first overland postal system, a human feedback loop that finally joined East and West. From St. Joseph, Missouri, the Pony Rider, one of a carefully selected company of "young, skinny, wiry fellows, not over 18, and willing to risk death daily," and a saddlebag full of letters charged across the prairie toward the stagecoach station in Seneca, Kansas, seventy-seven miles to the west.[1] Along the way, he and the mail changed horses six times; in Seneca, the rider himself finally rested, and handed over the saddlebag and its precious contents to the next rider on the team. Horses and riders relayed the mail steadily westward from station to station, traversing the 1,966 miles to Sacramento in ten days.[1,2]

The monoamine and peptide transmitters that deliver high-priority messages between brain and body are partnered by a delivery scheme that's the physiological equivalent of the Pony Express: the endocrine system. Hormones are the mail carriers in this biological relay, versatile chemical messengers equally at home in the brain and in the bloodstream.

The "neuroendocrine express" originates in small, short-axon cells clustered at the center of the hypothalamus. Peptides manufactured in these *parvocellular* hypothalamic nuclei—permissive signals known as *releasing factors*—are handed over to a plexus of delicate capillaries for delivery to the pituitary gland, the first way station in the endocrine relay.

Like a miniature plum, the pituitary dangles from a slender stalk of nerves and blood vessels at the interface between brain and periphery. A central furrow scallops the gland into two divisions. The hindmost, or posterior, lobe responds to direct commands from hypothalamic neurons by releasing the hormones oxytocin and vasopressin. In contrast, the anterior lobe waits for blood-borne signals, the releasing and inhibiting factors dispatched by the home station in the brain. As effortlessly as the Pony Express riders swept the mail pouch from one saddle to the next, the incoming and outgoing circulation swaps peptide for peptide, releasing factors for pituitary hormones, the next members of the hormone relay team. These go-betweens speed from the pituitary to secretory glands in the body: the thyroid, the adrenal, the gonads. Here, they pass on the neural command to secrete the familiar hormones thyroxine and cortisol, testosterone and estrogen. Figure 6.1 outlines the route from the brain to one of these end points, the adrenal gland.

The endocrine journey from brain to body ends in target tissues rich in hormone receptors. But it does not end at the cell surface. Unlike receptors for norepinephrine and serotonin, which rely on an intracellular relay of chemical reactions to carry the news of transmitter binding to the genes, many hormones can communicate with the nucleus directly. Their receptors *are* DNA-seeking transcription factors, meaning they can switch genes on and off without intermediaries.[3] Endocrine communication may be slower and more laborious than neurotransmission, but once the messages arrive at their destination, they proceed directly to the heart of the cell, smoothly connecting the environment to the genome.

The trip west, from St. Joseph to California, allowed westerners to keep up with critical developments in Washington as the country raced toward civil war. But eastward messages were equally important, an opportunity to reassure family and friends that the journey over the Rockies had been completed safely, a chance to share the exciting news of gold. Feedback from target organs and the pituitary back to the brain is also important in the endocrine relay system (see Figure 6.1). Estrogen and testosterone, cortisol and thyroid hormone sweep into the anterior pituitary with an update on hormonal activity in the body, and the power to fine-tune the production and release of pituitary hormones. Similarly, the intricate vasculature of the brain ferries the hormonal messages back to the hypothalamus, prompting adjustments in the secretion of releasing factors. Reciprocal interactions occur at the cellular level as well. High circulating levels of hormones spur a downward adjustment in the number and sensitivity of receptors on target organs, while declining levels urge targets to upgrade their receptor populations, much as activity at the synapse regulates neurotransmitter release and postsynaptic receptor density.

The brain has a voice as well, for control of the endocrine system extends

Figure 6.1

The hypothalamic-pituitary-adrenal axis. Corticotropin releasing factor (CRF) produced by the hypothalamus stimulates the secretion of the pituitary hormone ACTH, which in turn prompts the adrenal gland to release glucocorticoid stress hormones. Glucocorticoids and ACTH also work backward to regulate their own release. These feedback mechanisms represent an important avenue for control of the stress response.

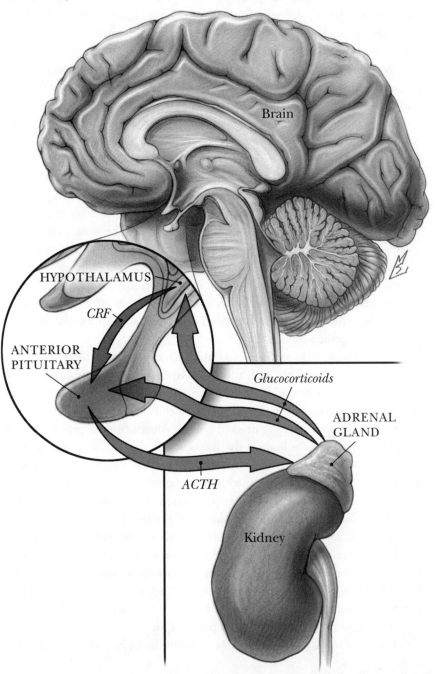

beyond the hypothalamus to include regions that process and interpret vital information about the environment. Among the most important are the base camps of monoamine projections—from the locus ceruleus, the raphe nuclei—that link these sources of norepinephrine and serotonin to the hypothalamus.[4] Thanks to these connections, the little cells of the hypothalamus are not only the central post office of the endocrine system but also the "transducers that convert neural information into hormonal information."[5]

Feedback, receptor regulation, and neural input endow the neuroendocrine system with the same flexibility enjoyed by the brain itself. Increases and decreases in endocrine activity—the hormonal equivalent of the processes that moderate synaptic communication between neurons—continually fiddle with the system to match response and the demands of an ever-changing environment. And events that alter neural estimates of safety, reward, and stability, processes monitored and regulated by norepinephrine and serotonin, change endocrine as well as neural function. Not only are our brains continually being rebuilt by our experiences, but so too are our bodies.

HORMONES, LIKE NEUROTRANSMITTERS, tempt us with another possibility of a simple explanation for violence, a "cause" we can blame, a break we can fix. But endocrine function is also a dialogue, a reciprocal process that reacts to behavior as well as drives it. Timing, personal history, and context play critical roles, while interactions between transmitters and hormones yoke the nervous system and endocrine system. The answers provided by hormones are rich, complex explanations that confirm the close association of the brain and the environment.

The Not-So-Gentle Sex

People are not born bad. But being born male seems to be a step in a violent direction. According to 1996 FBI statistics, 90 percent of the 18,108 individuals arrested for murder that year were men.[6] Antisocial personality disorder—the form of violence characterized by a sluggish neural alarm system and a conscience on permanent vacation—is three times more common among men than women.[7] Rapists, stalkers, and mass murderers are predominantly men.[7] Serial sexual killers, the most feared violent criminals, are, as far as forensic experts can tell, exclusively men.[7] Men are even more likely than women to kill themselves. Recent figures show that three-quarters of all suicide victims are men; in some age brackets, the suicide rate among men exceeds that among women by more than tenfold.[8]

Why are men so violent? Debate on this issue resurrects the age-old conflict between nature and nurture. Some contend that men learn to be aggressive, that they're the product of a culture that glorifies and promotes bad behavior. Others disagree. Men, they insist, are victims of hormones.

A room full of experts could easily be organized into two camps—on one side, culture advocates, on the other, those cheering for hormones, with a few chairs in the center aisle for the undecided. But the participant closest to the truth would be the individualist who picked up a chair and switched hourly from side to side, to emphasize that the relationship between hormones and behavior, like similar alliances between norepinephrine and fear, serotonin and aggression, is not linear, but circular. The social environment ties the endocrine system to the external, as well as the internal world; conversely, experience is meaningless until it is translated by the brain into the language of molecules.

"EVERYBODY KNOWS," Klaus Miczek laughs—with an ironic emphasis on *everybody*—"that testosterone causes aggression." He's conducting an armchair tour across the rocky terrain of hormonal influences on aggression, and this, his first slide, reiterates the widely accepted link between male hormones and male violence. Miczek, however, is not everybody. For him, this fact-finding mission will not follow a direct route from hormones to behavior, but a circuitous path from brain to experience and back again, familiar territory to neurobiologists studying the chemical foundations of aggression. A committed skeptic, he's titled the slide, "Myth."

Miczek sighs and leans back in his chair. "The problem is that there's a kernel of truth in this statement. That's what makes it alluring." Next slide. He points to the front page of an 1849 report by a German researcher named Berthold,[9] the source of what Miczek calls the "endocrine paradigm."

Berthold's approach was simple, his results convincing. He took barnyard roosters and castrated them to subtract testosterone, then compared their behavior before and after the surgery. Intact birds were noisy and belligerent. Castrates were as quiet and mild-mannered as the hens they supervised. They didn't crow, they didn't signify to their peers, and they didn't fight. When Berthold gave these roosters daily testosterone injections to replace the missing hormone, masculine behavior, including aggression, reappeared.

Berthold's observation, Miczek notes, has been repeated so many times, in so many species—from zebra finches to red deer, cattle to chimpanzees—that it's no wonder *testosterone* has become synonymous with male aggression.[9–12] He adds, "It worked in 1849, and it keeps on working." Wipe out male hormones, and lions turn into lambs. Add back the testosterone, then stand back and watch the fur (or feathers) fly.

Take the familiar laboratory rat. Male rats, like roosters, lose interest in bullying unfamiliar males if they're castrated. Hormone replacement therapy restores their aggression toward intruders.[11,13,14] Male-on-male aggression in mice appears even more hormone dependent. In a series of studies conducted in the early 1980s, British behavioral biologist Paul Brain demonstrated that castrating male mice significantly reduced not only the offensive aggression directed toward an intruder but also the defensive aggression precipitated by shock. Testosterone injections rekindled both kinds of aggression—and the higher the dose and the longer the treatment, the fiercer the fighting.[15]

Behavioral endocrinologist Joseph DeBold, who has collaborated with Klaus Miczek for over two decades, stresses that "a biologist—especially a behavioral biologist—would never reduce the interaction between hormones and experience to a dichotomy." He notes that even in the lowly rat, hormones are not everything. In a critical reexamination of "Berthold's law," DeBold and Miczek discovered that castration pacified only socially naive animals, rats with no prior fighting experience. Seasoned veterans, accustomed to trouncing intruders, remain aggressive even after their testosterone has plummeted to negligible levels.[16]

Male primates undo the knot tying aggression to testosterone levels even further. Some species *are* influenced by hormones. In squirrel monkeys, for example, the seasonal fluctuations in hormone levels that accompany breeding cycles—a "take-it-away-add-it-back" protocol designed by nature—correspond to peaks and troughs in the frequency and intensity of conflicts between adult males.[17] But in the congenitally hostile rhesus monkey, testosterone and aggression are no longer so closely linked; neither seasonally nor surgically induced changes in hormone levels alter aggressive behavior.[18,19]

By the time the hormone tour reaches the level of humans, the evidence in favor of a causal relationship is decidedly equivocal. For example, the most violent men are not necessarily the most hormonal. A 1976 study comparing plasma testosterone values in convicted sex offenders and nonincarcerated volunteers found, as predicted, the highest levels among the most violent offenders: rapists who had also beaten, stabbed, or shot their victims. But a follow-up study by the same group failed to replicate the earlier finding.[20,21] Two studies based on saliva or CSF levels of testosterone—more accurate measures of the amount of active hormone than plasma values—reported a significant association between violent behavior and testosterone levels among convicted offenders.[22,23] However, Canadian investigators who measured levels of nine steroid hormones, including testosterone, in men charged with a violent crime and those charged with nonviolent property offenses could not detect a single statistically significant difference between the two groups.[24]

Law-abiding men receiving testosterone to treat sexual dysfunction or endocrine disorders don't fall into violent rages.[11] And a recently completed study of anabolic steroids, athletic performance, and emotional behavior concluded that steroid use, contrary to locker room horror stories of "'roid rage," does not, according to study director Shalender Bhasin, "turn men into beasts."[25]

But there's another side to the testosterone story. Male hormones *are* an important factor in male sexual behavior—and in primates, sex and status go hand in hand. Could testosterone have more to do with the desire to win than the urge to kill?

SOCIAL HISTORY—sexual activity and, especially, fighting experience—has a profound impact on primate testosterone levels, coincident with an intricate social structure and a corresponding need for greater neural flexibility. The period of social instability that follows a change in group organization is one of the best times to observe the effect of experience on hormones. Figure 6.2 illustrates the ups and downs in testosterone levels corresponding to a turbulent seven months in the life of Ribot, a socially mobile rhesus monkey studied by researchers at the Yerkes Primate Research Center in Atlanta.[26,27]

Scientists Robert Rose, Irwin Bernstein, and Thomas Gordon shifted Ribot between a variety of social environments, ranging from his own private cage to a sixty-eight-member mixed-sex breeding colony. Each move prompted

Figure 6.2

Testosterone and social change: the story of Ribot. Hormone levels in this adult male rhesus monkey rose and fell in conjunction with changes in his social environment that altered his social status.

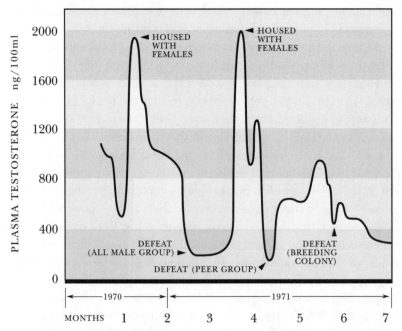

Adapted with permission from R. M. Rose, I. S. Bernstein, and T. P. Gordon, Consequences of social conflict on plasma testosterone levels of rhesus monkeys, *Psychosomatic Medicine 37* (1975): 50–61.

a change in Ribot's social status—and a corresponding change in his testosterone level. When the team transferred him to a compound housing thirteen sexually receptive females, his status and his testosterone levels soared. Ribot was less successful when pitted against other males in larger social groups. He lost fights—and status—in an all-male group, a mixed-sex peer group, and the large breeding colony; each defeat was followed by a drop in testosterone. In contrast, hormone values in the victorious males who defeated Ribot increased three- to fourfold over the values observed prior to the status competitions.

The outcome of each conflict was the critical determinant that regulated both testosterone and status. Winning, with its positive effect on social standing, stimulated testosterone; losing, on the other hand, not only resulted in a loss of rank, but also depressed hormone levels.

These changes reset response mechanisms to bias future behavior. Defeat "produces the fall in testosterone and that . . . lower level decreases the probability of aggressive action on the part of the subject."[26] Conversely, the rise in testosterone that follows a successful status contest set the stage for additional aggressive maneuvers to defend the newly achieved position and to improve rank even further.

Winning and losing also influence human testosterone levels. Syracuse University researcher Allan Mazur found that among a group of men who competed for a hundred-dollar prize in doubles tennis matches, those who won decisively experienced a postgame rise in testosterone levels, while their defeated opponents recorded a substantial drop.[28] Teammates who won their matches by only a narrow margin, neutralizing their feelings of success, experienced a loser's drop, rather than a winner's rise, in hormone levels. Similar increases and decreases have been reported in college wrestlers after winning or losing matches and in military recruits during the low-status period of basic training.[29,30]

In rats, quarrels over dominance often erupt into physical aggression. Primates, with a more sophisticated range of social communication options that includes facial expressions and eye contact, as well as posture, odor cues, and vocal interchanges, can often settle status contests before they escalate to a direct confrontation. Humans, who enjoy the added benefit of language, have the greatest opportunity of all to achieve superiority without bloodshed. But when more refined methods of allocating rank fail, men, who struggle as tenaciously over status as the males of other species, may well resort to violence to resolve the question. As a consequence, the relationship between male hormones and male aggression is intimately linked to the relationship between testosterone and status seeking. The interaction is nicely illustrated in a study comparing testosterone levels in aggressive prison inmates, nonaggressive offenders at the top of the pecking order, and mild-mannered prisoners who were followers rather than leaders.[31] As predicted, the aggressive prisoners had high hormone levels. But so did the socially dominant, nonaggressive group; in fact, the difference in testosterone between the two was negligible. Losing suppressed testosterone. But winning, by violence or attitude, had an equally positive effect on male hormones.

Among primates, status competitions add a flexibility to the social structure that encourages promotion of those individuals most capable of defending the group and preserving the peace. The hormone-reinforced outcomes of those competitions also stabilize the social order and discourage low-ranking individuals from initiating or perpetuating fights they cannot hope to win. In the external world, visible emblems of status—appearance, posture, privilege—offer a public record of social successes or failures. In the brain, high levels of testosterone and serotonin replace luxury cars and trophy wives as status symbols. Together, these chemical messengers keep score of the never-ending struggle between the status quo and the quest for advancement.

Boys Will Be Boys

Princeton University criminologist John DiIulio calls them "superpredators." They're the kids under eighteen who are thought to account for 25 percent of the violent crime in our country.[32] Between 1989 and 1994, arrests for violent crime (murder, rape, robbery, or aggravated assault) rose 46.3 percent among fourteen- to seventeen-year-olds, three to four times the increase observed in other age groups.[33] And although figures for 1995–1996 suggest that the violent crime rate for juvenile offenders finally may have entered a downward trend,[34] experts predict that the worst is yet to come, when the 40 million children currently under age ten reach adolescence in the inaugural years of the twenty-first century.[35]

While we contemplate the implication of these figures, however, we need to consider one final demographic detail. The faces of tomorrow's teenage killers belong overwhelmingly to today's little boys.

Men are nine times more lethal than women, and adolescent boys are already dangerous. Can it be mere coincidence that this is the same time that escalating levels of male hormones flood the body and the nervous system? Is puberty the culminating event in a hormonally driven process of sexual differentiation that begins edging toward aggression even before birth?

Studies in mice—a species in which prenatal hormonal manipulations are performed unwittingly by siblings—highlight the organizational role of sex hormones during brain development, an influence that will bias subsequent endocrine responsiveness. Female mouse fetuses positioned in utero between two males are exposed to more of their brothers' testosterone and less estrogen than females who develop next to a sister. As pups, these "2M" females have higher blood levels of testosterone; as adults, they boss other females and behave even more aggressively following testosterone injections. Males who spend the first weeks of their lives between two female fetuses, on the other hand, are less responsive to testosterone as adults.[36]

After birth, testosterone continues to oversee the process of sexual differentiation that began in the womb. Now hormones guide the first stages in the development of sexually specific behavior patterns—in males, for example, the propensity to attack other males rather than females.[37]

By the time mice and other rodents reach sexual maturity, the weapons are in place. Exposure to testosterone during the critical period of rapid brain development has titrated hormonal sensitivity and selected a battery of sex-specific responses suitable for use in threatening circumstances. Puberty finalizes mature patterns of social interaction: sexual advances toward females and the stereotyped behavior patterns that characterize adult aggression toward unfamiliar males.[11,38]

But now a second factor, context, begins to play an increasingly important role. The outside world takes the male behavior pattern defined by testosterone, a competence that might be called "aggressive efficacy," and specifies when, where, and how it should be used. As environmental information accumulates, testosterone shares the responsibility for regulating aggressive behavior with experience, acting as a facilitator and historian, not a dictator.

An experiment by psychologists Renee Primus and Carol Kellogg, of the University of Rochester, demonstrates how sexual maturity brings a new awareness of external cues, linking testosterone, perception, and response.[39] Female rats, paired with a stranger in a familiar cage or a new, previously unexplored enclosure, were equally sociable—sniffing, nudging, and grabbing—in both environments. Males, on the other hand, were much more sensitive to their surroundings. Beginning at puberty, male pairs placed in an unfamiliar cage tended to avoid one another, while pairs confined to a familiar environment interacted far more frequently, a difference prevented by castration at an early age. Testosterone, in other words, did not simply release or drive social behavior, but governed the ability to match key features of the environment with specific responses. A rat—or a human—who cannot make such distinctions is guaranteed to do the wrong thing at the wrong time.

Testosterone levels in boys begin to surge at approximately age ten, rise rapidly over the next several years, and finally begin to plateau at about age fourteen, when aggressive behavior starts winding up.[40] Coincidentally, this is also an age of rapid social change. Peer group rankings shift daily, with petty quarrels, athletic competitions, academic events, and girls offering ample opportunities to win or lose face. Conflicts with parents and other adults mark an urgent determination to muscle into the broader social structure, conflicts with higher stakes and a greater risk of defeat. Testosterone, now available in adult-strength doses, dutifully records each success or failure.

But is it the rise in testosterone that causes violent behavior in adolescent boys? Data collected over the past decade by psychologist Elizabeth Susman (currently of Pennsylvania State University) and colleagues at NIMH and the National Institute of Child Health and Human Development fail to support a relationship between antisocial behavior, including aggression, and elevated testosterone levels; in fact, these studies suggest just the opposite.[40-42] Delin-quent behavior, adjustment problems, and rebelliousness in adolescent boys aged ten to fourteen were associated with *lower,* not higher, testosterone levels. On the other hand, a Scandinavian study of fifty-eight Swedish boys between the ages of fifteen and seventeen (the last stage of puberty) did find a significant relationship between plasma testosterone levels and verbal, as well as physical, aggression.[43,44] The most interesting feature of this study, however, was not what it said about the relationship between hormones and aggressive behavior, but what it said about the relationship between hormones and perception. Testosterone was most clearly linked to aggression when the violent behavior was a reaction to an overture perceived as a threat. Boys with higher levels of testosterone were more impatient and irritable, readily frustrated, and quick to take offense. The tendency to start fights, rather than fight back, was less strongly correlated with hormone levels. Testosterone, the report concluded, was less likely to drive aggression than it was to magnify a boy's "readiness to re-spond vigorously and assertively to threats and provocations."[44]

THREAT PERCEPTION involves neurotransmitters as well as hormones, in par-ticular, the norepinephrine-containing neurons of the locus ceruleus that form the central branch of the global sympathetic response system. The power of this system lies in its flexibility, in its aptitude for tailoring the level of arousal to the level of risk posed by the environment. Each individual is born with an innate sensitivity to environmental stimuli—some more reactive, others less so. But the final working level of arousal is set by experience, by the nature of the inter-actions between the child and the outside world. This postnatal maturation process is reflected in the gradual development of the startle reflex.

"From a developmental point of view, the startle reflex goes through some major changes in early childhood," observes Robert Pynoos of the UCLA Neu-ropsychiatric Institute. Confront adults with a sudden high-pitched tone, and they jump, Pynoos explains. But warn them a fraction of a second earlier, and they're not so quick to startle. Two-year-olds don't habituate this way. They react even more intensely to the second stimulus, which, he notes, explains why his young daughter became more and more frightened during her first fireworks display. The tendency to overreact peaks between four and five years of age, then gradually declines.[45] However, reaction times don't fall to adult values until about age eight, just before pubertal hormones kick in.

Pynoos points out that the evolution of the startle reflex coincides with developmental changes in self-defense strategies. "If you put young children in a dangerous circumstance, they'll turn away from the danger and seek out protection. But about age eight, you start to hear the child fantasize about intervening actively: 'I wish I could have gotten the gun away from him.' It's a precursor to adolescence, when kids make real decisions about whether to intervene or not."

The hormonal changes of puberty therefore collide head-on with the first mature responses of the brain's alarm system, in the volatile context of an unstable social structure and frequent clashes with authority. But what happens when the alarm system has already been overloaded, when an abusive parent, daily exposure to a violent, dangerous neighborhood, or repeated harassment by a troubled sibling have made it clear that the world is a dangerous place? Pynoos and his colleague, Edward Ornitz, analyzed startle reflexes in school-age children suffering from PTSD and found that their startle responses failed to habituate (Figure 6.3).[45] "Instead of a ten-year-old pattern, their responses look like a five-year-old pattern," he says. As a result, these children continue to overreact to environmental cues, their catecholamine hazard-detection systems placed on alert by trauma. Instead of relying on adults to protect them, as a five-year-old might, they've got a ten-year-old's determination to solve their own problems. Tragically, the result can be an adult decision to use a gun.

Adolescence is a time when perceptual and response patterns formed by years of interaction between brain and environment are tested in adult society for the first time. When this process has gone well, risk assessment will be accurate, the need to resort to violence minimal. When it has gone poorly, gaps, misconceptions, and excesses will be clearly exposed—and violence may well ensue. Hormones may be the most obvious biological event of puberty, but adolescent violence, like adult violence, is more likely to be the product of a brain that has developed a misguided notion of threat than a brain fired up by testosterone.

The Thrill of the Chase

Fred S. Berlin, M.D., Ph.D., has his Baltimore office in a stately red brick row house only a few blocks from the apartment building I lived in as a graduate student. Meticulously restored and tastefully furnished, the exposed wood, the softly faded oriental carpets, the little watercolors of the Johns Hopkins Hospital suggest a Victorian rectitude not quite congruent with the electronic whine of the fax machine. This afternoon, I am the only visitor.

Dr. Berlin waits at the top of the second-floor staircase, a friendly man who laughs easily and quickly puts a newcomer at ease. That kind of accep-

Figure 6.3

Trauma delays development of the startle response. A normal eight-year-old (light bar, far right), like an adult, responds less vigorously to an anticipated stimulus. But in a child who developed PTSD after witnessing a shooting, the startle response is enhanced, rather than inhibited—just as it would be in a much younger child. Maturation of the startle response was delayed for two years by the trauma.

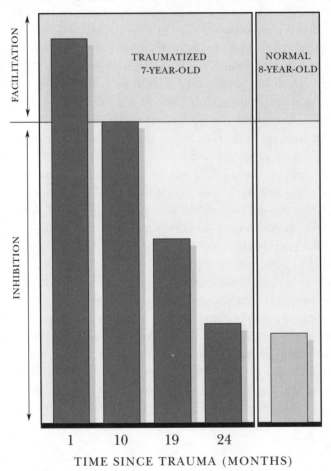

TIME SINCE TRAUMA (MONTHS)

Adapted with permission from R. S. Pynoos, A. M. Steinberg, and E. M. Ornitz, Issues in the developmental neurobiology of traumatic stress, *Annals of the New York Academy of Sciences 821* (1997), 176–193.

tance is important here, for this is not an upscale psychiatric practice dedicated to assuaging the anxieties of worried urban professionals. The men who pour out their hearts to Berlin are pedophiles and exhibitionists, rapists and sadists, voyeurs and obscene phone callers—men the rest of the world fear and loathe.

Berlin's twenty-year-old mission has been to care for and to understand these men, psychiatry's untouchables. Like many other physicians, his selection of this unorthodox specialty was inspired by a patient. "I was just a resident," he recalls, "and I had a man who came in and turned out to be a sexual sadist. He'd actually fashioned a club out of wood and had a chain on it, and he was fantasizing about assaulting his wife with it." The patient found this idea highly erotic. His wife did not. A psychologist as well as a physician, Berlin puzzled over the possible origins of this strange and terrifying behavior, and he was troubled by his recognition of a clinical void, by the realization that although "we had treatments for schizophrenia, depression, and a whole range of other disorders, we had virtually nothing when it came to sexual disorders. I saw an incredible need."

Malicious or merely repulsive, the sexually disordered trouble law enforcement officials, judges, legislators, and mental health professionals—and fascinate writers, talk show hosts, and television viewers. Perhaps, Berlin suggests, because sex is involved, no other aberrant behavior has more power to evoke moral outrage. Surely, many insist, no other form of male violence could be more unequivocally driven by male hormones.

If sexual aggression is an endocrine problem, removing the offending hormones must be the solution. This sort of thinking led the California legislature to enact tough new laws in September 1996 mandating "chemical castration"—periodic injections of the testosterone-lowering drug Depo-Provera—for child molesters convicted of a second offense.[46] Under the new law, even first-time offenders can be remanded for treatment if a judge deems their offense especially horrific. Assemblyman Bill Hoge, architect of the law, told a reporter from the *New York Times,* "There is no crime out there more heinous than child molestation. . . . Californians are fed up with those who prey on kids and we're going to jail them and make it so that when they return to society, they don't have the desire to do what they used to do."[46]

Legal advocates protest that the California law violates everything from the right to privacy to the right to refuse medical treatment. But as aggression researchers could tell them, the real problem is even more fundamental. Castration laws are not only legally questionable but biologically unsound, based on the belief that by manipulating hormones, we can control aggression as well as desire. For those like Fred Berlin who treat sex offenders, this is the most dangerous myth of all.

One of Berlin's favorite quotes is an observation by John Hughlings Jackson, the same nineteenth-century neurologist who cautioned against overly simplistic associations between anatomy and behavior: "The study of causes must first be preceded by the study of things caused." Berlin's research on the thing caused, sexual aggression, has convinced him that oversimplification is once again the root of widespread misconception—that deviant sex, like violence, takes a variety of forms. Mandatory chemical castration and other global solutions miss the mark, he believes, because sexual desire is not the underlying factor motivating all sex offenders.

For example, some sex offenders are psychotic or disoriented by drugs and alcohol.[47] As a result, their ability to distinguish a resistant partner from a consenting one is compromised. They may not fully understand that their actions are harmful to the victim, or they may be too disinhibited to worry about it. These individuals aren't likely to be good candidates for antitestosterone therapy. Both their violence and their sexual behavior are secondary to their mental health problems, and although they may well benefit from drug treatment, it's more likely to be therapy with medications that reduce delusions or treat addiction than hormone blockers.

For other rapists and child molesters, sex is a weapon, not an appetite. Some assault their victims impulsively, out of anger or on a dangerous whim. Especially likely to lose control when they've suffered a humiliating defeat—a job loss, a divorce—they channel their rage into a sexual format, but what they're really after is revenge, not satisfaction. Their behavior is actually impulsive aggression in which rape is substituted for a gun, as one rapist's description of his unpremeditated attack on a couple in a parked car illustrates:

> When the marriage broke up . . . I didn't know what I was gonna do. I just wanted to hurt somebody. I took the gun and ski mask out of the trunk, walked up to the two in the back seat naked, pointed the gun at them, told them to get out of the car, put the boy in the trunk of the car. The girl was standing there. She said, "What do you want?" . . . I didn't want, you know, sex. . . . It was I wanted her crying and upset like I was fantasizing about my wife.[47]

A second group of offenders more focused on violence than on sex are those motivated by antisocial tendencies. Again, the sex act itself is less important than immediate gratification. These attackers aren't necessarily angry with their victims; in fact, they may not care about them at all:

> I've been a person that whatever I wanted, I took. . . . I see this woman and I say to myself, "I want her," and I took her. . . . The motivation was selfish. . . . I wouldn't consider another human being's right to own property or to live without the fear of somebody intruding upon their reality. . . . I saw her and I just raped her in her apartment.[47]

Reducing testosterone levels in impulsive or antisocial sex offenders isn't likely to have much of an impact on their behavior because their principal problem is aggression, not sex. Put simply, when you take away sex from a rapist who's also a psychopath, you're still left with a psychopath.

Empirical studies support this contention. Trials of Depo-Provera and other testosterone-blocking drugs, involving hundreds of sex offenders, have demonstrated that although these drugs act like a bucket of cold water on sexual behavior,[11] they have little or no effect on antisocial aggression.[48,49] In fact, when recidivism rates during antitestosterone therapy are broken out into sexual and nonsexual (including violent) offenses, it's only the sex crimes that decline among the treated individuals.

Sex drives aggression for just one type of offenders: those motivated by intense, uncontrollable cravings for socially unacceptable sex acts and partners, disorders psychiatrists call "paraphilias."[47] The activities most men find pleasurable neither arouse nor satisfy them. Instead, individuals with paraphilias are erotically attracted to behaviors most people find repugnant: fondling children, exposing themselves, stealing women's underwear, beating and humiliating a submissive partner. They are not always vicious. They may actually profess to love their victims, oblivious of the hurt and trauma their "affection" causes.

Paraphilias are disorders of sexual orientation, and Fred Berlin emphasizes that orientation "is not a volitional choice." But experts disagree on how the brain and the environment intersect to make this choice for the affected individual. Some believe that specific sexual preferences may be present from birth. A higher-than-normal incidence of neurological deficits or chromosome aberration has been cited by others. And some paraphilias appear to be "imprinted" by sexual abuse or other early childhood experiences. Everyone does agree that once sexual desire has been cued to inappropriate outlets, redirecting orientation is incredibly difficult, even when the afflicted person is miserably unhappy with his lot.[50]

Berlin has treated hundreds of paraphilia-driven sex offenders, including the sadist wielding a club and chain, with Depo-Provera and believes that in many cases, the intervention has "turned these people's lives around." The intense urge to indulge the forbidden fantasy ebbs; as it does, the risk of offending declines. Long-term follow-up of more than six hundred sexually disordered men treated by Berlin or his colleagues at the Johns Hopkins Hospital with testosterone-blocking drugs and specialized group therapy showed that five years after starting treatment, more than 90 percent had successfully resisted the temptation to strike again.[51]

Sexual orientation may be fixed early in life, but sexual behavior itself is malleable. When a sex offender truly wants sex rather than revenge or cheap thrills, lowering testosterone levels can make a difference—not because it controls violence, but because it dissects out and regulates the sex drive clamoring for an outlet.

But what if sex and aggression can't be disentangled? What if the deviant fantasies require suffering, a dead body? When paraphilia takes a violent or homicidal turn, aggression engulfs sex, testosterone becomes irrelevant, and turning back is no longer possible.

TED BUNDY. John Wayne Gacy. Jeffrey Dahmer. There are violent, dangerous men, and then there are men who are not simply abusive or hostile or vengeful, but predatory. They are men for whom stalking and killing have taken the place of foreplay and for whom homicide is the definitive feature of their paraphilia.

Serial killers are rare but prolific. Numbering somewhere between fifty and five hundred individuals at any given time, authorities estimate that they may account for up to 10 percent of homicides each year.[7] Death tolls attributable to a single killer that total dozens of victims are not unusual. Not all are motivated by sex. But those who do murder in search of sexual satisfaction can be readily distinguished from the vengeful or the rapacious. Over time, their murders assume a unique, ritualistic pattern that is the telltale symptom of a paraphilia as well as the foundation of a psychological fingerprint, or profile, that can guide police efforts to track down the killer.

Forensic psychiatrist Robert Simon observes that serial killers "do not want a partner, they crave a victim."[7] Careful observation suggests that the term *predator* to describe these sex offenders is eerily accurate. Discrete and cunning, serial sexual killers don't start arguments with their targets, but stalk, lure, or ambush them. As labels like the "Hillside Strangler" and the "Boston Strangler" suggest, more than a few shun handguns, preferring instead to strangle or suffocate victims in the fashion of many animal predators. Some even eat their prey.

Lust murder is a strange form of predation, one in which the thrill of the chase and the sight of blood are as important as consumption. More than mere sadists, they are sexually deviant psychopaths whose thirst for arousal is so profound that they need murder and sadistic sex to slake it, killers so distorted they merit their own aggressive category: predatory psychopathy.

Behavior that has escalated to the point of spontaneity is not likely to respond to quick fixes. Simon observes that predatory psychopathy bears a suggestive resemblance to sensitization phenomena such as kindling. He notes that serial sexual killers tend to kill more frequently—and are progressively crueler to their victims—over time, as if the behavior was growing more and more attractive as the ability to control it was slipping further and further out of their grasp. The idea is speculative, Simon admits. But "because of the escalating pattern of killings, the relief that some murderers feel after the killings, the quickening of the cycle, and the out-of-control feelings, the kindling model seems to fit."[7] The sexual needs of predatory psychopaths are so inexorably enmeshed with their antisocial aggression that they have slipped beyond the reach of Depo-Provera. Many experts fear that they may be beyond responding to any intervention, chemical or otherwise.

The sex offenders California lawmakers hope to control with drugs include the impulsive as well as the deviant, psychopaths as well as pedophiles, drug addicts as well as sex addicts. If Berlin is right, some will lose their attraction to forbidden partners, but many will not overcome their attraction to violence. Whatever the law's political benefit for those eager to appear tough on crime, it unwittingly provides an opportunity to test the hypothesis that sex and aggression are biologically distinct events. The problem is that the outcome variable will be the lives of women and children.

A Feminine Perspective

The focus on testosterone, sexual predators, and male homicide rates has reinforced the idea that violence is a masculine prerogative. Men are aggressors. Women, on the other hand, are peacemakers—unless they are targets.

Women are raped, beaten, or attacked by someone they know seven times more frequently than men;[52] even as girls, they're two to three times more likely to be victimized by child molesters, rapists, or sexually abusive family members.[53] Trauma perpetuates fear. The most extreme example of social defeat, it also profoundly alters the likelihood of future aggression by resetting chemical parameters to favor defensive, rather than offensive, behavior. Once injured, the fear of retaliation constrains women's aggression. When the aggressor views leaving as an openly hostile act, as is the case in many violent marriages, fear may even prevent women from running away. Men—bigger, heavier, and quicker to pick up a gun—fight back. Women, all too conscious of their vulnerability, often have few options but to freeze.

Women, however, are not divided into pacifists and victims. As researcher Kathryn E. Hood, of the Department of Human Development and Family Studies at Pennsylvania State University, hastens to point out, "Women can and occasionally do murder and maim."[54]

Ask Darlie Routier. She's on death row, convicted of stabbing her five-year-old son Damon to death with a butcher knife.[55] Ask any of the forty-two other women currently awaiting execution for killing husbands, boyfriends, their children and grandchildren, police officers, strangers.[55] Or ask Donna Ratliff, an Indiana teenager who burned down her own house—while her mother and sister were inside.[56] Ask Jacqueline Williams, who wanted a baby so badly she helped her boyfriend murder a pregnant woman and slice the live fetus from her womb,[57] and Jillian Robbins, who opened fire on the lawn outside the Penn State University Student Union, killing one student and seriously wounding a second.[58] Ask mothers videotaped in the act of hitting, choking, or smothering their infants by researchers studying Münchausen syndrome by proxy—intentionally injuring their children in a play for attention.[59] Predatory serial sexual killers are men, but you could ask Aileen Carol Wournos what it is like to be a serial *enterprise* killer, robbing and murdering at least seven men foolish enough to offer her a ride.[7,55]

Dr. Hood is right. The question is not whether women are aggressive, but how and why.

EXAMINED in the proper context, females, including human beings, may not be less aggressive as much as they are differently aggressive. Women, for example, are less obvious about their aggression. To protect themselves from both social disapproval and painful retaliation, they learn to select methods and victims (e.g., children) less likely to provoke public scrutiny. Kathryn Hood notes that while adults insist that boys are more aggressive than girls, the two sexes, when questioned directly, rate themselves as equally aggressive.[54] Who's right?

A long-term investigation conducted by Hood and colleagues at the University of North Carolina suggests that neither side is exaggerating; boys and girls just come to choose different weapons.[54,60,61] The Carolina Longitudinal Study found that at age ten, both sexes tend to settle their differences openly, using fists, feet, and verbal insults. But by age thirteen, the girls have stopped shouting and punching. Their teachers believe they've matured into peaceful citizens. But the girls themselves reveal that what has changed are their tactics. They've replaced physical violence with emotional violence: ostracism, snubbing, gossip mongering, and backstabbing. Their aggressive behavior, in Hood's words, hasn't vanished; it has "gone underground, invisible to nonintimate peers and adult authorities." Because their aggression is "no longer a classroom management problem," they can act out freely without incurring punishment. The preference for covert methods over direct confrontation may persist into adulthood. In a 1994 study of aggressive behavior on the job, Scandinavian investigators found that working women were just as aggressive as their male colleagues. But like the Carolina teenagers, they were more likely to express their hostility by starting rumors or "forgetting" important assignments than by shouting at the secretary.[62]

Studies in animals also chart a different course for female aggression. Females fight over food and space, to punish insolence and to retaliate for attacks on a relative, but the patterns of their aggressive behavior often differ significantly from those of males. Among mice, for example, males are more hostile to strangers when they're separated from their peers at weaning and housed individually. Isolated females, on the other hand, barely notice an intruder. Their aggressive tendencies are strongest when they're group housed with four or five of their sisters and cousins.[63] Male aggression emerges like clockwork at puberty, while female aggression develops at a more leisurely pace. In a study comparing the development of aggressive behavior in male and female mice from a selectively bred strain, fighting between males escalated sharply at about 46 days of age, when the animals reached sexual maturity.[63,64] Female aggression, however, remained dormant for another six weeks, when it doubled at 90 days, then increased steadily over the life span, peaking at the hoary age of 270 days.[63,64] Given repeated opportunities to fight, older females grow even fiercer, attacking more quickly and more often until their aggression equals that of their male contemporaries. And mouse mothers dispel any lingering doubt that females can fight when necessary. In fact, nursing mothers are the most aggressive mice of all, attacking an unknown intruder up to twice as often in a single encounter as a fully mature male.[64]

FOR BOTH SEXES, the cumulative exchanges between the individual nervous system and personal experience that determine behavior are recorded in the language of neurochemistry. In men, the chemical account includes a history of the reciprocal exchanges between testosterone and fighting experience. It would be surprising if hormones did not contribute to emotional record keeping in women as well. But which hormones? Do violent women have man-sized testosterone levels to match their warlike behavior?

Some experts might say yes. Psychologist John Dabbs, one of the investigators who detected a significant association between hormone levels and violence in male prisoners, found evidence of a similar link in violent female offenders and, more recently, among a group of violent, drug-abusing young adults that included girls as well as boys.[65,66] But outside prison walls and delinquent populations, the bond between testosterone and female aggression begins to disintegrate. Testosterone levels in adolescent girls (testosterone rises in females as well as males at puberty) bear no relationship to their aggressive or defiant behavior toward parents and peers.[40,41] Women suffering from endocrine disorders that drive testosterone to levels as much as 200 percent of normal values have more hair and lower voices, but not quicker tempers.[11] And although women who take male hormones along with estrogen after menopause swear it enhances their libido, they don't report a sudden upswing in violent behavior.[11]

Maybe the problem isn't male hormones specifically, but sex hormones in general. Perhaps testosterone has responsibility for regulating male aggression—albeit in a complex and environmentally sensitive fashion—while estrogen, the female counterhormone, oversees women's aggression.

Women themselves are often quick to blame a bad temper on their hormones—more specifically, on their monthly menstrual periods. Along with physical symptoms ranging from headaches to weight gain, the hormonal trough that precedes menstruation is thought to mark a dangerous period of irritability, anxiety, and potentially murderous mood swings. Evidence in favor of a link between the menstrual cycle and aggressive behavior first surfaced in the 1960s, when researchers analyzed monthly cycles in female offenders and determined that nearly half had committed their crimes just before or just after the beginning of menstruation.[67] Subsequent studies profiled a premenstrual syndrome (PMS) thought to explain episodic hostility as well as physical discomfort in affected women—and even advanced as a legal defense by some.[68,69]

As is the case for testosterone, however, the case for PMS—and the changes in estrogen and progesterone presumed to underlie its physical and behavioral manifestations—as an explanation for female aggression is far from ironclad. From the start, the argument has confused aggressive behavior, crime, and incarceration rates. The majority of subjects in the original studies hadn't committed violent crimes, but minor property offenses.

In addition, there's the question of exactly how premenstrual hormones exert their effects. A Dutch study examined the emotional responses of premenstrual women to anger-provoking situations and compared them to the responses of women tested midway into their monthly cycle.[70] The premenstrual group, especially those who reported symptoms of PMS, *did* react more angrily to provocation. But just like school boys with high testosterone levels, the "hormonally driven" aggression of these women had all the telltale signs of an overreaction to threat: changes in blood pressure, cortisol levels, and feelings of anger. The researchers concluded that the hormonal fluctuations associated with the premenstrual period "have the effect of making women more susceptible to responding emotionally to negative life events," that is, that their primary effect is to change the perception of threat rather than directly fueling aggressive behavior.

Finally, the shifts in mood and hostility associated with PMS appear to be suggestible. A recent study by Hood pitted college-age women with diametrically opposed viewpoints in a series of four weekly debates.[54] Videotapes of the encounters revealed that women who were contentious in the first debate were equally outspoken in all of the others. Diary records of their moods and emotions over the same thirty-day period were also consistent; their feelings of anger, frustration, depression, or anxiety varied little from week to week. But asked at the end of the observation period to "describe the changes during their most recent menstrual cycle," they invariably reported "feeling aggressive" in the week prior to menstruation. The women were convinced that PMS had made them belligerent. Their actual thoughts and actions showed otherwise. As the investigators concluded, "Women's beliefs about their own biology . . . do not reflect their own behavior."[54]

PREMENSTRUAL SYNDROME doesn't necessarily live up to its bad reputation. But estrogen is still a surprisingly important dialect in the chemical language of aggression. During adolescence, in fact, it makes a more presentable contribution to teen angst than testosterone. The research team from NIMH, the National Institute of Child Health and Human Development, and Penn State University that found lower testosterone levels in aggressive adolescent boys also recorded angry, defiant, or manipulative behavior by teenage boys and girls while they worked with their parents on a problem-solving task, then correlated the incidence of aggression with hormone levels. The results in boys were not significant. But for the girls, hostile outbursts, especially toward their mothers, were associated with higher levels of estrogen.[71]

Estrogen's impact on female aggression has also been tested directly. Elizabeth Susman and coworkers at Penn State compared the incidence of fighting and other aggressive behaviors in boys and girls who received hormone therapy to treat delayed puberty.[72] Aggression escalated rapidly in girls who received estrogen, even at the lowest dose prescribed. Boys, in contrast, had to take a higher dose of testosterone for a longer time before their aggression increased; at low doses, they were significantly *less* aggressive than estrogen-treated girls.

Estrogen may even be the principal hormonal player in male aggression. In mice, it is more potent than testosterone at reversing the antiaggressive effect of castration. From a biochemical point of view, this is not as radical as it seems, for testosterone is really only a hormonal way station, an intermediary readily converted by a brain enzyme into estrogen. As a result, testosterone's influence on aggression may actually be mediated in part by female hormones. Estrogen, Susman suggests, may do double duty at puberty, facilitating conflict in both boys and girls.

The violence of men and the violence of women differs in form, substance, and chemistry. But it is the result of the same biological process. Regardless of sex or age, behavior is not simply driven by a single hormone or by one neurotransmitter, but evolves out of the interlocking relationship that links perception, interpretation, and response, the circle that joins the brain and the surrounding world.

Threat: A Hormonal Call to Arms

Conflict is stressful. Whether an individual is a perpetrator or a victim, the aggressive encounter activates neurochemical mechanisms dedicated to crisis management. The sympathetic nervous system and the catecholamine transmitters norepinephrine and epinephrine constitute an arm of this emergency response system. But the task is too important to be left entirely in the hands of one agency, no matter how competent. Managing stress effectively requires a partnership between the nervous system and the endocrine system, between catecholamines and the glucocorticoid "stress hormones" secreted by the inner core of the adrenal gland.[73]

However, glucocorticoids have the authority to do more than fight fires. Chemical power brokers, they barter with brain monoamine pathways for a voice in regulating the stress response. Their contribution defines the critical boundary between coping and overreacting, between a nervous system capable of withstanding stress and a system spiraling out of control.

Everybody knows that hormones facilitate aggression. But only some are thinking about the right hormones.

DOCKED AT its home base in the hypothalamus, the wellspring of the endocrine stress response—corticotropin releasing factor (CRF)—waits for the announcement of a threatening event, which it delivers without delay to the pituitary. The aroused pituitary dispatches the next hormone in the series, adrenocorticotrophic hormone (ACTH). The circulatory system carries ACTH to the adrenal gland, which, in turn, secretes glucocorticoids, along with norepinephrine and epinephrine.

The structure of CRF was long one of the best-kept secrets in biology. Scarce and fragile, isolating enough of it to verify its identity from the complex jumble of brain proteins proved as difficult as drawing a single intact hair from a tightly woven braid.[74–76] But for such an elusive peptide, CRF has a long reach. Its actions are not limited to a private conversation between hypothalamus and pituitary. Using fluorescent antibodies to CRF or radioactive probes that detect expression of the CRF gene, neuroanatomists have found CRF-containing cells in the prefrontal cortex, amygdala, hippocampus, raphe nuclei, and locus ceruleus—"all the correct places," observes leading neuroendocrinologist Charles Nemeroff, chair of the Department of Psychiatry at Emory University School of Medicine, "to mediate stress and emotion."

Like other hormones, releasing factors, and peptides that have wandered out of the hypothalamus and into the brain, CRF moonlights as a neurotransmitter. Activated neurons containing the peptide release CRF; across the synapse, it binds to specific receptors, initiates a cascade of biochemical reactions inside the cell, and precipitates physiological and behavioral expressions of stress.[77] CRF increases blood sugar, respiration, and heart rate; slows digestion; and kills interest in leisure-time activities like eating and sex. It intensifies startle responses and facilitates fear conditioning. When tiny doses of CRF are injected directly into the brain, rats become hyperactive and defensive; dogs, more vigilant and aggressive; and monkeys, nervous and clingy.

CRF and catecholamine-containing neurons enjoy a special and reciprocal intimacy. Noradrenergic inputs from the locus ceruleus to the hypothalamus feed directly onto CRF-producing cells; these fibers "turn on" CRF release.[4] Going the other way, CRF neurons projecting to the locus ceruleus return the favor, driving up firing rates and increasing plasma levels of norepinephrine and epinephrine.[77] Mutually reinforcing, CRF and norepinephrine collaborate to ensure that no challenge—to survival, well-being, or peace of mind—goes unmet.

Once the brain has categorized an event as threatening, glucocorticoids and catecholamines coordinate the call to arms. The sympathetic nervous system is the headstrong member of this partnership. It rouses the body to action, energizing the individual to flee or fight. Glucocorticoids protect and direct.[78,79] They deploy immune cells to the lymph nodes or to wound sites, while corraling wasteful autoimmune responses. Their firm management prevents other volatile responses to the stress of injury—inflammation, swelling, fever—from spiraling out of control. In the brain, glucocorticoids remind stress-charged monoamine systems to stop, look, and listen. By means of strategically placed receptors in the limbic system and cerebral cortex, they check the surge in norepinephrine synthesis and release provoked by threat and promote the moderating influence of serotonin. They blunt startle responses, slow fear conditioning. Without the endocrine system, the stressed organism might slowly self-destruct; with glucocorticoid intervention, the impact of stress is cushioned.

But glucocorticoid secretion is itself a double-edged sword, capable of destroying the very system it is charged with protecting.[78–80] If the endocrine side of the stress response is not shut down itself after the immediate threat has passed, glucocorticoid excess can fuel the growth of atherosclerotic plaques, choke critical immune responses, whip insulin secretion into a frenzy, leach calcium from bone, even assault delicate neurons.

Feedback mechanisms are one way of containing stress hormones. Like other neurohormonal circuits, the relay between CRF, ACTH, and the glucocorticoid hormones (often referred to as the _"hypothalamic-pituitary-adrenal,"_ or _HPA, axis_) is regulated by hormonal input at the level of the brain as well as the pituitary. An HPA axis liberated from this self-control mechanism—for example, by surgical removal of the adrenal glands—allows the hypothalamus and the pituitary to do as they like; predictably, CRF and ACTH levels soar.[75] Postsurgical glucocorticoid injections reinstate control, restoring the balance between the brain, the pituitary, and the periphery.

The brain itself adds a second level of control. The catecholamine inputs to the hypothalamus that drive CRF release can also turn down CRF stimulation of the HPA axis (binding to alpha receptors is the "on" signal; binding to beta receptors, the "off" signal).[78] Serotonin, enhanced by stress hormones, responds by inhibiting further glucocorticoid activity.[81] The body's internal clock, timed by neurons in the suprachiasmatic nucleus of the hypothalamus, superimposes its daily rhythm on hormone secretion. In a way not yet completely understood, the steady beat of this pattern makes it easier for neural control mechanisms to shut off the stress response.[82] Above all, the limbic system—particularly the hippocampus, a structure rich in glucocorticoid receptors—monitors the progression of HPA activation, poised to wire the hypothalamus and shut off the fractious hormone at its source.[82]

LEGEND HAS IT that the young king Arthur was asked by his mentor, Merlin, whether he preferred the sword Excalibur or its scabbard. Without hesitation, Arthur voted for the sword.

"Then you are a fool," retorted Merlin. "The scabbard is worth ten swords. While you have it in your possession, no wound can kill you. He who lives by the sword can also die by it, but with the scabbard, you will emerge unscathed from even the fiercest battle."

The ability to respond to threat—the powerful sword wielded by stress hormones—can be life saving. But the ability to contain this response—the sheath fashioned by internal self-control mechanisms that curb catecholamine and glucocorticoid activation—is life preserving. Should these containment mechanisms weaken or fail, the protection afforded by the scabbard is lost. The loss uncouples the brain from the external world and issues an open invitation to behavioral catastrophe.

The Stress of Life

In the story, King Arthur's precious scabbard is stolen by the enchantress Morgan le Fay. In real life, control of the stress response can also be lost in a single calamitous moment—or it can slip away gradually, eroded by repeated exposure to threat, social pressure, loss, and uncertainty. Individually, each small insult has a negligible impact. But over time, the nervous system finds it harder and harder to rein in the stress response, as it edges progressively closer to losing control entirely.

The wear and tear of living in a group, of forging and sustaining friendships, outmaneuvering enemies, maintaining status—"the stress of life," as pioneer Hans Selye called it—can be a ready source of the kind of repetitive pressure that steadily shapes the perception of threat and the corresponding physiological response. Over time, the progressive reinforcement of individual response patterns acts collectively to stabilize the social organization and keep the group together—but at a physiological cost to some members. The payoff is security; the price, stress.

Not surprisingly, those who struggle along at the lower strata of society pay a higher price than those at the top. Take the less successful members of one of nature's most status-conscious primates, the olive baboon. Every year, Stanford University researcher Robert Sapolsky, an authority on the neurobiological consequences of stress, shelves his laboratory duties for several months and travels to the Masai Mara National Reserve in Kenya, where he studies the hormonal ups and downs of these baboons in their native environment. His observations demonstrate the physical toll of social stress in a species that needs to devote little time to finding food or fending off predators, leaving them, Sapolsky observes, with "hours each day to devote to generating social stressors for each other."[83]

Subordinate male baboons get no respect but plenty of grief. Routinely forced to hand over food, female companions, and sleeping places, subordinates are at the mercy of their superiors, who also like to beat up underlings as a way of playing down their own failures. For the unfortunate subordinates, all this abuse adds up to chronic, inescapable stress—and persistently elevated cortisol levels.[83]

Sapolsky and other stress experts believe that the high cortisol levels associated with "subordination stress" reflect stress response mechanisms that have escaped containment.[83] Neurobiologists can catch a glimpse of such runaway responses after a challenge dose of the synthetic glucocorticoid dexamethasone. When the feedback mechanisms that normally regulate cortisol secretion are intact, the extra steroid prompts the hypothalamus to suspend glucocorticoid release temporarily. Nonsuppression (the term used if the body ignores the dexamethasone turn-off signal and continues to pour cortisol into the bloodstream) is a sure-fire sign that the containment floodwall has been breached.

Sapolsky anesthetized wild baboons, dosed them with dexamethasone, then measured their cortisol responses over a six-hour period—and found that subordinate animals with the highest baseline cortisol values showed the weakest response to the drug.[83] Chronic subordination stress had frayed the physiological safety net that should have checked hormone secretion. The weapon that inflicted this damage, Saplosky suggests, may actually have been the very feedback mechanism designed to contain it. Studies in rats show that repetitive stress not only boosts corticosterone levels but also downregulates neural glucocorticoid receptors, in a misguided attempt to compensate for the excess hormone. As a result, the brain underestimates circulating glucocorticoid values and fails to alert the pituitary that it's time to slow down. The consequence is a self-perpetuating and self-defeating cycle in which stress elevates cortisol, high cortisol levels reduce neuronal sensitivity, unresponsive neurons lose additional control over the HPA axis, and the unsupervised adrenal gland continues to secrete too much cortisol.

But freedom from social stress can't be won simply by clawing your way to the top of the social ladder. Even dominant animals can be "stressed out" if individual differences and past experience have shaped a coping style that emphasizes quick reflexes over confident management. And intimidation isn't the only way of securing temperate cortisol levels and intact endocrine feedback mechanisms—the "executive health benefits" that are the physiological hallmark of social success.

Among Sapolsky's baboons, some dominant males do have the take-charge attitude you'd expect from upper management.[83–85] Confident of their superiority, they react only to clear instances of insubordination—and then they act swiftly and decisively. Because they "pick the right fights," they rarely lose, but when they do, they relieve the stress of defeat by thrashing the nearest available inferior. Not surprisingly, their resting glucocorticoid levels are comfortably low, and their endocrine systems respond briskly to dexamethasone, quickly and efficiently downregulating cortisol secretion.

Other males succeed by being loyal friends rather than strong leaders.[83-85] They're the troop members everyone wants as a grooming partner, the good sports who relieve weary mothers by entertaining the youngsters. These animals also enjoy low glucocorticoid levels—evidence that social stress can be reduced by affiliation as well as aggression.

For some dominants, however, life at the top can't bring peace of mind.[83-85] If they notice a rival, they drop whatever they're doing to fret and pace—even, Sapolsky notes, if the enemy happens to be sound asleep under a tree twenty feet away. Directly confronted, however, they hesitate, waiting to be hit first before fighting back. When they win, they still need to shore up their confidence by promptly attacking a subordinate. These touchy, suspicious animals have significantly higher cortisol levels than their self-assured colleagues, despite their comparable social standing.

The critical factor governing baboon glucocorticoid levels, in other words, is not status but safety. When a male baboon sees a threat behind every bush, he can't help but feel the steady pressure of stress, pushing for responses more likely to be impulsive than decisive, hostile rather than authoritative.

Sapolsky concludes, "There are lots of different ways to be a dominant male." But only some are conducive to the health and well-being of the stress response system.

THE SOCIAL ORGANISM, like its component members, is always changing. Infants mature into competitive young adults. A newcomer joins the group. A top-ranking individual sickens and dies, sparking dissent over who will take his place. One alliance coalesces, another collapses. Fortunately, such upheavals usually occur sporadically, giving everyone time and space to accommodate. But sometimes social change is cataclysmic. When a community doesn't merely regroup but recreates itself, conflict is inevitable. And where there's open warfare, there's sure to be stress.

Primate researchers can easily create anarchy simply by putting several individuals who don't know one another together in the same cage. Recent experiments carried out at the National Institute of Child Health and Human Development, for example, tracked behavior and stress responses in one of the world's smallest primates, the common marmoset, during community formation.[86] Groups of three male and three female marmosets, previously housed as heterosexual couples, were simultaneously released into large enclosures furnished with such primate amenities as branches, hiding places, and perches. But the six unfamiliar residents had little time to explore their new neighborhood. Chaos reigned, as both male and female monkeys jostled, threatened, and bullied their way toward a stable social organization. Aggressive behaviors, including direct attacks, increased tenfold over the number observed prior to the move. Cortisol levels soared, and feedback responsiveness to dexamethasone plummeted.

Six weeks later, peace had been restored. As the community stabilized and stress declined, aggressive encounters and cortisol values receded to baseline levels. But a silent reminder of the tumult remained, a lingering impairment in feedback sensitivity that was a reminder of how even a brief period of social instability had brought the stress response system a step closer to losing control.

ADOLESCENCE is an excellent example of the consequences of social upheaval for humans, as teens challenge their parents for authority and struggle to find their place in an unstable and contentious peer hierarchy. That some overreact to the stress of transition is hardly surprising. Their struggle is reflected in higher levels of aggressive or defiant behavior—and a hormone profile characteristic of an organism under stress.[40,71,87] The greater the reactivity to stress, as reflected in a brisker cortisol response to a distressing or unpleasant event (e.g., having an experimenter draw blood and conduct a physical examination) the greater the incidence of behavior problems.[87] Stress can even suppress an adolescent boy's rising testosterone levels, suggesting that male sex hormones influence behavior at the discretion of the environment.[40,87]

The outcome of stressful encounters in adolescence may have long-term implications for behavior, physiology, and neural development. For example, male hamsters thrashed daily by an aggressive adult during puberty were more submissive toward peers when they were adults—but more aggressive toward younger or weaker animals.[88] Subordination stress also increased the density of serotonin-containing fibers in the hypothalamus, a structure and a chemical pathway known to participate in the generation of aggressive behavior.

Most adolescents muddle through somehow. Their hostility fades as they find a niche in the social structure; their stress hormone levels rise and fall in concert with the ups and downs of everyday life. But hostile, impulsive adults have as much difficulty adapting to the stress of life as most adolescents, and become as enraged at every minor setback or petty annoyance as a teenage boy on the rampage. Their social life may not be chaotic to begin with, but their impulsive aggression manages to generate its own disorder on a regular basis.

Cardiologists know these people well. They originally lumped the chronically angry along with a larger population well known for their impatience, competitiveness, and irritability—people said to have "Type A" personalities. Researchers first discovered a link between Type A personality traits and an increased risk of heart disease in the 1960s.[89] But subsequent studies demonstrated that not all Type A individuals were equally likely to succumb to cardiovascular disease. Some seemed to stabilize early in the course of the disorder; ultimately they fared no worse than patients who had more agreeable personalities. Scientists began to reconsider the earlier data, attempting to zero in on the specific traits in the original Type A profile most closely associated with the risk of coronary artery disease.

Their results surprised many, including, no doubt, underachievers who had rationalized their indolence as "heart smart." Ambition and workaholism were not the culprits. Of all the Type A traits originally described by cardiovascular researchers, hostility turned out to be best predictor of future heart disease.[90] Corporate executives could breathe a little easier. Only hard-driven, intense people who also yelled at secretaries, fumed at the line in the copy room, and raged over the trivial mistakes of coworkers—individuals some researchers labeled "Type H"—faced a greater chance of developing clogged arteries and dying of heart attacks.

The Type H personality does its damage to the cardiovascular system by gradually uncoupling the stress response from control mechanisms. Each angry outburst calls in a false alarm to the stress response system that floods the body with norepinephrine, epinephrine, and cortisol. Blood pressure surges, straining and fraying the inner surface of blood vessels. Plaques of fibrous, fat-filled tissue scar the injured vessel walls. Over time, the repetitive damage to the vessels progressively impedes blood flow, and the repetitive activation of the nervous system progressively erodes the accuracy of risk assessment. With each alert, damage to the vasculature and the miscalculation of threat increase. Ultimately stress responses may escape containment altogether, leading to persistent hypertension, relentless suspicion, impulsive reactions, more anger—and even more stress. This vicious circle may end not in the suffering of victims but in an early death for the aggressor.

THE ULTIMATE RAT race is Robert and Caroline Blanchard's visible burrow system. Like marmosets, rats forging a new community get acquainted by fighting over who's in charge. Again, within a few days, a stable social arrangement emerges, with one male firmly in place at the top of the hierarchy, the other males demoted to subordinate status, and the females on the sidelines.[91,92]

Life as a subordinate rat is even more miserable than life at the bottom of a baboon troop. A few subordinates cope by attaching themselves to the dominant animal; these lackeys shadow his every move, on the surface and in the burrows, even sleeping in the same chamber. Most avoid him as much as possible. Either way, they suffer—physically, behaviorally, and emotionally—from the consequences of chronic, uncontrollable social stress, the same sort of torture, Robert Blanchard points out, endured by people who can't easily escape an aggressor because they're married to him, related to him, or need to keep the job he supervises.

Subordinates sleep less, move less, and eat less than dominants; even if they're removed briefly from the colony every day and force-fed, they can lose as much as a quarter of their original body weight in a few short weeks. Their resting glucocorticoid levels are two to four times higher than those of the dominant rat, their testosterone levels significantly lower. Unlike marmosets, the formation of a stable hierarchy doesn't normalize stress hormone values, meaning that subordinate rats are habitually exposed to excess levels of corticosterone (the rat equivalent of cortisol). Confined to the burrow system in this state for long, they often die.

Errol Yudko, a postdoctoral fellow in the Blanchards' lab, put an end to their ordeal before that happened. He removed subordinate rats from the visible burrow system after two weeks to measure the biochemical and behavioral toll that stress had taken on them. He recorded plasma corticosterone and testosterone levels. To learn if chronic stress had changed their ability to respond to a new threat, Yudko restrained each rat in a Plexiglas tube for an hour and noted the glucocorticoid response to the stress of confinement. Finally, he compared their behavior to that of dominant rats in a battery of test situations designed to represent varying levels of threat.[93]

The stress response systems of most subordinates were still raring to go, even after weeks of debilitating social stress. Restraining these rats sent corticosterone levels soaring. Like all other rodents, they were visibly distressed when confronted with life-threatening evidence of a predator—a handful of bedding from a cat's cage, a glimpse of the cat from across the room. But they were also preternaturally fearful in more benign situations. When Yudko reached into the test cage to pick up a dominant rat, for example, it was as likely to sniff, peer at his hand, or even bite as it was to run away. Subordinate animals, on the other hand, froze or attempted to hide; when he finally did get a hand on them, they struggled frantically to escape. Flipped on their backs, they cowered rather than immediately righting themselves. And when they were transferred to an "open field"—a large, uncovered enclosure that's the rodent equivalent of a dark parking lot—they shrank into the corners, reluctant to place as much as a paw into the exposed central area.

Some subordinates, however, played by a different set of rules. Yudko could often spot these mavericks from their very first moments in the burrow system—they were the ones who were the most aggressive. Unfortunately, they were also the ones who took the worst beating from the dominant. And when aggressive subordinates lost, they took it even harder than the rats that seemed to know their place right from the start. Quick to fight, defeat transformed them into supersubordinates. They spent the rest of their two weeks underground, pressed into the shadows as if the dominant were about to pounce on them at any moment. Even after they were rescued, they continued to behave as if they were about to be annihilated. Compared to the other subordinates, these rats were even more desperate to evade capture, even slower to get up if they were rolled over, even more reluctant to step out onto the open field.

But their endocrine systems had given up. Even after an hour trapped in a cramped plastic tube, their cortisol levels were nearly normal.[92]

HUMANS WITH DEPRESSION have a lot in common with the subordinate members of animal communities. They eat less, sleep less, lose weight, avoid challenges, give up in the face of crisis. Their cortisol levels are elevated, and as many as 60 percent fail to respond to what ought to be a cortisol-suppressing dose of dexamethasone.[94] But that's only the beginning. Depression links the "giving-up" response to environmental cues associated with the precipitating episode, often related to a stressful life event. Subsequent exposure to these reminders—situations resembling the original event in some way—may provoke a new episode of depression, further remodeling the brain and endocrine system and increasing the risk of still more episodes in the future. The cycle gradually dissociates depression from its original source; ultimately, it may escape environmental influences altogether to become a self-perpetuating, recurrent condition for up to 40 percent of the depressed population[95]—a sobering example of what can happen when the stress response gets out of control.

Logically, it would seem that no one ought to experience a greater rise in plasma cortisol levels than individuals suffering from PTSD. But researchers have found just the opposite: rape victims, Holocaust survivors, and combat veterans with PTSD actually have lower-than-normal levels of cortisol and a greater-than-normal response to even low doses of dexamethasone.[96–99]

In PTSD, as in the Blanchards' supersubordinate rats, the stress response system appears to have checked out. But the situation is actually far more complicated. When trauma expert Rachel Yehuda and a research team at the Mount Sinai School of Medicine tracked the pattern of cortisol release over the course of the day in a group of combat veterans with PTSD, they discovered that the stress response system was disrupted but far from destroyed.[100] Compared to normal volunteers, PTSD victims had a more pronounced diurnal pattern of cortisol release. Although overall their cortisol values were lower, they were also more volatile, fluctuating dramatically over the twenty-four-hour observation period. In other studies, the group also found that PTSD patients had more than the usual number of glucocorticoid receptors and they secreted more CRF—the starting point of stress response activation.

Yehuda and her colleagues concluded that the entire HPA axis has not shut down in PTSD, as the low cortisol levels might suggest; it has actually shifted into overdrive. The smallest blip in the environment triggers a stress response, but because of the extra receptors, feedback is also operating at peak efficiency. As a result, each slight rise in cortisol shuts the system down as quickly as stress turns it on. Stretched as taut as a rubber band, the stress response system oscillates at every reminder of the original trauma, pulsing cortisol with each quiver.

Cortisol responses are also depressed in one other group of individuals with abnormal reactions to stress—the antisocial, those whose hearts don't miss a beat at the thought of hurting another person. Studies of recidivistic violent offenders, adults with antisocial personality disorder, and antisocial adolescents have all documented statistically significant reductions in cortisol.[23,101–103] Among teenage mothers, low concentrations of CRF during pregnancy are associated with a higher incidence of antisocial behavior in the first stressful weeks following delivery.[87] Even school children displaying the earliest signs of attitude—defiant behavior that doesn't respond to punishment—have lower cortisol values than their prosocial peers.[104]

Psychopathy, it would appear, may also be a stress response disorder—but one in which perception and response react too weakly rather than too vigorously. With both the sympathetic nervous system and the HPA axis in low gear, antisocial individuals are left practically impervious to threat. No warning bell of anxiety or disgust sounds when they're about to commit an atrocity. Conflict is no longer stressful, so they can fight and not feel a thing. Appearing to lack a conscience, what the antisocial may really lack is the biological machinery necessary to warn them that they're edging toward disaster—an operative stress response system.

THEFT IS ONLY ONE way to lose a scabbard. It can be left behind, surrendered in battle, swept away in a raging river. Control of the stress response system can also be lost in many ways. But whether the end result is depression or impulsive, hostile aggression; PTSD or the total breakdown of antisocial personality disorder; the root cause is always a misunderstanding between the brain and the environment, a discrepancy, Robert Sapolsky concludes, between "how much threat is really out there and what a person perceives."

A Question of Balance

The black mare stands three inches higher than me at the shoulder; she is so tall that even with a mounting block, I still need a leg up. It's been twenty years since I last rode a horse. My hands are cold, and I've forgotten everything I ever knew about balance.

Like a beginning skier or middle-aged skater on a new pair of rollerblades, I'm more concerned with falling than technique. But balancing by brute force is hard on the body. Knees are asked to do the work ordinarily assigned to strong legs, ankles resist instead of flexing, back and shoulder muscles take over until confidence catches up. No wonder strains and sprains from overuse are such common injuries among weekend athletes.

My right knee—and the horse's mouth—will pay the price for my unsteadiness. For now, though, the ground seems hard and very far away, and the fear of falling wins out over the prospect of pain.

"STABILITY IS EVERYTHING. It is more important than economy," wrote the great physiologist Walter Cannon, the investigator who first recognized the importance of balance to survival, the equilibrium he called homeostasis. Yet Cannon acknowledged that, paradoxically, the key to balance is not constancy but flexibility. Quoting the nineteenth-century French physiologist Charles Richet, he observed, "In a sense, it [the living organism] is stable because it is modifiable."[105]

Adaptation is the essence of homeostasis. Nonetheless, after Cannon, the term came to mean something more like stagnation. Homeostasis did not merely stabilize physiological parameters, such as blood pressure or glucose levels, but held them hostage to an immutable set point—a "normal" value to be maintained at all costs. In keeping with this emphasis on regularity, the textbook analogy for homeostasis became the thermostat. The home owner set a temperature; then fan and furnace upheld it, regardless of the weather, the season, or the time of day.

But what happens when two people disagree on the optimal indoor environment? Interpersonal stability will demand more flexibility. When she's home, the set point is a balmy 70 degrees—shirt-sleeve temperature. When he's home, the thermostat goes down at least ten degrees and sweaters are in order. In a real home, the steady state is not so steady. In fact, it fluctuates within a range of permissible values, back and forth, up and down, according to the time of day, the schedules of the occupants, and the capacity for compromise.

Real bodies are also flexible. Thanks to the brain's responsiveness to the challenges posed by the environment, both current and anticipated, the balance points for the nervous system—the cardiovascular system, the immune system, even basal metabolism—are not static, but can fluctuate to meet the changing demands of a dynamic world. Blood pressure, for example, follows one set of rules during sleep, another during an uneventful workday, and a third during a two-hour argument with a relative. Stress researchers call this elasticity *allostasis,* meaning "stability through change"—and it allows the organism to gear up or shift down temporarily without losing balance.[78–80,106]

But there's a price to pay for flexibility. The new steady state permitted by allostasis may maintain balance by asking organ systems to perform in extraordinary ways, at the limits of their standard operating range. In the short term, the surge in demand can be accommodated. But when chronic or severe stress allows stress responses to escape containment, the demand is relentless. This pressure—Rockefeller University stress expert Bruce McEwen calls it "allostatic load"—wears on body and brain, steadily grinding away the ability to cope.

"If allostasis is the process of adaptation, then allostatic load is the price of adaptation," explains McEwen. He likes to compare the long-term impact of stress to the difference between balancing a seesaw with two five-year-old children or with two five-ton elephants (Figure 6.4). Both pairs balance, but the elephants place a greater burden on the seesaw. Long term, this balancing act can mean injury or illness—and an overreactive or underreactive nervous system. Forced to operate at a capacity for which they were never designed, the reciprocal processes that regulate neurotransmission and neuroendocrine function during stress overshoot, break down, or oscillate frenetically (Figure 6.5).[80] The dysregulation surfaces as the misperception of threat and emotional instability, typical of a family of maladaptive reactions that can be characterized as "stress response disorders."

Depression, impulsive Type H aggression, PTSD and antisocial personality disorder are very different expressions of a common failure to assign the correct emotional valence to memories, thoughts, or external events; as a result, stimulus and response are mismatched. All have in common a loss of control over the neural and endocrine components of the stress response; as a result, all are associated with abnormal glucocorticoid levels, altered endocrine feedback mechanisms, and the dysregulation of norepinephrine and serotonin function. What differs is the specific pattern of the neurochemical changes, the charac-

Figure 6.4

Homeostasis Versus Allostasis.

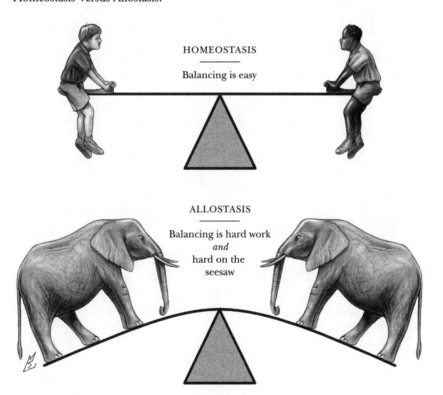

HOMEOSTASIS

Balancing is easy

ALLOSTASIS

Balancing is hard work
and
hard on the
seesaw

Figure 6.5

A stress response system that has not been taxed by chronic or severe stress reacts briskly to a challenge, but is quickly brought back under control (top panel). A system asked to do too much, too often, on the other hand, may escape normal containment mechanisms. Overwhelmed by allostatic load, it oscillates, locks in, or gives up.

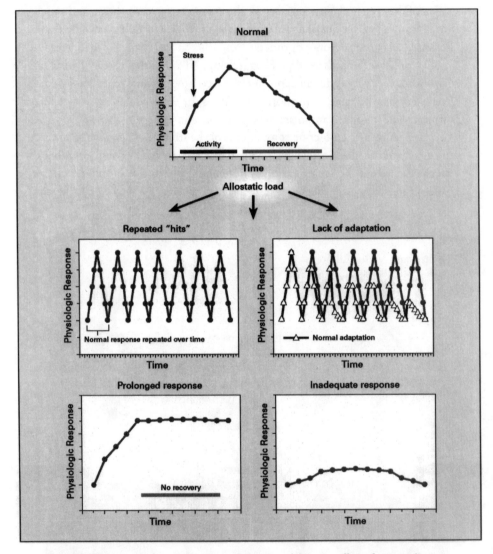

teristic behavioral profile, and the nature of the allostatic load inflicted by each disorder; these patterns and profiles are outlined in Table 6.1. In depression, the inability to shut down the stress response exposes the brain and body to chronically elevated levels of cortisol. In PTSD and impulsive aggression, it is the frequency of "hits"—and the wear caused by repeated peaks and troughs in glucocorticoids—that loads the nervous system. And in the antisocial personality, load has become irrelevant, the system, unresponsive.

This is why bad neighborhoods, bad homes, and bad relationships breed violence—not because of a willful deterioration in moral character but because of a steady deterioration in the ability to cope. As stress wears away at the nervous system, risk assessment grows less and less accurate. Minor insults are seen as major threats. Benign details take on a new emotional urgency. Empathy takes a back seat to relief from the numbing discomfort of a stress-deadened nervous system. Surrounded on all sides by real and imagined threats, the individual resorts to the time-honored survival strategies: fight, flight, or freeze. The choice will be of enormous importance to society, but it will matter little to the nervous system. All are stressful, and all will perpetuate the steady erosion of mental and physical health, the fatal attraction to unacceptable responses.

Stress powerful enough or persistent enough to overload the nervous system can also sensitize it. Over time, depression, flashbacks, and angry outbursts often seem to pick up steam, accelerating in frequency and intensity as if each episode pushed the reactivity of the nervous system a notch higher. Like

Table 6.1. Key Features of Stress Response Disorders

	ALTERED NOREPINEPHRINE FUNCTION	ALTERED SEROTONIN FUNCTION	SENSITIVITY TO ENVIRONMENT	CORTISOL LEVELS	SIGNATURE BEHAVIOR
Depression	Yes	Yes	Overreactive	Increased	Withdrawal
PTSD	Yes	Probably yes	Overreactive	Decreased; altered pattern of release	Fear; changes in responsiveness and memory
Impulsive hostile aggression	Probably yes	Yes	Overreactive	Increased	Rage
Antisocial personality disorder	Yes	Not known	Underreactive	Decreased	Lack of empathy

kindled seizures, excessive reactions to threat, real or imagined, may gradually assume a life of their own, until they reach a kindling-like spontaneity, an automatic response to external cues cursed with an emotional significance they were never meant to have. NIMH's Robert Post believes adaptation that's gone overboard may reflect the impact of stress at a cellular level. He and his colleagues have shown that stress can fiddle with the genome, cranking up levels of gene-regulating transcription factors that can switch on genes coding for proteins critical to brain function—the enzymes that build neurotransmitters, receptors, second messengers, structural proteins.[107,108] Stress, in other words, doesn't just transiently alter catecholamine levels or endocrine function and then move on; it actually reconstructs the brain, with long-lasting consequences for neural function and behavior.

IN BOTH ANIMALS and humans the long-term damage done by stress surfaces as a compulsion to behave in unproductive and even harmful ways. For example, a subordinate rat stressed by confinement with a belligerent dominant male in Robert and Caroline Blanchard's visible burrow system doesn't check for predators by poking his head out of the burrow for a few quick seconds, like any sensible rat, but sticks his neck out—and freezes.[92] He may stand transfixed for more than half a minute, in full view of everyone on the surface. "It's craziness," marvels Caroline Blanchard. "If there *was* a cat there, it would have eaten him in that time. But he can't let it go and move on."

Bruce McEwen puts it a different way. "People say 'stress makes you stupid.' But what it really does is limit your options." Responses mounted under duress tend to repeat themselves until, with time, the aggressive, the traumatized, the depressed seem to forget that they have other options.

Forgetfulness is a reasonable consequence of stress, for of all the brain regions receptive to glucocorticoids, none is more exquisitely sensitive to the wear and tear of stress than the hippocampus, a transitional neighborhood between the limbic system and the temporal cortex. The bridge between memory and emotion, the hippocampus is the holding area where the cortex parks short-term memories while deciding if they're worth the space in permanent storage, as well as the database for the information about context that is so crucial to emotional memory.

With one of the highest densities of glucocorticoid receptors in the brain,[82] the hippocampus is bound to be "ground zero for stress hormone sensitivity," explains Robert Sapolsky. Under ordinary circumstances, this sensitivity is a blessing, for glucocorticoid activation of the hippocampus is a critical element in the feedback control mechanism that contains the stress response. When excessive or repetitive stress overwhelms containment mechanisms, the double-edged sword of glucocorticoid hormones can be wielded against hippocampal neurons.[78,80,82,109,110] Worse, the damage may not be obvious until it's too late. Subtle, pernicious assailants, unbridled glucocorticoids may wear out the cells by releasing the powerful stimulant neurotransmitter glutamate, divert the glucose that feeds them, undermine defense mechanisms that shield them from metabolic wear and tear, deprive them of essential growth factors.[110] The endangered cells slowly weaken, more vulnerable to environmental insults with every stressful episode.

"You don't have to shoot a neuron in the head," says Sapolsky. "You just have to keep kicking away at its defenses, sap just enough of its energy to make it queasy and lightheaded. Unfortunately, today may turn out to be the worst day of that neuron's life—and then it's in trouble."

Glucocorticoids can also forgo covert operations and launch a frontal attack on vulnerable hippocampal neurons, shriveling dendrites on the cells like leaves on a scorched plant. Fortunately, if stress ends and glucocorticoid levels are finally contained, the injury can be reversed. But if stress continues to drench the hippocampus in excess cortisol, the beleaguered neurons may die.[111]

Some investigators believe that the toxic consequences of cortisol exposure may explain the reduced hippocampal volume observed in PTSD patients. Similarly, a recent MRI study of depressed patients also found evidence of hippocampal atrophy—and the extent of the atrophy was correlated with the duration of the depression and, presumably, the duration of exposure to the elevated cortisol levels that characterize this stress response disorder.[112]

Are these changes evidence of permanent damage—wholesale cell death— or a sign of struggling neurons in mortal danger? The animal data suggest that prompt intervention to lower glucocorticord levels can reverse stress-related injury, but time is of the essence. Even an optimistic interpretation would admit that atrophy is a warning that a neurobiological clock is ticking, that even if injured cells are still alive, it is only a matter of time before they wake up to the worst day of their lives.[109]

By progressively remodeling brain areas that participate in emotional memory, stress may slowly obliterate positive memories, leaving behind only the painful ones. More constructive coping responses are lost, and the brain fixates on an increasingly smaller portfolio of counterproductive reactions. With fewer and fewer alternatives, violence, depression, and fear stop being options and become a way of life.

Seven

OUT OF THEIR MINDS

They say that those who forget the lessons of history are con-demned to repeat them. Fortunately, the network organization of the brain provides excellent protection against such potentially catastrophic forgetfulness, routing strategic environmental data through two decision-making circuits: a rapid-response loop centered on the amygdala and a slower-acting loop that circles through the cerebral cortex as well as the limbic system. What this cortical circuit lacks in urgency, it makes up in precision, adding the detail necessary to a fully informed opinion on the state of the world, a sophis-tication essential to sound emotional judgment.

The detail comes from memory and experience—the lessons of personal history—retrieved, organized, and mirrored back to the amygdala from the cortex. Familiar faces and voices are placed in context. Unfamiliar situations are scanned for the smallest familiarity linking them to the past. Possible re-sponses are rated according to their success or failure in previous encounters of the same sort. A confirmed threat will second the amygdala's decision to act de-cisively. But when the cortical analysis suggests that the "threat" is not a cause for alarm—that the scraping sound you hear is the family dog scavenging cat food, not an intruder—the cortex reasons with the amygdala. Cortical projec-tions returning to the lateral nucleus of the amygdala damp down intense emo-tions to reflect the updated picture of reality made possible by acquired knowledge.

A brain that cannot connect past and present or that has forgotten how to adapt is in danger of losing touch with reality, of getting lost in an ever-changing world. And without a firm attachment to reality, false alarms can look suspiciously like real threats. As a result, the analytical and reasoning skills of the prefrontal cortex are as essential to accurate risk assessment as the security check performed by brain catecholamine pathways and the emotional valuation carried out in the amygdala.

THE PREFRONTAL CORTEX, one of three cortical areas known as association cortex, is the showpiece of the primate brain. In humans, it spans nearly one-third of the cortical surface,[1] a busy, bustling, neural marketplace trading in ideas, perceptions, and interpretations.

Speed is of the essence in the prefrontal cortex. With so much information to process and exchange, it can't always afford to wait around while messages are systematically relayed from protein to protein, receptor to genome. For high-speed transmission, it turns instead to a group of common amino acids that do double duty at the synapse. Their receptors govern channels that admit charged particles, or ions; the inward flow sends an on or off signal to the postsynaptic cell in only milliseconds.[2] In contrast, second-messenger-driven protein fiddling can take minutes to complete; getting down to the level of genes, several hours.[2,3]

On signals are the province of the amino acid glutamate, better known outside the neuron for its important role in protein synthesis.[4] The chief targets of glutamate-containing cortical neurons are the distinctive neurons known as pyramidal cells, easily identified by their wedge-shaped cell bodies.[5] Pyramidal cells are dispatchers.[6] Their axons form the road out of a given sector of prefrontal cortex, directing local messages addressed to other areas of cortex, as well as long-distance messages traveling to distant corners of the central nervous system. Glutamate influence on pyramidal cells therefore directly controls cortical output.

Gamma-aminobutyric acid (GABA)—an amino acid best known to botanists before it was discovered in the mammalian brain in the 1950s[4]—is the major inhibitory amino acid transmitter. GABA is popular with the interneurons that shuttle messages over short distances within the cortex. Many synapse on pyramidal cells, allowing GABA, as well as glutamate, a significant measure of control over outbound information.[5,7]

Responsibility for tempering the energetic cortical neurons using GABA or glutamate transmitters falls to the more deliberate norepinephrine, serotonin, and dopamine pathways coursing into the prefrontal region from their base camps in the brain stem. Together, these monoamine inputs form a neural highway patrol, monitoring the flow of traffic, warning speeding neurons to slow down, hustling laggards.

Dopamine projections to the cortex specialize in the prefrontal area, and concentrations of dopamine here are among the highest in the cortex.[8] Elegant anatomical studies show that the targets of these projections include both pyramidal cells and the ubiquitous GABA-containing interneurons. Using antibodies to dopamine and the powerful resolution of the electron microscope, Patricia Goldman-Rakic, a neurobiologist at the Yale University School of Medicine, observed dopamine synapses dotting the spines of pyramidal neurons in monkey and human prefrontal cortex;[9,10] and Francine Benes, of Harvard University and the Laboratory for Structural Neuroscience at nearby McLean Hospital, used two fluorescently tagged antibodies—one directed against dopamine, the other against GABA—to demonstrate the fingers of dopamine fibers clutching the cell bodies of GABA-containing interneurons (Figure 7.1).[5]

By virtue of these connections, dopamine axons have elbowed their way straight into the middle of prefrontal cortical action. But to be heard, they have

Figure 7.1

By directly or indirectly regulating the activity of pyramidal cells in the prefrontal cortex, dopamine plays a critical role in that region's most important task: working memory.

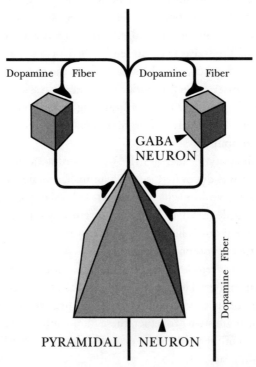

Adapted with permission from F. M. Benes, Is there a neuroanatomic basis for schizophrenia? An old question revisited, *Neuroscientist 1* (1995): 104–115.

to follow the rules: no shouting and no whispering; too much or too little dopamine is detrimental to cortical function.[11] When dopamine activity charts a middle course, it preserves a balance of power between excitatory and inhibitory influences in the prefrontal cortex. Outside this window, dopamine is at best ineffective; at worst, seriously disruptive.

Individuals contemplating the best way to explain the mysterious workings of the prefrontal cortex—and the problem-solving, analytical processes it supervises, thought processes that cognitive neuroscientists call "executive functions"—need space, a large flat surface where they can organize their reference materials. The brain facing a confrontation also needs space to spread out its reference materials while it decides how to react. This desktop of the here-and-now, a component of executive function known as "working memory," has commandeered a swath of prefrontal cortex to serve as a sort of mental scratch pad, where the brain can juxtapose current sense data and mental images of similar experiences, jot down the pros and cons of various response options, free-associate, muse about the past, and run through various what-if scenarios.[12–14] One response emerges from this split-second mind-mapping exercise, an option that stands out above all the others. Working memory files the spare responses, and directs brain centers that initiate and control voluntary movement to carry out the chosen course of action.

"Working memory is critical to comprehension," explains Goldman-Rakic. "As a result, it's a core process necessary for executive function." The essence of working memory—and the reason it's critical to comprehension—is its ability to simultaneously compare past and present, the raw data brought to you live by perception and images from the brain's archives. These recollections—Antonio Damasio calls them "dispositional representations,"[15,16] meaning that they're reconstructed, like digitized photographs, from fragments coded in the firing patterns of related neurons—are not just dusty old artifacts. As Damasio points out, they are animated by emotion, linked to the inanimate features of the memory and reactivated when it is summoned to the center stage of working memory. Thanks to this emotional overlay, remembered images, like current images, have meaning and value.

The ability to manipulate representations, as well as concrete information, means that our behavior can be guided by ideas, concepts, and plans rather than depending solely on environmental cues.[13] Representational knowledge, in other words, is what adds the element of reason to risk assessment. It is the door that admits moral beliefs, intellect, and learning. In addition, it is the mechanism that invests these abstractions with meaning. Peace is attractive because the image of behaving nonviolently can be retrieved along with pleasant inspirational feelings, and when the memory of hitting someone resurfaces coated with a thin slime of guilt, we're not likely to use force in that situation again.

If working memory goes wrong, reason and comprehension go out the window. When the workspace contains incomplete, irrelevant, or erroneous information, or when the associations between environmental data and representational knowledge are fuzzy, the brain may reach irrational and even bizarre conclusions. If working memory is disconnected, the brain has only the rough outline of reality sketched out by the limbic loop of the emotional network to go on. If it is damaged, representational thinking will be lost.

The brain deprived of cortical input must rely on first impressions, whether they're right or wrong. Predictably, patients with damage to the prefrontal cortex, like Damasio's Elliot, act on impulse. Without working memory, they are "environmentally dependent," reacting to events rather than interpreting them.[17]

However, the prefrontal cortex doesn't have to be struck with a blunt object to fall down on the job. Chronic or extreme stress, for example, can enter erroneous information about the environment into representational memory. Disease can interrupt retrieval, analysis, or output. Events that interfere with chemical transmission, especially the precision essential to optimal dopamine function, can confuse working memory.

When early warning mechanisms misfire, risk assessment mechanisms miscalculate, and fact-checking mechanisms fail, no defenses remain to patrol behavior. Excessive and inappropriate responses are the inevitable result.

"The Prism of Psychosis"

Nancy Schultz looked at John du Pont and saw a traitor who repaid her husband's friendship with three 44-caliber bullets.[18] First District Attorney Joseph McGettigan, who prosecuted the du Pont case, saw "an arrogant person who took the most arrogant act a person can take . . . separating the soul from a person's body."[19] But defense attorney Thomas Bergstrom and defense psychiatrists Phillip Resnick, Robert Sadoff, and William Carpenter urged the jury to look at John du Pont and see him "through the prism of psychosis."[20]

During their first meeting, du Pont told the psychiatrists that a clone created by government agents or enemies of the Buddhist church had killed David Schultz.[21,22] Then he changed his mind, and insisted that Schultz had been killed by either Republicans demanding campaign contributions or the CIA, in retaliation for a failure to turn over secret videotapes stolen from the Russians. At various points in the interview, du Pont claimed to be the Dalai Lama, the Christ child, the president of Bulgaria, a Russian czar, heir to the Third Reich, and owner-guardian of the Shroud of Turin. The psychiatrists' conclusion—that du Pont was not only seriously mentally ill but so delusional he was incompetent to stand trial—won the millionaire his three-month reprieve from justice at Norristown State Hospital. Later, at du Pont's murder trial, Resnick, Sadoff, and Carpenter agreed unanimously that John du Pont was insane, a paranoid schizophrenic so enslaved by his bizarre delusions that he was incapable of distinguishing reality from fantasy, much less right from wrong.[23]

Resnick hypothesized that in the months before the murder, du Pont's obsessive fear that he was the target of a conspiracy had steadily escalated, following a fire at his Foxcatcher estate. A minor head injury, incurred in a fall at Schultz's home, and a glimpse of a homemade bottle-rocket launcher Schultz had created for his children convinced the millionaire that the wrestler was at the heart of the plot. David Schultz was actually a government-designated assassin preparing to kill him. If du Pont wanted to stay alive, he needed to kill Schultz first.

The jury didn't buy Resnick's theory. They acknowledged that du Pont's thought processes were abnormal. As an alternate juror put it, "There was definitely something wrong with this guy." What they could not accept, however, was that fear, rather than malice, had driven du Pont to kill. The juror continued, "I think he was mentally ill but he knew what he was doing at the time, that he knew he was killing a man and that it was wrong."[24]

One of the less attractive consequences of a mind that can work with ideas is that it can create its own monsters. It can conjure a stalker from a man's footsteps, turn a shadow into a thug with a gun. These terrors represent the possible, and so we tolerate them, much as we humor a child's irrational fear of the dark. But conspiracy theories? Russian spies hiding in the walls? An American millionaire, the Dalai Lamai? These sound suspiciously like a bad joke. We know that our minds can play tricks on us, but the alternative reality of madness is incomprehensible.

"Du Pont wasn't angry at David Schultz, nor did he hate him. But he was fearful of David Schultz. And that fear was not based on reality," argued Bergstrom in his closing argument. "That's the hard part. We're used to things we can see and touch. We're not used to delusions. We're not used to psychosis."[19]

MORE THAN 2 MILLION Americans suffer from some form of schizophrenia, the archetypal chronic mental illness.[25] Few are cosseted millionaires. In fact, thanks largely to four decades of steady pressure to release the mentally ill from psychiatric hospitals, perhaps as many as a quarter-million are homeless,[26] and countless others drift in and out of acute treatment facilities, halfway houses, and family homes, where they are seen as a threat and a burden. When they behave in bizarre or alarming ways, they often end up in prison.

Schizophrenia usually rides into the victim's life on the coattails of adolescence, although an onset as late as the mid-forties is not unknown.[27] The disease follows a waxing and waning course, full-blown episodes alternating with periods of relative quiescence. With age, it gradually loses its grip over some, while others progress from psychosis to a dementia as incapacitating as Alzheimer's disease.

The classic schizophrenic is often portrayed as a disheveled, incoherent madman who hears voices and talks back to them. In reality, the disease can present with a bewildering array of symptoms, ranging from vivid hallucinations to a mute passivity. Patients may be withdrawn and apathetic or wildly agitated, unresponsive or overemotional. Moods, thoughts, and speech patterns are strange and inappropriate, but the exact form of these distortions is highly idiosyncratic. The patient may be disoriented, unmotivated, incoherent, even stuporous.

Psychiatrists have attempted to impose order on this chaos by grouping patients with similar symptoms into subtypes. Catatonic patients are withdrawn to the point of total rigidity, assuming statuesque postures from which they refuse to be moved. Disorganized schizophrenics are slovenly, aimless, seemingly lost in their own eccentric worlds.

Paranoid schizophrenics—the group to which Resnick, Sadoff, and Carpenter assigned John du Pont—are convinced that someone is out to get them. *Paranoia*, a term used by Hippocrates to describe the delirium that accompanies a high fever, can be loosely translated from its Greek origins as "beside oneself."[28,29] In popular usage, the term has come to denote suspiciousness, a mistaken belief that you're being followed, watched, duped, manipulated, or threatened. Not everyone who's paranoid is schizophrenic. The ranks of the chronically persecuted include "many of life's least lovable character types—the bigot, the injustice collector, the pathologically jealous spouse, the litigious crank, the fanatic."[30] These garden-variety paranoids keep civil court dockets full, the op-ed page humming, local politics contentious, extremist religious groups financially solvent. Their complaints are unwarranted but not irrational, and they function satisfactorily, if rather unhappily, in society. Obnoxious but rarely dangerous, their paranoia can be considered a personality disorder rather than a psychosis.

The mental health community gets more interested when an individual's persecution complex starts to interfere with his or her daily life or the rights of others. Complaints that the neighbors spend their days gossiping about you are one thing. But if you start believing that the neighborhood is mobilizing a group effort to have you evicted, your suspicion has ballooned into a delusion—an assertion that's not only unlikely to be true but improbable. Psychiatrists call a paranoid belief system so extreme that it flies in the face of obvious evidence to the contrary a delusional disorder. Delusional individuals are not necessarily incapacitated or psychotic; many hold jobs, form relationships, attend to personal hygiene. In fact, if it weren't for their questionable beliefs on certain topics, many would seem perfectly normal. And their delusions, while fantastic, are not totally impossible. Unlikely as it may be, the neighborhood *could* have a collective grudge against you, or as the saying goes, "Just because you're paranoid doesn't mean they're not out to get you."

If you're convinced the neighbors are planning to abduct and dismember you because they're aliens from a hostile planet, your delusions have lost touch with the real world. This is the boundary of paranoid schizophrenia. These individuals assign special powers and influence to the outside world. They display the disordered thinking and emotional derangement that are the cardinal symptoms of schizophrenia, and their behavior has become sufficiently disorganized to disrupt all aspects of ordinary life. They may hear voices commanding them to take protective action or hallucinate an incipient attack by an innocent bystander. Their efforts to "protect" themselves from such invisible threats often turn violent.

The odd behavior of schizophrenics frightens people. But determining whether that fear is well founded is more difficult than you might think. Studies of prisoners or psychiatric patients can be misleading; with the passage of laws barring involuntary commitment and the demise of state hospitals, the mentally ill are unusually likely to be "hospitalized" in prison, while exceptions that still permit commitment for individuals deemed "dangerous" have overstocked psychiatric hospitals with violent patients. Recent studies have turned to community survey data to overcome these biases. Based on information from the Epidemiologic Catchment Area Survey, the most extensive evaluation of the prevalence of mental illness in the general population, a 1990 study by psychiatrist Jeffrey Swanson, of the Duke University Medical Center, reported that rates of violent behavior among the seriously mentally ill (including schizophrenics) were as much as five times those observed among people who did not have a psychiatric disorder.[31] Compared to the general population, schizophrenics were nine times more likely to have fought with someone during the past year—and twenty-one times more likely to have used a weapon in doing so.[31]

But not all schizophrenics contribute equally to these inflated rates of violence. In fact, most are more likely to end up as victims of violence than as perpetrators. According to a new study underwritten by the MacArthur Foundation, the majority of mentally ill individuals discharged from psychiatric hospitals were no more violent than the rest of us. Patients who abused drugs or alcohol after discharge, however, *were* significantly more violent—rates of violence in this group were as much as five times higher than the rate observed in the general population.[32] The patient in the throes of an active psychotic episode is also more likely to be violent, especially when that crisis involves paranoid delusions that someone is threatening them or trying to control them.[33–35] Believing that an individual living on your property is a government assassin, as John du Pont did, would probably qualify.

In paranoid schizophrenia, the disconnect between perception and reality results in a gross overestimate of threat. As his anxiety mounts, the paranoid individual grows increasingly "beside himself" in his distress, and the risk of serious violence increases. Ultimately, he may feel compelled to "get them before they get me." The resulting violence is not malicious but self-defensive, a pathological exaggeration of the normal fight-or-flight response.

Recent data show that delusions of threat or persecution—false beliefs that Columbia University researchers Bruce Link and Ann Stueve call "threat/control-over-ride" symptoms—were twice as likely as other psychotic symptoms to lead to assault.[35] Worse, when paranoid patients are violent, they're more lethal. Of all mentally ill individuals, they're the most likely to commit murder.[36]

Unfortunately, paranoid schizophrenics are more likely than the catatonic or the openly hallucinatory patient to be out and about in the community. Older, on the average, than patients with other forms of the disorder when symptoms first make an appearance, people with paranoid schizophrenia may retain a toehold on real life, drifting from menial job to menial job—or shielding their eccentricity behind an extravagant inheritance.[27] Their dangerously inflated sense of threat, combined with just enough residual motivation and organization to act on their delusions, means that while they may not be able to plan their day, they can plan acts of revenge against their persecutors, and they're more than capable of figuring out how to use a gun. As a result, paranoid schizophrenics keep their psychiatrists and law enforcement officials awake at night, the one group of mentally ill individuals people have cause to fear.

The menacing theme of paranoid delusions is what motivates the aggressive response of the individual with schizophrenia. But how do such bizarre misconceptions come to life in the first place? How does a brain that is apparently functional for fifteen, twenty, even thirty years suddenly lapse into one that identifies its owner as the infant Jesus, sees colleagues as conspirators, hears voices muttering threats from inside the walls?

When it comes to reasoning, planning, remembering, reality checking, one region of the brain stands head and shoulders above the rest: the prefrontal cortex. It's here that the brain adds reason to emotion, analysis to perception. And it's here that researchers have gone to search for the answer to the complex and difficult problem of schizophrenia.

THE 1950S WERE a tough time to be a mother. Give your baby too much attention, and you'd surely spoil her; hover, and your overprotectiveness would make her neurotic; ignore her, and you'd turn her into a schizophrenic. According to the experts of the day, schizophrenia was caused by aloof, rejecting mothers who didn't really want to be parents—or partners, either, for that matter. The "refrigerator mother" and the "communication-deviant" couple were part of a larger conceptual framework of family interaction theories that placed the blame for psychosis directly at the feet of parents.[37]

At the same time that clinical investigators were busy condemning mothers, pharmacologists were unwittingly redeeming them. French researchers discovered that a group of compounds originally synthesized as antihistamines could reduce the delusions, hallucinations, and bizarre thought patterns that separated schizophrenics from the outside world. In 1957, the first of these drugs, chlorpromazine—marketed in the United States under the trade name Thorazine—was introduced to the medical community. The era of biological psychiatry had begun.

How could a drug accomplish what years of psychoanalysis or family therapy could not? The secret of Thorazine's neural magic, investigators learned, was its effect on dopamine. The delusion-killing efficacy of Thorazine and other antipsychotic drugs, known collectively as neuroleptics, correlated with their power to block dopamine receptors.[38] The conceptual jump from mechanism of action to a mechanism of psychosis was obvious. Schizophrenia was caused not by bad parenting but by bad chemistry, by hyperactive dopamine neurons that flooded synapses with too much transmitter. Neuroleptics intervened by shielding dopamine receptors from this excess stimulation, effectively lowering activity until it fell back within the permissible bandwidth.

Today, neuroscientists know that the origins of schizophrenia go beyond dopamine, and they've added GABA, glutamate, serotonin, and several peptides to the list of schizophrenia's neurochemical targets.[39] They know that a dopamine excess is not the "cause" of schizophrenia, any more than excessive testosterone is the "cause" of aggression. But drugs that block dopamine receptors are still the only known antipsychotics. Drugs that increase dopamine transmission, on the other hand, can precipitate a psychosis as convincing as the real thing. Dopamine may not be the only guilty party in schizophrenia, but its fingerprints are all over the disorder, a trail of guilt that leads from the brain stem, along the mesocortical dopamine pathway, straight into the prefrontal cortex.

SCHIZOPHRENIA RESEARCHERS may no longer blame parents, but some still suspect that the disorder originates in the first years of life. The culprit, they believe, is a brain that has done a shoddy construction job, a developmental error that results in a cortex assembled incorrectly.

Normal cortex is made up of six layers, readily distinguished by the shape and type of neurons in each. The fibrous top layer, layer 1, contains axons that stretch laterally to crosslink cortical regions. Layers 2 and 3 are home to small neurons; deeper layers—especially layer 5—to large pyramidal cells. During development, the deeper layers "set" first, their large cells migrating out of a germinal bedrock known as the subplate zone. Upper-layer neurons leapfrog over them and shimmy up radial supporting fibers to their second-story destinations; as they depart, evidence of the subplate zone melts away, leaving behind only a subterranean network of milky, myelin-coated axons that shuttle messages in and out of the cortex.

Four separate anatomical studies have now found that in the prefrontal cortex and entorhinal cortex (a major way station connecting prefrontal cortex to the hippocampus)[13] of the schizophrenic brain, the small, last-born neurons never arrive.[40–43] The top-most cortical layers are depopulated, while lower layers are choked with extra cells. Some immature cells even linger in the white matter below the cortex, the last vestige of the subplate. Instead of assembling inside out, into a traditional six-layered cortex, cortical neurons in the brain of the future schizophrenic have improvised a novel architecture (Figure 7.2).

If you've ever assembled one of those confounded pieces of furniture that come with hundreds of screws; several pieces that look right, but don't quite

Figure 7.2

(A) Normal brain. (B) Entorhinal cortex from a patient with schizophrenia, showing obvious disorganization in layers II and III, and in layer III, a patchy appearance as well.

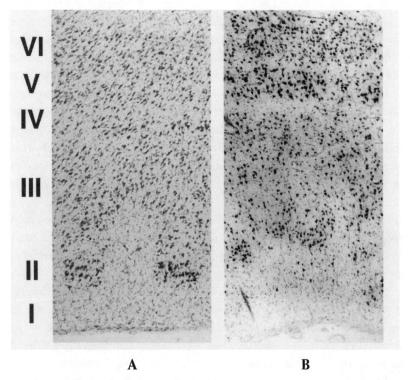

A   B

Photo courtesy of Dr. Steve Arnold, University of Pennsylvania. Reprinted from S. E. Arnold, Hippocampal pathology, in *The Neurobiology of Schizophrenia*, ed. P. J. Harrison and G. W. Roberts. Forthcoming from Oxford University Press, Inc. Used by permission.

match; and instructions that don't quite make sense, you know how easy it is to accidentally join a right side to a right side, or to insert a critical tab into its slot incorrectly. Of course, you usually don't realize you've done something wrong until you turn the piece over two hours later. Similarly, the architectural creativity of the late-maturing schizophrenic prefrontal cortex is typically subtle enough to escape notice during the childhood years. Then, in adolescence, age, or stress, or increased demand, or hormones—no one knows for sure—may finally ask things of the misaligned prefrontal cortex and disrupted cortical-limbic system circuitry that they are unprepared to deliver.[44] Thought, reason, and personality come unhinged under the insupportable pressure.

Another anatomical study, a detailed microscopic analysis of brain tissue taken from schizophrenic patients after their death by McLean Hospital's Francine Benes, may explain how dopamine could be involved in schizophrenia without causing the disorder.[5] Compared to sections of cortex taken from normal controls, cortical tissue from schizophrenic brains had significantly fewer of the GABA-containing interneurons targeted by incoming dopamine axons. The dopamine neurons themselves were spared. But the loss of their partners may lead to an imbalance that simulates a dopamine excess.

From a thinking point of view, a frontal cortex that's architecturally and chemically unsound is a frontal cortex that's unqualified to carry out that most important of cortical responsibilities: working memory. Perceptual information may still arrive intact. But because of the disrupted circuitry, relevant historical and contextual information from representational memory may not. Strange associations, fragments of unrelated memories, free-floating ideas drift in and out of the workspace. Nothing matches. As a result, first impressions can't be verified, emotional data are misinterpreted, sensory data aren't recognized. It's as if you sat down to write a book chapter and found your desktop cluttered not with papers, pens, and medical journals, but with shoes, rocks, a doll from your childhood, a broken teacup. Just try to generate a coherent product from that mess.

An inability to match past to present, and internal concepts to external data—a deficit that cognitive neuropsychologists Daniel Kimberg and Martha Farah have described as a "weakening of the associations" between the materials assembled on the working memory desktop[45]—means that schizophrenic thought has "a strong focus on the irrelevant and a weak focus on the relevant."[13] Schizophrenics, in other words, make a lot of mistakes in interpreting the world around them, mistakes that show up as delusions, hallucinations, strange—even dangerous—conclusions. Worse, they have no way of catching their errors and correcting them before they result in unacceptable behavior.

Without working memory there can be neither insight nor reason; without reason, there can be no self-control. A man in this bind could readily believe a friend had turned into an assassin, could easily fail to understand what a horrible mistake it would be to kill him.

THE AMERICAN PUBLIC, quick to accuse the mentally ill of every imaginable evil, is paradoxically reluctant to accept that some violent behavior is a consequence of serious mental illness. Our sense of righteousness demands that people take responsibility for their actions, especially when those actions appear capricious, arrogant, or cruel. Understandably, we want to keep the gateway between prison and psychiatric hospital as narrow as possible.

But the time for someone to take responsibility is *before,* not after, paranoia has degenerated into violence. Treating everyone with a chronic, debilitating mental illness as a potential killer is unfair and unreasonable. But ignoring signs of an impending breakdown in control isn't the answer either. The sobering lesson of the du Pont case is that we ignore a progressive deterioration in threat calculation at our own peril.

Children Who Do Too Much

A good horse's mouth is as sensitive as the human ear. A good rider's hands speak with a clear, emphatic voice. Pair a responsive horse and an articulate equestrian and the ride is a conversation, the bit a telegraph, the taps, pauses, pulls, releases issued from hand to mouth a calm sequence of clearly spoken requests: "a little faster here, bend your neck more and don't cut that corner, now turn left, slow down, we're done."

Beginning riders don't talk; they shout. They lean on the bit, using the reins instead of their back and hips to balance. Every time they lose a stirrup, turn, or shift gaits, they slam the horse in the mouth. Their hands clench up or daydream when they should be giving directions.

All of this jabbing, poking, and tugging adds up to a lot of meaningless chatter. The horse's delicate mouth is trammeled by noise. Yet he's still confused, for only a fraction of the torrent of signals coming from the rider actually mean anything. The real commands are buried in a cacophony of irrelevant detail.

Lesson horses ultimately stop paying attention. They develop a "hard mouth," and a reputation for being unresponsive, inattentive, headstrong, even defiant. In reality, they're overwhelmed, unable to differentiate the rider's intentions from her inadequacies. Until the rider learns to talk, the horse has no good reason to listen.

Some children have a hard time paying attention, too. You have to call them eight times for dinner. You send them to their rooms for jumping on the couch, and as soon as they're back downstairs, they're jumping again. They're the kids who talk out of turn, spill the milk, get out of their seats, fidget, disassemble the hall closet, smash heirlooms, cut in line, "forget" directions. Their parents wear out long before they do.

They're kids who can make a shambles out of school and family life. They pick fights, annoy classmates, torment siblings. They can't share. They forget that you can get killed running into the street. Adults can't manage them, and other children can't stand them.

When the failure to listen becomes dangerous, destructive, or incompatible with learning, it's clear that something more than youthful high spirits is at play. That's when parents seek help and child psychiatrists label the unresponsiveness "attention deficit–hyperactivity disorder" (ADHD). ADHD is not simple misbehavior; it's no more willful than the insensibility of a horse enduring a first-time rider. Attention fails when there's too much to attend to, when the needle of the signal is lost in the haystack of perceptual noise. An overwhelmed brain has no priorities, and without priorities, responsible actions are impossible.

ADHD troubles parents, educators, and child psychiatrists because about half of children who can't pay attention are also behind the door when the rules of conduct are handed out.[46–48] In this subgroup, inattention festers into an oppositional, hostile defiance that may net them a trip to the nearest mental health professional. If they're not careful, it may land them in the nearest juvenile detention facility. According to Tony Rostain, of the Philadelphia Child Guidance Center, "It's the kids with the oppositional form of ADHD who develop a real problem with aggression. If you look at kids who present for treatment here in the clinic, probably more than three-quarters fall into this category. Inattentive or hyperactive kids who are pleasant socially and get along with others can be managed. The aggressive kids are harder to manage."

Long-term follow-up suggests that the concern parents and physicians have about hyperactivity is not unfounded. Oppositional ADHD in children can be a prelude to worse behavior in adolescence and adulthood. At least twenty reports linking early hyperactivity and attention deficits to later antisocial or violent behavior have appeared in the clinical literature.[47,48] Other studies have documented an increased risk for substance abuse and suicide.[48] Even when it doesn't end in criminal behavior, ADHD that persists into adulthood is a social handicap that can damage work and personal relationships.

Some children with the aggressive form of ADHD are hostile and impulsive. "These kids are hypervigilant to the extreme," explains Rostain. "They can't calm down enough to pay attention. They're easily frustrated, on a hair trigger." Others, less emotional, have more in common with the underreactive, "low-arousal" aggressor. Skin conductance is lower. Adrenaline and cortisol secretion respond sluggishly to stressors. Rostain says these children "aren't vigilant enough. They are stimulus hungry, and are overactive as a way of achieving a normal level of arousal."

The combination of hyperactivity and low arousal may be especially ominous. In a long-term study of social development and adjustment, Swedish researchers charted sympathetic nervous system activity (as a function of epinephrine release in response to the stress of an arithmetic test) and the incidence of hyperactivity in a group of seventy boys at age ten and again at age thirteen.[47] For each boy, the physiological measures were then compared to the incidence of criminal offenses, including violent crimes, over a twenty-year period.

Boys who committed crimes as adolescents but turned over a new leaf when they reached adulthood were rated as more hyperactive than boys who had no recorded offenses. However, their epinephrine levels—before and after stress—were, if anything, slightly higher than levels in the law-abiding group. In contrast, boys who continued their criminal career as adults were significantly more hyperactive than either the controls or the juvenile offenders—and their epinephrine levels were significantly lower. Nine of the eleven persistent offenders had both high hyperactivity scores and low epinephrine excretion values. The combination of ADHD and a lethargic stress response system, the authors propose, may seriously compromise social learning, culminating in a failure to internalize and obey social rules and conventions.

IN ADDITION to supervising working memory, the prefrontal cortex plays a critical role in the regulation of attention.[49] To carry out this task, it networks with the posterior parietal region of cortex and two midbrain nuclei, the superior colliculus and the pulvinar, which are also involved in the processing of visual information. Cognitive neuroscientists believe that these posterior elements of the attention system have responsibility for the spatial orientation of attention—disengaging from the current focus of attention, moving the focus, and reengaging it in the new location.[49,50] According to Steven Pliszka, an ADHD expert at the University of Texas Health Science Center in San Antonio, responsibility for identifying, interpreting, and comparing the new focus to information in memory is then "handed off" to the prefrontal cortex. There, working memory takes a good look at all of the available information spread across the workspace (including the attention system) and reaches an executive decision on how to respond. Figure 7.3 illustrates the neural circuits mediating attention.

The groundwork for this entire chain of events is laid by the central arm of the global sympathetic nervous system, the norepinephrine-containing neurons of the locus ceruleus.[49] When nothing much is happening, locus ceruleus neurons idle at a low background volume. But when a significant event demands recognition and response, a burst of activity from these cells activates posterior and frontal cortical attention systems, sharpening their sensitivity to

Figure 7.3

"Paying attention" requires the coordinated effort of several brain regions. Alerted to a significant event by the global sympathetic nervous system (1), posterior nodes of the attention circuit redirect the focus of attention (2, 3). Responsibility for interpreting information about the new stimulus is then "handed over" to the prefrontal cortex and working memory (4). Dopamine activity shields working memory from irrelevant information (5) and locks in the selected response (6).

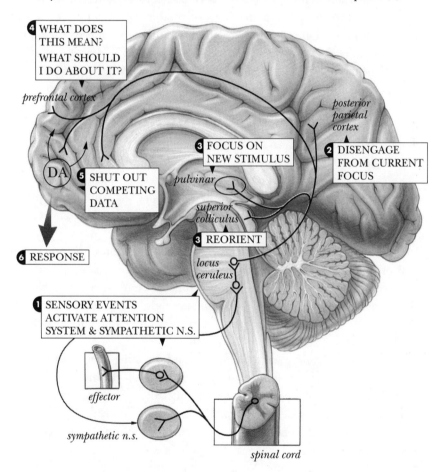

Adapted with permission from S. R. Pliszka, J. T. McCracken, and J. W. Maas, Catecholamines in attention-deficit hyperactivity disorder: Current perspectives, *Journal of the American Academy of Child and Adolescent Psychiatry 35* (1996): 264–272.

novel stimuli. Once information about the potential emergency is forwarded to the prefrontal cortex, dopamine chimes in. By temporarily damping down cortical sensitivity to the steady influx of new data, dopamine activity creates a quiet space in which working memory can choose the best response from the available options.

Imagine you're headed across a dark parking lot toward your car when you notice a man approaching from the opposite direction. Suddenly you shift from planning what you're going to have for dinner to sizing up the stranger, who commands all of your senses. Inside the brain, this disengage-shift-reengage sequence is coordinated by the posterior attention system, the alarming nature of the situation brought into sharp focus by norepinephrine. Body and brain are ready to react. Before you decide what to do, however, you'll need to decide whether this guy is a real threat or another commuter on the way to his own car, his own uncooked dinner. And to do that, you'll need the frontal cortex, meaning that the orienting circuit must hand off its description of the scene to the cortex. Working memory holds this information on-line, while it retrieves the additional emotional, contextual, and factual information necessary to interpret the situation. Perhaps the man's trench coat and laptop computer suggest that he's just another businessman; the cortex decides you can relax and proceed to your car. Perhaps he's scanning the lot with a frown on his face and doesn't seem to notice you. The cortex decides he's probably looking for his car, but maybe you'd better hurry up and keep a watchful eye on him. But perhaps his furtive demeanor and the way he's eyeing you match existing data on parking lot attackers. In this case, you may well decide to look for reinforcements, search for a security phone, even scream.

To carry out any of these options, you'll need to suppress the others and focus on the task at hand. If continuing to your car seems appropriate, you don't need to cry for help. If you need to run, you can't waver. Now the inhibitory action of dopamine suspends processing of new input, and the selected response is locked in. You get in your car and drive home or mobilize security in time to catch a dangerous predator.

Problems in the early stages of alerting and orienting could lead to difficulty in focusing attention. Because the activity of the norepinephrine-containing neurons of the locus ceruleus is crucial for initiating the orienting sequence, some experts have suggested that a malfunction here that alters the background firing rate could be a major factor in ADHD.[49] If locus ceruleus neurons chatter away even when there's nothing to talk about, the cortex becomes overwhelmed by noise. Nothing in the sensory stream stands out, and the prefrontal cortex stops paying attention.

Other investigators, like Russell Barkley of the University of Massachusetts Medical School, believe that the root of the problem in ADHD may lie with an underactive prefrontal cortex.[51,52] Their evidence comes from PET studies in adolescents and adults with ADHD that show a drop in glucose utilization—a reduction in brain activity—in the prefrontal cortex. In addition, they note that patients with damage to the prefrontal cortex are impulsive, distractible, emotionally unpredictable, and socially compromised—classic symptoms of ADHD.

Dopamine and its inhibitory action are critical to frontal cortical function. Drugs that increase dopamine activity, such as Ritalin and dextroamphetamine, improve attention and reduce hyperactivity.[53] Put the two together, and it would seem reasonable to suggest that if ADHD results from a malfunctioning prefrontal cortex, what's not performing up to par are dopamine neurons. The problem in ADHD, in other words, might be a mirror image of the problem in schizophrenia, where dopamine influence exceeds normal boundaries. Once again, an unwelcome change in behavior can be traced to dopamine activity that falls outside the narrow window associated with optimal function. The reduction takes the prefrontal cortex out of the decision-making loop. According to Barkley, that loss relinquishes the control of behavior to external events rather than internal constraints. Behavior is no longer guided by the cortex, leaving the individual environmentally dependent, tied to the events of the moment in reflexive fashion.

Because the cortex feeds back to the locus ceruleus, where it's supposed to keep the background firing rate in line, the problem goes full circle. The meaning of the event is lost; the problem of differentiating what's important and what's not, what's really threatening and what's not, grows. If the attention deficit is accompanied by a reduction in sympathetic activation—low arousal—the brain is left with few resources for assigning emotional significance to people, places, events. Without emotional significance, any response is acceptable, including, if the situation seems to call for it, an aggressive response.

Too Much Fun

The misconception that someone's out to get you is more likely to end in violence than any other type of delusional belief. But if you weren't born with a rearranged frontal cortex, don't worry. You can easily recreate the same lethal delusions without resorting to pathology. All you need is time, money, and a steady supply of cocaine.

Cocaine—or, more accurately, the alkaloid form known as crack—gets much of the blame for the tidal wave of violent crime that crushed urban communities during the 1980s; conversely, its steady fall from the top of the pharmacological charts gets a share of the credit for the recent downward trend in violent crime rates. Affordable, readily absorbed when smoked (thanks to the immense amount of permeable surface offered by the lungs), and more addicting than powdered cocaine, crack added thousands of new consumers to the recreational drug market. An expanding pool of buyers attracted an influx of suppliers—and an abrupt end to the traditional ways of doing business. Clandestine trading gave way to gangs and guns, street warfare, soaring homicide rates.

Cocaine has had an especially nasty reputation since it was first introduced to the American public in the 1880s.[54,55] The rapid appearance of cocaine in home remedies, tonics, and beverages—including Coca-Cola, "a drink offering the advantages of coca without the danger of alcohol"[54]—was followed almost as rapidly by disturbing reports of the drug's addictive potential and its dangerous effects on the heart. Cocaine had adverse effects on behavior as well, as documented by the renowned German pharmacologist Ludwig Lewin, among the first to describe symptoms of cocaine-induced delusions that led to violence: "Mental weakness, accompanied by psychic irritability, bitterness toward others, faulty judgment, suspiciousness, erroneous assumptions, a warped interpretation of facts, unjustified jealousy . . . lead the subject . . . to a state of delusion. . . . An unfortunate man who had ingested three grams or more of cocaine armed himself against imaginary enemies. . . . Another one broke dishes, furniture, and beat up a friend."[55]

By the turn of the century, cocaine was so firmly linked to addiction and violent behavior that President William Howard Taft declared it "the most serious drug problem America has ever faced."[54] One by one, state laws restricted access to the drug to licensed physicians, culminating in the sweeping control measures of the 1914 Harrison Act. Cocaine, forbidden even in patent medicines, became the most tightly controlled psychoactive medication in America, less acceptable and even more feared than opium.

Cocaine abuse plummeted. The public lost interest in cocaine, and the medical community forgot the violent paranoia it could ignite. But the drug did not vanish. It merely withdrew, waiting patiently to be rediscovered in the 1970s, during America's "second cocaine epidemic."

At first, the resurgence of "the great white plague"[55] was viewed as little more than a social annoyance—pharmacological recreation for those who had outgrown marijuana and were bored by alcohol. Slowly, however, reports of the same disquieting side effects noted a century before—rapid addiction, cardiovascular abnormalities, and a paranoid psychosis so similar to schizophrenia that emergency room physicians found it nearly impossible to distinguish the two—began to resurface. With the introduction of crack, reports of toxic effects multiplied, and the association between cocaine and aggressive behavior grew from anecdotal to legendary.

Police reports, government survey data, and hospital-based surveys of patients treated for cocaine-related emergencies all confirm that the relationship between cocaine and human aggression is more than just folklore.[56-58] Users themselves agree that hostility, paranoia, and aggression are not an infrequent feature of their drug experience. One study of freebase cocaine abusers, for example, reported that 62 percent described themselves as paranoid while on the drug, and 28 percent admitted to violent thoughts and actions, including an attempted homicide.[59] Among a sample of more than 450 cocaine abusers calling a Chicago drug abuse hot line, 42 percent said that the drug made them angry, more than 80 percent said they experienced feelings of paranoia and suspiciousness, and nearly half admitted to violent crimes ranging from fistfights and arguments to rape and homicide.[60]

The common denominator in these studies is paranoia, the user's tormented belief that he's being surrounded, followed, or marked for elimination. Ludwig Lewin himself would not have been surprised to hear researchers conclude that cocaine-related violence "may in part be a defensive reaction to irrational fear."[55] As any psychiatrist who has seen paranoid schizophrenia in action can tell you, where there's a fear of personal harm, even if it's delusional, a violent response is an ever-present danger.

CRACK COCAINE use may have declined, but in some parts of the country, another stimulant, methamphetamine, has enjoyed a renaissance. From a home base in the Southwest, a wave of methamphetamine abuse roiled up in the early 1990s and surged steadily eastward.[61–63] In San Diego County, for example, once known to speed freaks as the methamphetamine capital of North America, the number of hospital admissions related to methamphetamine doubled between 1991 and 1994, deaths attributable to amphetamine overdoses during the same time period more than tripled, and the annual amount of the drug seized by authorities skyrocketed from 1,409 pounds to 13,366 pounds.[61] Women have been drawn to methamphetamine in unprecedented proportions,[62] and the number of high school seniors who have used the drug in the past year has doubled in the past six years.[64]

"Crank" costs about the same as powdered cocaine, but a single hit can last over a hundred times longer, making it far more cost-effective. It doesn't have the same shady connotations as crack, and you can probably buy some in your own neighborhood. Amphetamine is the stimulant of Middle America, homey, almost respectable, a tool of the trade for cross-country truck drivers, college students, dieting housewives. As one researcher told a *New York Times* reporter, "Crank is a drug of the times, for people who don't have enough time."[62]

Gang warfare isn't a big feature of the methamphetamine trade, but speed is still synonymous with violence. Some experts, in fact, have claimed that "amphetamine use is more likely to lead to violence than any other drug."[65] In communities where speed has enjoyed a renewed popularity, arrests for violent crime have soared; spectacular incidents of extreme violence—like a New Mexico father who beheaded his teenage son in an amphetamine-induced rage, then threw the head out the window of his van[66]—follow in its wake. Even when they're trying to get straight, methamphetamine-abusing patients have a reputation for aggressiveness; drug treatment centers report that they're even angrier, more hostile, and more likely to lash out than patients being weaned from crack cocaine.[67]

Cocaine and methamphetamine, like all other psychoactive drugs, tinker with brain chemistry in ways that are bound to have behavioral consequences. Nonetheless, social analysts—and the drug-fearing public—have been reluctant to attribute drug-related violence to drugs. The biology-free perspective insists that the violence of drug abusers is largely economic (fueled by the need for money to buy drugs) or social, kindled by territorial disputes, questionable business practices, and revenge.

But addictive drugs are more than contested commodities. Devious con artists, they charm their way into chemical pathways mediating reward, mood, and risk assessment. Once inside, they turn nasty, exploiting, reengineering, and finally enslaving the cells that admitted them. The brain that has been hoodwinked by drugs is a brain with a new and jaundiced view of the world, a brain in which pharmacology has transformed the nature of the interaction between the nervous system and environment.

Purely social explanations for drug violence fall short because they fail to include the brain. They do not explain why methamphetamine and crack cocaine are worlds apart socially yet depressingly similar in their behavioral effects. They do not explain delusions of psychotic intensity. And they do not explain how a select group of chemical substances came to be important enough to kill for in the first place.

PLEASURE IS the secret to the power of addictive drugs. When drugs show up to party, they head straight for the dopamine-containing pathways that track rewards and punishments. There, they'll stage a celebration the brain won't soon forget. Unfortunately, they're also exuberant, sloppy party guests, making a tremendous mess and then interfering with every effort to clean it up. Mess is not conducive to healthy dopamine function. If transmitter activity is to keep to the critical middle ground, dopamine synapses must be kept in pristine condition, clear of excess neurotransmitter that might shift the system into overdrive. The crucial task of waste removal is the responsibility of specialized proteins known as transporters, biological ionizers that trap dopamine at large in the synapse and pump it back into the presynaptic cell for recycling.

Narcotics have their opiate receptors. Antipsychotic drugs latch onto dopamine receptors. Yohimbine, the drug that provokes panic attacks in PTSD sufferers, begins its reign of terror by hooking up with receptors intended for norepinephrine. The central dogma of modern neuropharmacology might be summed up as "a place for every drug and every drug in its place." Cocaine, however, proved more reluctant than most others to reveal its point of entry. The cocaine receptor remained a mystery until 1987, when Michael Kuhar, then at the National Institute for Drug Abuse's Addiction Research Center and currently chief of the Neuroscience Division at the Yerkes Regional Primate Research Center, discovered that the appeal of cocaine and related drugs correlated with a fatal attraction for the dopamine transporter.[68] By an unhappy coincidence, nature has shaped a cleft in the transporter protein just the right size and shape to accommodate a cocaine molecule. When the drug slips neatly into its docking space, dopamine is locked out, the synapse is flooded with "levels of dopamine never seen in nature," says Kuhar, and the balance essential to proper dopaminergic action goes out the window, replaced by a breathless euphoric rush that gives new meaning to the concept of reward.

Methamphetamine can't gain a toehold on the cocaine binding site, but it still finds a way to block dopamine reuptake. Just to make sure the volume's turned all the way up, it also encourages the presynaptic cell to release even more dopamine into the synapse.[69] And unlike cocaine, the methamphetamine experience isn't over in an hour; because amphetamines are cleared more slowly from the brain, the drug continues to fiddle with dopamine transmission for eight or more hours.

Stimulants don't limit their rabble-rousing to dopamine pathways. Less choosy than some other drugs, they're happy to block transmitter reuptake at norepinephrine or serotonin synapses as well.[69] But it's their predilection for short-circuiting waste removal at dopamine synapses, especially those of the mesolimbic pathway that end in the nucleus accumbens (see Chapter 5), that has caught the eye of researchers. That's the critical feature common to drugs people are willing to kill for—cocaine, amphetamines, opiate narcotics—and that differentiates them from ho-hum drugs—antibiotics, for example.[70] You name it—if the drug is addictive (even if it's a legal high like alcohol or nicotine), you'll find it disorganizing dopamine function in the nucleus accumbens.

The dopamine reward system was designed to serve as an internal parent, delivering a chemical payoff for behavior such as feeding or mating that accomplishes a biological goal or ensures survival.[71] Allowing drugs like cocaine access to these essential reinforcement mechanisms is as foolhardy as going on vacation and giving a group of rowdy teenagers the keys to your house. The interlopers take over and convince dopamine neurons to reward drug taking as enthusiastically as they endorse eating and sleeping.[70,71] Abuse escalates into addiction. The addict *needs* this party to last and last, and he's willing to do whatever it takes—rob, cheat, bully, even murder—to ease the anguish of wanting, to overcome anyone foolish enough to stand between him and the drug that has come to seem essential to his survival.

Once you're hooked on stimulants, your life is in danger of degenerating into one long party, a string of intermittent binges during which you smoke or sniff or inject the drug until every last crystal is gone. If you want to stop, you'll find that exiting the party circuit is a herculean task; even with strong support, your chances of being drawn back into the cycle are much greater than the chances you'll be able to stay out of the loop for good.

The repetitive bingeing will surely alter your social and professional life. With time, you'll also begin to notice some changes in the drug experience itself.

Your current highs, for example, will pale in comparison to the vibrant, ringing euphoria that memorialized your introduction to cocaine or amphetamine. When you run out of drug—or vow to take one more stab at quitting—you don't feel relief or pride; you feel tired and dejected and hostile. Bingeing again, your irritability wanes, but now you're increasingly edgy and nervous, even panicky. If you're unlucky—and a recent survey suggests that as many as two-thirds of cocaine addicts are[72]—your anxiety will devolve into frank paranoia, complete with vivid delusions that you're about to be arrested, pursued, or attacked.

Your world has changed, all right. While you were having fun, your favorite stimulant was busy rearranging not only your life, but your brain.

As WALTER CANNON noted, "Stability is everything." Reciprocity and balance are essential features of neurotransmission, the result of mechanisms that adjust pre- and postsynaptic elements—transmitters, receptors, second messengers, transcription factors—to maintain a dynamic equilibrium. In this circular system, it is impossible to disrupt one element without affecting the others. If transmitter levels go down, postsynaptic sensitivity will be turned up—by increasing the number of receptors, improving the efficiency of message transfer from the outside to the inside of the neuron, or temporarily shutting off activity-dampening autoreceptors. If transmitter levels go up, receptors and effectors gear down.

If stimulant consumption becomes a way of life, cocaine and methamphetamine will arrogantly muscle their way into the dialogue between dopamine neurons and their targets over and over again. And with time, the repeated blockade of transporter proteins not only sets off periodic tidal waves of dopamine, but instigates more profound and long-lasting changes on both sides of the synapse.[70,73–75] From the level of the beleaguered dopamine receptors, the message of addiction is passed from protein to protein along the cascade of biochemical reactions that link receptors and genes. The steady cry for help recruits transcription factors, and transcription factors switch critical genes on and off.[76] Structural and functional proteins encoded in those genes reshape the neurons to cope with the demand of chronic drug use (Figure 7.4). Ultimately, drug-induced remodeling surfaces as new patterns of drug response and drug craving that are likely to come as unwanted surprises to the substance abuser.

One of the first is the discovery that the longer you indulge, the less fun drug taking becomes. The loss of sensitivity—pharmacologists call it tolerance—is the result of the brain's self-defensive efforts to moderate the cacophony of supranormal dopamine levels. To stay happy, the tolerant addict needs more and more drug, more and more money, and perhaps more victims to supply it.

Tolerance is only one consequence of the brain's heroic attempt to balance the stimulating effects of drugs like cocaine. During the excitement of stimu-

Figure 7.4

Chronic exposure to drugs like cocaine induces a variety of adaptations in mesolimbic dopamine neurons, including changes in the levels of critical signaling proteins, changes in gene expression, and even changes in neuronal structure. Drug-induced remodeling surfaces as tolerance, craving, and behavioral sensitization.

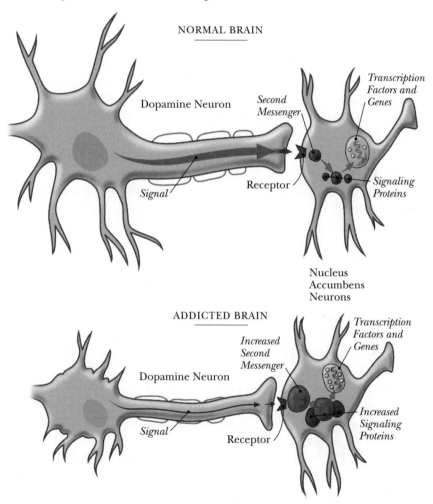

Adapted with permission from D. Beitner-Johnson, X. Guitart, and E. J. Nestler, Common intracellular actions of chronic morphine and cocaine, *Annals of the New York Academy of Sciences 684* (1992): 70–87.

lant intoxication, it's difficult to recognize that overstimulated monoamine pathways are actually standing on the brakes to restore some semblance of order. Removing the drug may stop the flood of transmitter, but it won't immediately stop the system from overcompensating. What was once adaptive is now problematic, too damped down, the glow of stimulant euphoria replaced by an irritable, gloomy cloud. The synapse has escaped containment. Like the stress response overflowing its natural boundaries, the resulting dysregulation leads to behavioral symptoms of imbalance—depression, anxiety, hostility and overt aggression—cardinal symptoms of the early stages of withdrawal.[77]

Finally, everything but the fond memory of cocaine fades. With tolerance out of the way, the lasting damage cocaine and methamphetamine do to the nervous system can finally be seen. The most obvious is a recurrent, obsessional craving for drugs, often sparked by the mere sight of people, places, or things associated with drug use. And the most surprising is sensitization. Tolerance allows the brain to cope with outrageous dopamine levels by trying to ignore them. Sensitization—an increased sensitivity to some drug effects after repeated exposure—encourages the brain to carry a grudge, much as amygdala neurons can collect the grievances imposed by a series of electric shocks.

Electrically stimulated cells finally blow off steam in the form of kindled seizures. Repeated exposure to cocaine can also provoke seizures. But more important (from a social point of view), it provokes paranoia, pushing neurons further and further until they blow off steam in the form of psychotic—and potentially murderous—delusions.[72,78]

Pharmacologically, sensitization is the opposite of tolerance; in fact, it's sometimes called "reverse tolerance." Instead of needing progressively larger doses to achieve euphoria, the cocaine-sensitized abuser needs progressively smaller doses to precipitate delusions. There are also more subtle differences. Tolerance is nearly universal. But only a subset of stimulant users progress to paranoia, suggesting that some individuals are more vulnerable to sensitization than others, even if they consume less drug less frequently.[78]

Susceptible abusers first notice a nagging anxiety. The anxiety worsens with every binge; once paranoid symptoms appear, they grow more threatening each time the brain encounters cocaine.[78] And in those individuals, sensitization outlives even the most protracted period of sobriety; relapses after years of abstinence can still cause delusions. Some researchers believe the adaptations that underlie sensitization may be permanent.

Every time the brain en route to sensitization and paranoia is exposed to stimulants, more dopamine courses through the limbic system and the prefrontal cortex. So it's no surprise that the resulting behavior resembles that associated with the dopamine imbalance that's a trademark of paranoid schizophrenia, complete with its lethal edge. A 1991 survey found that nearly 40 percent of paranoid cocaine abusers admitted for inpatient treatment had armed themselves with a weapon; 6 percent had actually shot at someone they believed to be an attacker.[72]

STRESS DISORDERS and substance abuse are coconspirators; catch one and you're likely to nab the other as well. PTSD, for example, is twice as common among substance abusers as it is in the drug-free population.[79] Studies estimate that between 60 and 80 percent of combat veterans who seek treatment for PTSD also abuse drugs and alcohol.[80,81] Conversely, 70 to 80 percent of patients in substance use recovery programs report that they were physically abused as children.[82]

People who have been beaten, shot, raped, terrorized, or abused turn to psychoactive drugs not to feel good but to avoid feeling bad, to find diversion, soothe an overactive nervous system, muffle despair and rage, add some of the sparkle back to life. To many who treat traumatized patients, therefore, stress-induced substance abuse is a pain problem. To a neurobiologist studying stress, however, it's a traffic problem. Sensitization to stress and sensitization to stimulants follow the same route to work—the mesolimbic and mesocortical dopamine pathways that mediate reward and regulate reality checking.[83,84]

Stimulant drugs and environmental stress are such convincing stand-ins for one another that Seymour Antelman, of the University of Pittsburgh, refers to cocaine as a "pharmacological stressor."[85] Rats primed with multiple small doses of cocaine are significantly more frantic in response to the stress of foot shock than animals whose dopamine neurons have not been sensitized by stimulants.[83,86] And rats sensitized by stress (five consecutive daily sessions in a test cage where they received twenty minutes of intermittent shocks) released more dopamine into the nucleus accumbens and were significantly more hyperactive after a challenge dose of cocaine than rats that had not experienced stressful shocks.

Foot shock is an uncommon adverse event outside the laboratory, but subordination stress—the outcome of losing an aggressive encounter—is part of daily life for any social animal, and it can also alter the reinforcing properties of stimulant drugs. Margaret Haney, a former student of Klaus Miczek and currently of the Substance Use Research Center at Columbia University, allowed socially naive controls and rats defeated in four successive encounters with aggressive residents the opportunity to self-administer brief pulses of cocaine. Defeated animals dosed themselves with significantly more cocaine than the unstressed rats—and they did it twice as fast.[87] Stress made cocaine even more attractive and the compulsion to use it even more powerful than usual.

The brain is an easy target for the malignant effects of trauma and stress. And one of the more distressing neural consequences of surviving an assault or enduring years of abuse may be an increased vulnerability to the allure of drugs and an even greater sensitivity to their most socially destructive effects.

America's Favorite Drug Goes on Trial

Say "violence" and "hormones" in the same sentence, and people think "testosterone." Pair "violence" and "drugs," and they think of crack and speed, warring gangs and desperate addicts. But the drug with the strongest connection to violent behavior is sold in stores, not on the street; it's cheap, easy to obtain, socially acceptable, and perfectly legal. That drug, of course, is alcohol.

"No drug is more intimately related to aggression than alcohol."[88] Klaus Miczek has a slide featuring this quotation, taken from a 1990 review by Iowa State University researchers B. J. Bushman and H. M. Cooper. But just below is a second quote, this one taken from a contemporaneous review by British behavioral biologist Paul Brain: "Alcohol is *not* an aggressive drug."[89]

"It's stunning, absolutely stunning," observes Miczek. "Two people looking at the same data and reaching completely opposite conclusions. Everyone wants to know who's right."

Personal experience and statistics are on the side of Bushman and Cooper. Everyone knows someone who "can't hold his drink," a person you need to treat gingerly after he's had a few. Experts agree that alcohol can be real trouble; they estimate that more than half of all violent crimes are committed by a perpetrator under the influence of alcohol.[90–92] Nonviolent crime is also associated with alcohol, but individuals arrested for a violent crime have higher blood alcohol levels.[92] Among murderers, the rate of alcohol consumption before the crime has been reported to be as high as 83 percent.[65] In the community at large, alcohol turns up in the same places as domestic violence, child abuse, suicide, and lethally aggressive driving with a suspicious frequency.

But numbers alone can't convict alcohol. Many social scientists would side with Paul Brain, contending that the epidemiological evidence, while voluminous, is no more than circumstantial; simply because alcohol was present at the crime scene doesn't mean it's responsible. Alcohol-related violence, they might argue, is a "nurture" phenomenon, the result of cultural beliefs that permit or even encourage aggressive behavior in drinkers, environments that give people license to act out when they're drinking. Alcohol, in other words, is ruined by bad company.

Who's right about alcohol and aggression? You should know by now that there's no one right answer. The response to alcohol is a joint effort between pharmacology and experience, chemistry and expectation. The right question to ask is not, "Are drinkers more aggressive?" but "when?" "with whom?" and "how?"

As long as they're not behind the wheel, most social drinkers aren't a danger to anyone. But every party has a few wet blankets, drinkers who would rather pick a fight than start a conversation. That's the first key to understanding the biology of alcohol: recognizing that it either loves you or hates you. When the chemistry is bad, trouble is never far behind.

"It takes all kinds—and some of them aren't," my father used to grumble. Klaus Miczek has learned that one of the features that differentiates the kind from the unkind is their response to alcohol.

"When we give a group of, say, a hundred mice a fixed dose of alcohol, some fight less than they did before. Some mice look the same. But perhaps ten of them really stand out. They go berserk on alcohol."

Miczek and his colleagues have examined dozens of groups of mice, and in each and every one, some mice become so mellow on alcohol they may stop fighting altogether, while a small proportion—ranging from 10 percent to as high as 30 percent—turn vicious.[93,94] "If you averaged the incidence of fighting," notes Miczek, "you'd probably come to the conclusion that alcohol doesn't do anything. And your conclusion would be right—except for this minority who overreact."

"Overreact" is an understatement. After even a small dose of alcohol, the number of attack bites and threats by AHA (alcohol-heightened aggression) mice increased by more than 200 percent.[95] Tolerance is not in their vocabulary either. Tested repeatedly over a three-month period, AHA animals fought after drinking alcohol every time.

Why does alcohol make monsters out of some mice? No one knows. "They're not the big bullies," explains Miczek, "and they're not the meek, inhibited ones, so disinhibited by alcohol that all hell breaks loose. They are not the ones who run around the lab getting into all kinds of trouble—including fighting—either. They're not the most active at reproducing."

AHA mice fight more, but not differently; their encounters follow the same gap-burst pattern characteristic of rodent aggression, with appropriately mouse-sized spaces between bursts. They weigh the same as their alcohol-resistant counterparts. They can hear, see, and smell equally well. Pharmacokinetic studies show that they absorb and break down alcohol at a similar rate. Neurochemical assays haven't yet uncovered any differences in transmitters or receptors. Perhaps fortunately for other mice, they don't even breed true; the offspring of AHA mice are no more likely to be alcohol-sensitive than the offspring of normal mice.

The aggression of AHA mice remains a mystery, a critical mystery. If we knew the reason for their sensitivity to alcohol, we might begin to have an idea why alcohol is such an incendiary compound for some humans. Equally important, it would be useful if we had an accurate estimate of how many "AHA" people exist. For all the boxes of statistics we've collected, we simply don't know how many people in the general population—not prisoners, but average citizens—turn violent after drinking. And it's what we don't know that keeps hurting us. Intoxicated criminals are only the tip of the iceberg. Hidden in the crowd of social drinkers are an unknown proportion of people for whom alcohol is an explosive substance, people who are an accident waiting to happen.

NEUROCHEMICALLY, ALCOHOL is almost as sociable as some drinkers. Perhaps its most important partner is the inhibitory neurotransmitter GABA. Just as cocaine has its own docking site on the dopamine transporter, alcohol has a reserved space on the GABA receptor, where it acts to boost the effect of the transmitter.[94,96] In addition, like other addictive substances, alcohol fraternizes with mesolimbic dopamine neurons.[70,97] Alcohol-induced increases in the activity of this pathway prompt adaptive changes similar to those prompted by cocaine and methamphetamine, changes that compose the siren song of the bottle.

Alcohol also enjoys a chummy relationship with serotonin. Exploratory studies in the late 1970s first noted that cerebrospinal fluid drawn from young male alcoholics contained significantly lower levels of the serotonin metabolite 5-HIAA than fluid taken from nondrinking cohorts. Later studies showed that the reduction was limited to one special group of alcoholics: men who started drinking early, following in the footsteps of their alcoholic fathers— men who were also impulsively aggressive. In their landmark study of violence, impulse control, and serotonin, Markku Linnoila and Matti Virkkunen observed that nineteen of twenty-one arsonists—the definitive crime of poor impulse control—had an alcoholic father and that these men had the lowest CSF 5-HIAA values,[98] an observation confirmed in subsequent studies by these researchers.[99,100] Alcohol, Linnoila and his colleagues hypothesized, temporarily relieves the emotional malaise associated with a disrupted serotonin system, but at the cost of disorganizing the system even more. As a result, the impulsive drinker may feel better, but he's more likely than ever to act first and think later.

Only impulsive aggression is correlated with reduced levels of serotonin metabolites; nonimpulsive alcoholics have normal levels of 5-HIAA,[99,100] suggesting that the effect of alcohol may be piggybacked onto a more complex relationship between serotonin dysregulation and impulse control. John Evenden, a British researcher for the pharmaceutical firm Astra Arcus, has further dissected the relationship between alcohol and impulsive behavior.[101] Evenden subdivides impulsive reactions into three types of errors: errors in *preparation* (not all relevant information is taken into account before making a decision), errors in *execution* (the responder quits before the goal is reached), and errors in *estimating consequences* (a quick but less valuable outcome is chosen rather than delaying gratification to get a bigger reward). Only the last is affected by alcohol. After consuming alcohol, rats given a choice between pressing a lever that immediately dispenses a single food pellet or a second lever, which delivers five pellets—but only after a delay—overwhelmingly vote for the immediate reward. Like impulsive people, the intoxicated rats can't wait for the future. They live for the moment.

Given that the inability to delay gratification is also a cardinal symptom of injury to the prefrontal cortex, it's tempting to speculate that alcohol may act, in part, by disrupting frontal cortical function. The intoxicated individual loses context and becomes "environmentally dependent," responding in a knee-jerk fashion to the here-and-now, without waiting for the voice of reason to offer a second opinion. When that spontaneous reaction includes a gun, alcohol-fueled impulsiveness can easily be lethal.

YOU DON'T HAVE TO be an alcoholic to be a violent drinker. In fact, with alcohol, less is more—more aggression, that is. Rats and monkeys given low doses of alcohol are significantly more aggressive toward intruders and subordinates. But at high doses, the sedative effects of alcohol take over, and the same animals attack less often than controls.[93,102]

A little goes a long way in humans as well. For example, among college students paired with an experimenter's accomplice in a competitive game, subjects who received a low dose of alcohol punished the confederate more severely for losing (by administering electric shocks) than subjects who received higher doses.[103]

"Chronic drinkers are a huge problem for society—but not because they're a fantastic threat," observes Klaus Miczek. "When it comes to violence, the problem is the binge drinker, the Sunday-afternoon-football drinker, the evening drinker." The exception is the alcoholic who has superimposed substance abuse on another problem—a personality disorder, a stress response disorder, schizophrenia. For these chronic drinkers, adding alcohol to the mix fans the flames, increasing the probability of violence.

But in general, when it comes to drinking and fighting, many of us worry too much about the wrong people. We cross the street to avoid the frightful, scruffy drunk loitering in a doorway. The kind of drinkers we're likely to encounter in a favorite night spot, on the road, and at home are the ones who should send shivers up our spines. Alcohol is full of surprises, and more than a few are nasty.

DARLENE DECIDED it was time to get back into rehab after her boyfriend held a loaded gun to her head for three hours.

"He told me, 'Get out of the room, I'm going to kill myself. If you don't get out I'm going to shoot you first.'" She stayed. "I didn't leave," she rationalizes, "because I knew if I walked out of that room, he *would* kill himself. I was willing to be a sacrifice. I did things then I would never have done when I was sober."

Darlene has been at this halfway house for about six weeks now. She insists that she has "finally found serenity." She still looks as if every drop of vitality had drained out of her body through a hole in her feet, her face so pale it's translucent.

"When I was drinking, I guess I didn't place a very high value on my life," she concludes. "I guess if he'd have pulled the trigger, that would have been okay by me."

Shelley, too, got into dangerous predicaments when she drank. "I've been in violent situations with men all my life—and it was always related to drugs and drinking. It would start off with some petty shit, some stupid comment like, 'Why the hell are you making that for dinner?' Next thing I'd know it would be a brawl. He'd try to overpower me; I'd fight back; in the end he'd win. Next day neither of us would remember what in Christ's name we were arguing about."

Perpetrators aren't the only ones who drink and take drugs. Victims can also be substance abusers, and their drug taking has an equally significant impact on the probability, intensity, and duration of violent behavior. In fact, at least half of all *victims* of violent crime are intoxicated at the time of the attack.[91]

Alcohol, like stress, makes people stupid. It trims danger down to size, encourages drinkers to talk when they should listen, to experiment with foolishly provocative behavior. The one thing it doesn't change is the probability of winning a fight. Victims may think they're more intimidating after a few drinks; they're actually just more vulnerable.

If you were a rat and a cat had just strolled through your living quarters, the last thing you should do is make your presence obvious. Freezing in place would be a wise decision, less likely to attract attention. Moving is riskier, and checking out the space where the cat was last seen is downright foolhardy.

But if you're a rat who's downed a small amount of alcohol, that's just what you'd do. Robert and Caroline Blanchard have studied the reaction of rats and mice to threat by placing a cat briefly into a compartment opening onto the rodent's cage.[104] Normal animals stay still—and as far away from the cat chamber as possible. Animals treated with low doses of alcohol, on the other hand, saunter back and forth across the cage, rear, sniff, actually venture into the cat compartment to check things out. No sober animal would ever act this recklessly.

Alcohol also increases self-defensive aggression. Wild rats, confronted with a grasping human hand, typically try first to run away, freeze if there's nowhere to run, and attack only if cornered. At low doses of alcohol, however, they attack immediately, seemingly oblivious of the risk posed by starting a fight with someone fifty times their size.

Under the influence of alcohol, rats take chances that can have lethal consequences. Humans similarly compromised by alcohol end up as victims of violence, usually after they've done things they'd never risk doing sober. Some, like Shelley, may egg on an opponent, or decide to fight when backing down is clearly a smarter option. Others, like Darlene, often can't see that it's time to get out until the trigger is pulled.

Home Alone

Women just can't win when it comes to substance abuse. They're no less vulnerable to the brain-remodeling effects of drugs than men but much more vulnerable to an attacker. Thanks to long-lasting effects on the stress response system, female drug users who are abused or assaulted are at risk for developing stress response disorders and of being victimized again in the future. When they finally decide to quit, drug-using partners threaten and harass them.

Women are big losers in the drugs-violence equation. But the biggest losers of all are their children.

No one knows for sure how many babies try drugs before they're even born. Their mothers aren't inclined to be honest, and current drug tests miss many of those exposed early in pregnancy. We do know that as many as 3 of every 1,000 infants born in the United States will have consumed enough alcohol to carry the physical stigmata of fetal alcohol syndrome (FAS): facial deformities, heart defects, abnormalities of the joints and bones, poor growth.[105,106] Epidemiological studies estimate 10 to 15 percent have sampled cocaine,[107] including a surprising 6 percent of those born in communities far from crack-ridden inner-city neighborhoods.[108]

For these children, the earliest encounters between the nervous system and the outside world can be an act of vandalism. Alcohol, for example, can have a devastating effect on the brain, as well as the body, during development. Just a few days of alcohol exposure is enough to kill immature brain cells.[106] In the worst case, exposure can wipe out enough neurons to stunt brain growth. Cells that survive wander as haphazardly as a drunk driver weaving across the road; they miss turns, lose their way, never arrive at their designated locations. Muted dendrites and confused axons fail to connect. In rats, prenatal exposure to alcohol sensitizes dopamine neurons, priming them to overreact to alcohol later in life.[109] In humans, chemical and functional derangement leads to a decrease in IQ scores, other cognitive deficits, behavioral problems, poor coordination and muscle tone, and hearing disorders.[106]

Fetal alcohol syndrome is a leading cause of mental retardation in the United States and other Western nations.[110] Attention deficits, problems with memory and problem solving, and impulsiveness are also common in children with FAS.[111] Even children exposed to levels of alcohol too low to cause FAS may still have trouble learning, paying attention, sitting still—a constellation of deficits known as fetal alcohol effects (FAE).[105,106]

A condition that blunts reasoning ability and disrupts attention mechanisms spells trouble for social learning, regardless of whether it has a direct effect on behaviors like aggression. So it's no surprise that children with FAS or FAE have been characterized as aggressive and defiant.[112] But there's no straight line from a mother with a drinking problem to a child with a behavior problem. Some mothers who drink heavily during their pregnancies have normal babies, while some who consume only one or two drinks a day have children with all the behavioral signs of FAS or FAE. The reason is that prenatal drug abuse involves more than alcohol consumption. Mothers who drink (or abuse drugs) are careless in other ways that can impair brain development. They are less likely to receive prenatal care or to pay much attention to their own health and nutrition, and they often abuse other drugs in addition to alcohol, placing their infants at even greater risk. Timing and duration are also important.[106] Binge drinking is often more destructive than regular consumption. Drinking during the first trimester of pregnancy is risky, but even a single bout of drinking during the third trimester, when the brain is especially vulnerable to the toxic effects of alcohol, can wipe out significant numbers of cells. The higher the peak blood alcohol level is, the greater is the risk of brain damage. It's not enough, in other words, to say that the nervous system has encountered a drug in utero. You also have to specify when, how long, and how much, as well as what other prenatal trials the brain may have been asked to endure.

CRACK COCAINE was supposed to create a generation of monsters. Eight or nine years ago, experts feared that "crack babies" would start out as underweight, irritable, agitated infants—and grow into aggressive, overreactive children.[113,114] Media reports took these concerns to extremes, suggesting that children exposed to cocaine in the womb would be permanently brain damaged, unable to give or receive affection, without conscience or self-control, devoid even of "the intellectual development to have consciousness of God."[114]

If you asked people on the street, they'd still tell you there's no question that infants exposed to cocaine today will be tomorrow's criminals. But recent research shows that there *are* questions. On the plus side, we now know that crack babies are not irreparably ruined. On the downside, we've learned that the brain exposed to cocaine during development does not escape unscathed, but changes in subtle, covert ways that may well have long-term implications for behavior.

"Neural development is a series of choices," says Pat Levitt of the University of Pittsburgh School of Medicine. Studies in animals by Levitt and others suggest that neurons—like drug-curious adolescents—can be coaxed into less-than-optimal decisions by cocaine. By interfering with the action of serotonin, cocaine addles synapse formation in the cortex.[115] By interfering with dopamine, it allows growing dendrites to run wild, snaking around and through the prefrontal cortex like overgrown vines.[116] It disconnects dopamine receptors from their second messengers, monkeys with transmitter levels, adds and subtracts neuronal components to remodel the infant brain just as it remodels the brains of adults.[117–119]

And what about human children? No one knows about their dopamine receptors, but work by Ira Chasnoff, of the University of Illinois College of Medicine, suggests that the after-effects of cocaine's interference in brain development can be seen years later and that the drug specifically targets social behavior, attention, and stress responses.[120] Chasnoff identified ninety-five children exposed to cocaine in utero and compared their cognitive and social development over the first six years of life with that of seventy-five children from the same inner-city community who had no drug exposure. "Looking at IQ, at four, five, or six years of age, we find there's no difference between the two groups on global cognitive functioning," says Chasnoff. "But we do find a significant difference between the two groups on behavior, with increases in aggressive behavior and impulsive behavior in the drug-exposed kids." Home environment, not drug exposure, was the best predictor of future IQ—and the single most important predictive factor in that environment, according to Chasnoff, is whether the mother continues to use drugs. On the other hand, drug exposure—independent of the environment, which had little impact— was the best predictor of future behavioral problems. Both mothers and teachers reported that cocaine-exposed children were aggressive, easily distracted, and more likely to be depressed or anxious than their drug-free peers.

Prenatal drug exposure is not pharmacological determinism. A drug-realigned brain is still sensitive to the power of the environment to pull and shape and carve, to recreate physiology, to mold perception and reaction. But a child who starts life already scarred from battling the prenatal environment has obstacles to overcome, has evidence that the world is a dangerous place.

What happens before birth, however, is only a starting point. The real trouble for many drug-exposed children begins *after* birth, when they leave the hospital in the arms of a drug-addicted mother. Women desperate enough to use drugs while they're pregnant aren't likely to find the new responsibilities of motherhood a compelling reason to stop. As a result, the biological impact of maternal substance abuse doesn't stop in the delivery room, but continues to pose a relentless insult to the developing brain.

An addicted mother's attention is focused on drugs, not her child, and her ability to cope with stress is compromised by the long-term effects of substance abuse. Unfortunately, a drug-exposed baby who's slow to respond or learn and needs endless clinic visits to manage long-term health problems, doesn't pay attention, throws tantrums, and gets into everything as soon as he can crawl is just the sort of high-demand child guaranteed to outstrip her limited emotional resources. Emotionally disorganized child and overreactive mother egg each other on, each negative interaction adding to the chemical record of disaster, the perception of threat and the risk of violence spiraling higher and higher. The more impaired the relationship, the greater the risk it will become a violent relationship.[121]

Women stand in the middle of the circle that joins substance abuse and violence. When they abuse cocaine and alcohol during pregnancy, they get another life off to a bad start. And when they keep on using drugs after the birth, they take a bad situation and make it worse—that is, provided they're around to have an influence. Drug-addicted parents have an uncanny knack for winding up in hospitals, prisons, or rehabilitation centers rather than at home. Some die. Some abandon their children. Many lose their children to foster care.

Thanks to crack cocaine and the resurgence of methamphetamine, children are losing mothers at an unprecedented rate. One in three female inmates are serving time for drug-related offenses. Three-quarters are mothers.[122]

The loss of a father may be grievous to a young child. But biologically, the loss of a mother is devastating.

COLUMBIA UNIVERSITY'S Myron Hofer is one of a growing number of researchers who are examining the biology of maternal separation for clues about how loss shapes future behavior and physiology. Like Klaus Miczek, he's concerned with asking the right question. When it comes to mothers and children, he believes that question is not how the abandoned infant responds, but why; as he puts it, "In maternal separation, exactly what is lost?"[123]

The answer begins with the discovery that the separation response is not one response but many. In rats, for example, every physiological response of an isolated pup is a reaction to one lost function of the absent mother: a more rapid heart rate, for example, to the change in ambient temperature, an increase in corticosterone to the physical sensation of the mother's touch. Together, "the elements of the lost interaction with the dam . . . turn out to be regulators of the infant's developing nervous system."[123] In an undisturbed nest, these regulatory processes are embedded in the mother-infant relationship, linking biological regulation to the sight, smell, and feel of the mother. The loss of the mother is an "escape from regulation," an uncoupling in which immature behavior and body functions are left to fend for themselves. In a few short minutes, the world goes from safe to perilous, organized to chaotic. Responses to future loss or stress will never be the same.

"The early environment programs the nervous system to make an individual more or less reactive to stress," says Michael Meaney, a biologist at McGill University. Mothers, Meaney believes, play a crucial role in this programming process. "The presence of the mother ensures that stress responses are minimized and growth is maximized." He has shown that a baby rat separated from its mother for no more than six hours a day during the first two weeks of life grows up to be an adult rat with a ruined stress response system, poorly contained by feedback inhibition and hypersensitive to distressing events like restraint and novelty—changes that may persist throughout its life.[124,125]

Meaney concludes, "If parental care is sufficient, the nervous system seems to conclude that the world isn't such a bad place to grow up. But when care is inadequate or unsupportive, the system may decide that the world stinks—and it better be ready to meet the challenge."

Or, in the words of psychiatrist Gary Kraemer, "The brain of the unattached infant cannot regulate behavior in the usual way."[126] He and his coworkers at the University of Wisconsin's Harlow Primate Laboratory have extended studies of maternal separation and behavior to a primate model: the rhesus monkey. Infant monkeys separated entirely from other animals, including their mothers, grow up to be pathological, socially clueless adults. However, it's not clear how much of the derangement is due to the loss of a mother and how much is due to social isolation. To block out the confounding influence of isolation, Kraemer's group removed infants from their mothers and reared them with peers.[127,128] The young monkeys spent the first six weeks of life in a well-appointed nursery, surrounded by plenty of toys and socialized by daily play dates with peers; at six weeks, they were moved to a cage with two of their playmates. Then, at eight months, the team shook up the social environment, reassigning peer-reared infants to cages with two unfamiliar partners and at the same time removing a comparable group of mother-reared infants from their home cages to three-monkey peer groups.

Peer-reared infants showed neurochemical differences—at first higher, then lower norepinephrine levels, for example—in the very first months after maternal separation.[127,128] But it's the social challenge of getting to know new peers that really demonstrates the impact of losing your mother. Mother-reared monkeys reacted to the stress of the transition with a classic fight-or-flight response: norepinephrine, cortisol, and ACTH levels shot up. Stress hormone levels in peer-reared monkeys, on the other hand, did not change. Their behavior, however, did; they were significantly more disruptive under stress than the mother-reared cohorts.[126,128]

Kraemer's group concluded that mother monkeys are as crucial to the organization of social behavior as mother rats are to the organization of physiological functions. The infant uses the emotional power of the mother-infant bond to internalize the rules of social conduct (presumably as representational memories); without her, social behavior, as well as adaptive biological and behavioral responses to social stress, fail to develop correctly.

Sluggish reactions to stress are a cardinal feature of antisocial aggression. Of course, that doesn't mean that all antisocial individuals grew up without the benefit of a mother or, conversely, that a mother's death or disappearance predicts a life of violence. But it may be more than an unhappy coincidence that the number of youthful "superpredators"—children without a conscience—has grown in tandem with the drug-related disappearance of mothers.

PERTURB DOPAMINE function, and you'll not only reset reward mechanisms and stress responses, but upset the balance of power in the prefrontal cortex. Addle cortical function and you'll disrupt working memory. Disrupt working memory and you've compromised thinking. And if you can't think clearly, all the moral instruction in the world won't do you any good—you'll have no more access to it when you need it than you have to the farthest reaches of space. If we want people to behave reasonably, we have to ensure that the neural mechanisms governing reason are up and running.

Eight

LEADING THE CHARGE

In spring, rain muddies the landscape; in high summer, the thick, sullen heat steams the outline of trees into a hazy olive blur. But October in Pennsylvania's Bucks County brings clean edges, a diamond-hard light that magnifies the contour of every leaf and paints tawny stripes on towers of cornstalks.

The sleek black and white coats of the three border collies are as crisply outlined as the intersection of the blue sky and the blaze of trees marking the far edge of the meadow. Perhaps the intense clarity of the landscape suits a dog that epitomizes canine perceptual acuity. Their dark, curious eyes catch trainer Eve Marschak's slightest movement; their ears snap forward at her whistle, even if they're on the other side of the field.

A "come bye" or "away to me" from Marschak launches them in a great arcing loop to approach their sheep from behind. Low to the ground, they snake out their heads and glare at the flock with a primitive ferocity. The sheep eye them back. A slight, sharp lunge, and the sheep break and run. Now the dogs will urge them along toward the trainer they consider their pack leader, smoothly balancing the flock along an imaginary straight line by rapid switches in position and speed that counteract any attempts by the sheep to flag, drift, or bolt. But it was not always a trainer or a shepherd at the receiving end of the drive. In the early days of the relationship between man and dog, the human pack leader was a hunter, and the goal of herding was a shared meal.

A recent analysis of DNA samples from sixty-seven different breeds of dogs, ranging from the toy poodle to the Saint Bernard, found that even those that seem to share little more than a bark and a bite can claim the wolf as a common ancestor.[1] When you watch a border collie creeping across the meadow, its chin nearly flat to the ground, you can still see that wolf. The intimidating glare the collie uses to motivate the sheep reminds observer Donald McCaig, author of four books on herding dogs, of "the wolves' tactic of selecting a victim in a herd by catching its eye and asserting dominance before the attack run." And the fleet sweep that carries the dog around the flock—the outrun—recalls the "head-'em-off-at-the-pass" maneuver characteristic of pack members "quick enough to outrun escaping prey and turn it back into the jaws of the pack."[2]

Surely herding, a behavior conceived by wolves and refined by generations of selective breeding, must be under genetic control. Why else would border collies that have never seen a sheep run circles around a group of milling children or stalk a band of sparrows? Why else can some rough ("Lassie"-type) collies and shelties, old English sheepdogs and German shepherds—breeds more often found in the show ring or a backyard than on a farm—still be seen "gathering" a ball rather than retrieving it, or struggling so obviously to reconcile heeling and herding as they run alongside a human jogging companion?

Yet something about herding transcends genetics. If border collies were built according to a standard program, fewer novice breeders would be surprised by the disappointing progeny of champions that "do not produce the way they work, because they are, as trainers say, 'man made.'"[3] Puppies with sterling pedigrees would never mature into dogs unfit for work because they are disinterested, distractible, overly aggressive, or afraid of sheep. If sheepdogs were born, not made, stock owners wouldn't need experts like Eve Marschak.

In a small training pen just off the main barn, Marschak is molding seven-month-old Turn into a sheepdog. Turn illustrates the raw native talent border collies bring to the herding task. You can't question her interest; she's so eager to have at the sheep that she must be brought into the pen on a leash. Her intelligence and tractability are already apparent. At an age when many household pets have not yet mastered "sit" or "stay," Turn, who has had no formal obedience training, drops willingly on command and never hesitates when she hears the whistle calling her into Marschak's welcoming arms. She can even show you an acceptable outrun, her childishly rangy form suggesting a shaggy greyhound as she gallops around the startled sheep.

But Turn also illustrates the limitations of heredity. Despite her eagerness and keen interest, she has a long way to go before anyone would call her a first-rate herding dog. She can collect the sheep into a loose knot, but once they start to move, her baseline skills are no longer enough to keep them from scattering or zigzagging haphazardly across the pen. And her intelligence may prove to be a liability as easily as an asset. Watching Bessie, an older dog, rough up a belligerent row of ewes corralled for routine medical treatment, Turn discovered how to manage sheep by biting them. Now she compensates for her inexperience by nipping, a major sin that will immediately disqualify her in a herding trial.

Turn was not born with balance so perfect it could draw a straight line across an October meadow; that part of herding will have to come from the outside, not the inside. Over the coming months, Marschak will do much of the actual work as she blocks, redirects, and corrects Turn's movements to shape her speed and agility into the polished technique of a full-fledged sheepdog. She'll do most of the thinking, until Turn's cleverness matures enough to keep her out of trouble.

An eye for movement and an urge to gather may be bred in the bone. But the path from puppy to herding trial winner also depends on the trainer—the most important factor in a young dog's environment—to choreograph these latent talents into the intricate dance between dog and sheep. The course of the trainer's approach to the dog depends in turn on the animal's response to her overtures; over time, she will hone the training environment to select those techniques that achieve the best results with this particular dog. As the dog matures, new skills must be learned, and new aspects of its innate nature emerge (an unexpected stubborn streak, a strong preference to work to the right side rather than the left side) to challenge the trainer, who must update her teaching technique to accommodate these new strengths and weaknesses. Dog and trainer act and react, with genes providing the tools and raw material to frame the herding behavior, and the trade-off between genes and environment, the force that determines whether potential will materialize into an effective sheepdog.

Even genetically identical individuals may turn out differently from each other if they are reared in different environments. A perennial fancier who takes three cuttings from the same prize plant, for example, may find that while the one planted in full sun blooms prolifically, a second planted against a wall with a northern exposure performs indifferently, and a third planted in the shade of a large maple never makes it through the first growing season. Geneticists can even quantify this fluctuation, in the form of a table or graph called a *norm of reaction,* that describes the change in outward appearance, behavior, or survival of genetically identical individuals as a function of systematic alterations in environmental conditions.[4] Norms of reaction offer scientists mathematical and visual confirmation of what gardeners and dog breeders know from experience: observable traits are built from the outside in as much as from the inside out.

Newspaper accounts of advances in behavioral genetics—with daily reports of genes for shyness and novelty seeking, handedness and alcoholism, sociability, sexual orientation, even aggression—often seem to imply that behavior is no more than the read-out of a genetic program. No wonder few aspects of the be-havioral biology of aggression make people more nervous than the suggestion that violent, antisocial, or criminal behavior might have roots in the genome. Yet the opposing viewpoint—the *safe* viewpoint—that behavior and heredity are independent events can't be correct either. Behavior is not grafted onto the nervous system as a cultural afterthought. It begins and ends in a brain built ac-cording to recipes filed in the genome. Without genetic guidelines, neural ar-chitecture would be chaotic, neurophysiology lawless, behavior impossible, neurochemistry haphazard.

Heredity adds its voice to the dialogue between the brain and the world, but not in the simple-minded, deterministic way that many outside the labora-tory fear. Genes do not herd behavior to a prearranged end point. Surprisingly malleable, the genome is open to interpretation throughout development, and the true impact of genetic factors on a complex behavior, whether it be sheep herding or human aggression, can be fully understood only when heredity is positioned in the context of environment, timing, and personal history.

White Lions and Violent Men

The reason people often misunderstand the role of genetics in human biology is that the basic ideas are deceptively simple. Even a visiting parent roped into Ca-reer Day (such as myself) can easily demonstrate to second graders the orderly patterns of inheritance first described by Gregor Mendel. She just needs to marry the science to one of the Philadelphia Zoo's favorite attractions: their lions.

Merlin, the zoo's current Lion King, commands the usual amount of re-spect from visitors. But his female companions, Jezebel and Vinkel, are the lions everyone really comes to see. Like every other lion that zoo visitors know from books and television, Merlin has a tawny coat. Jezebel and Vinkel, how-ever, are white.[5]

Coat color is one of the features that make up a lion's *phenotype*—the indi-vidual physical and physiological characteristics that can be observed or mea-sured. By comparing the phenotypic features of parents and their progeny (choosing as his model organism the common garden pea), Mendel hoped to take the guesswork out of plant breeding.[4] Through a combination of careful ob-servation and selective breeding, he succeeded in establishing pea plant lines that "bred true" (the offspring of a cross between any two members of the line pro-duced plants identical to their parents) for seven pairs of physical characteristics: flower color (purple or white), flower position (axial or terminal), seed color (yel-low or green), seed texture (wrinkled or smooth), pod color (green or yellow), pod shape (smooth or pinched), and stem height (long or short). And he also dis-covered that in each pair, one of the two options held the upper hand. For exam-ple, a cross between a yellow-seeded plant and a green-seeded plant usually resulted in yellow-seeded progeny. However, to dominate did not mean to obliterate. When one yellow-seeded plant from a yellow-green cross was paired with another, three-quarters of their offspring, on average, produced yellow seeds like their parents, but the remainder produced green seeds. The parent plants, although yellow-seeded themselves, had retained and forwarded the directions for producing green seeds to some of their offspring. Mendel proposed that yellow seeds and green seeds represented a pair of discrete, heritable units, one of which—yellow—overshadowed its green partner. A yellow-seeded plant with one yellow parent inherited either two yellow units—one from each parent—or a dominant yellow unit from one parent and a silent green unit from the other. Green-seeded plants, on the other hand, occurred only when both parents contributed a green unit.

Tawny fur is the leonine equivalent of yellow pea seeds. White fur, like green seeds, takes a back seat to the tawny option.[6] Of course, today, instead of "heritable units" we have the discrete snippets of DNA known as genes. Every lion cub inherits two genes for coat color: one from its mother and one from its father. Tawny dominates white, so if at least one of the two coat-color genes is a tawny, it will take charge, and the cub will have a tawny coat. White, on the other hand, is a *recessive* gene; it gives in to a tawny gene every time. Only a cub that has inherited a white gene from both parents can have white fur.

Jezebel and Vinkel are white because they have two recessive white fur genes. In genetic terms, they are *homozygous*. Merlin, on the other hand, is *heterozygous;* he has tawny fur but also has a silent white gene. Homozygotes breed true, producing cubs just like themselves. Heterozygotes breed surprises. In March 1996, tawny Merlin and white Vinkel became the proud parents of triplets.[7] Tombo and Tonyi were tawny coated, like Merlin. But their sister, Nakanda, sported the same striking white fur as her mother. She had inherited two recessive white genes: one from Vinkel and the other from her heterozygous father.

On average, about half the cubs born to a white lion mated to a tawny heterozygous lion will be tawny—but about half will be white. And while most of the cubs born to a pair of heterozygotes will be tawny, about one-quarter, on average, will have white fur instead—animals bearing one recessive white gene from each parent. This simple "Mendelian" pattern of trait inheritance is illustrated in Figure 8.1.

In a matter of minutes, even eight-year-old children can use these Mendelian principles to predict coat color for generation after generation of lion cubs. They'd be accomplished geneticists—if only all traits were determined by a single pair of genes, one dominant, one recessive, and if only every gene pair stuck to its phenotypic knitting, keeping its nose out of other traits' affairs.

Mendelian genetics, it turns out, is just an introductory course. Anyone who

Figure 8.1

Mendelian inheritance in lions. (A) The result of crossing a tawny heterozygote (Ww, where W is the dominant gene) and a white homozygote (ww).

A TAWNY HETEROZYGOTE   A WHITE HOMOZYGOTE

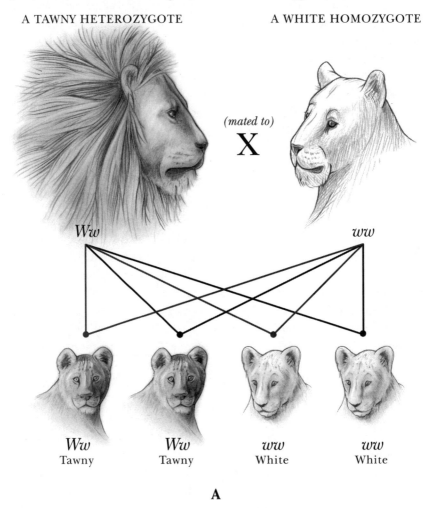

A

wants to be an expert at genetic analysis needs a higher education. To begin, many genes refuse to be limited to a single responsibility. The influence of these genes isn't confined to coat color or stem height, but ripples out to touch multiple traits, properties as complex as behavior, as fundamental as survival.

"WHAT KIND of dog *is* that?"

Nearly everyone asks that question the first time they see my collie. Children brake their bicycles and stare. Passing drivers pull over to the curb for a closer look. Even confirmed dog haters have been known to cross the street,

Figure 8.1 (continued)

(B) The result of crossing two heterozygotes.

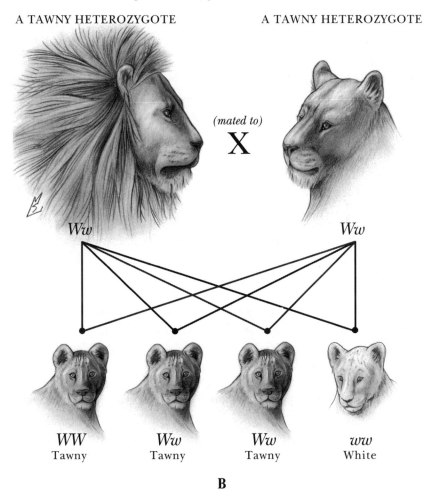

A TAWNY HETEROZYGOTE   A TAWNY HETEROZYGOTE

(mated to)

X

Ww   *Ww*

WW   *Ww*   *Ww*   *ww*
Tawny   Tawny   Tawny   White

B

their curiosity outweighing their apprehension. They're confused because they see a dog that's shaped like television's Lassie—the same bushy fur, the delicately tipped ears, the refined, pointed muzzle, the same compassionate expression—except that she's colored like a wolf. Shaula's silver coat, splashed with patches of black, hints of tan, and a swath of white ruff, is known as "blue merle," the most striking and least common of the three recognized collie color patterns.[8]

Blue merles are uncommon because they require careful attention to breeding. The merling gene is a second coat color gene that controls color intensity, diluting most of the black fur in what would otherwise be a tricolor coat to gray. Pairing a blue merle bitch (like Shaula's mother, Caralan Dream

Weaver) and a tricolor will give you a litter containing both blue merle puppies and tricolor puppies.[9] But why not mate two merles and get a whole litter of silver-streaked puppies?

Crossing a merle to a merle gives pale blue puppies known as "double dilutes." Crossed back to a tricolor, double dilutes throw litters consisting entirely of "the most spectacular blue merle puppies you can imagine," according to collie breeder Pat Buerger, who's responsible for the Caralan line—that is, if they survive puppyhood. The majority of double dilutes never get a chance to reproduce because they're born deaf, blind, or dead.

The merling gene influences several phenotypic traits, coat color, development of the visual and auditory systems, and survival. Researchers call such versatile genes *pleitropic* genes.[4,10] Pleitropy forces collie breeders to look beyond coat color, with its obvious appeal to potential buyers, and factor in the chances of selecting for a potentially lethal gene. And in human genetics, the existence of pleitropic genes warns both researchers and outside commentators to look beyond the most dramatic or most obvious trait affected by a gene, to tally all of its effects on the phenotype before jumping to conclusions about gene function.

THE TAWNY LION–white lion pattern of inheritance is a pattern critical to human health. Approximately six hundred human diseases are inherited in the same fashion as white fur and green seeds, including many of the most infamous genetic disorders: cystic fibrosis, muscular dystrophy, Tay-Sachs disease.[11] In these conditions, one gene, which comes in two forms—a "healthy," dominant version and a recessive form that's the troublemaker—decides the fate of each new member born to an affected family. The recessive variant, like white fur in lions, can travel silently from generation to generation, surfacing when two heterozygous individuals reproduce. Children who inherit two healthy genes are unaffected. Because one healthy gene can usually compensate for its nonfunctioning partner, children who, like their parents, are heterozygotes also avoid calamity (although they may pass the deadly gene on to their own children). A child unfortunate enough to inherit two recessive genes, however, has no fallback position. You can't make a functional protein without an error-free copy of its genetic recipe, and you can't make a healthy human being without a working version of all his or her component proteins.

Two faulty genes add up to one defective protein. When that protein plays a role in many physiological functions, its loss may be felt throughout the body. The resulting phenotype is a battery of signs and symptoms that reflect the gene's widespread influence. If you concentrate on one at the expense of the others, you may not only botch a critical diagnosis but also misjudge the very nature of the pleitropic gene.

Cystic fibrosis, the most common Mendelian-pattern genetic disease in Caucasians,[11] is an example of a disorder caused by a genetic defect with wide-ranging consequences. The gene itself contains the recipe for a protein known as *cystic fibrosis transmembrane conductance regulator* (CFTR), found in the se-cretory cells that line the small airways of the lungs, the interior of the stom-ach, the sweat glands, the pancreas, and the gallbladder.[12] CFTR molecules form a network of pores in the outer membrane of these cells that control the flow of chloride ions—the "chloride" half of sodium chloride, or salt. When the pump is defective, chloride cannot be piped into or out of cells. Cellular secretions become sticky or salty, setting the stage for physiological disaster.

The dysfunctional pump is the primary defect. But the clinical presenta-tion—the phenotype—reflects the wide-ranging secondary consequences of losing a protein critical to the healthy functioning of many organs.[11,12] Cystic fibrosis is famous for its assault on the respiratory system, where the secretory defect chokes bronchial passages with a thick mucus that's a fertile breeding ground for opportunistic bacteria. But the disease isn't confined to the lungs; it extends to other tissues that rely on the pump, from the mucus-secreting glands of the stomach and intestine to the enzyme-secreting cells of the pan-creas. Sticky secretions clog ducts, trapping insulin and digestive enzymes; ex-cess salt leaches out of the body whenever the victim sweats. As a result, an affected child has difficulty digesting food, tolerating heat, or maintaining fluid balance after an episode of vomiting or diarrhea. He may develop recur-rent sinusitis, nasal polyps, gallbladder disease, ulcers, intestinal obstructions, or diabetes. The tips of the fingers grow large, or "club"; the reproductive sys-tem may not grow at all.

Cystic fibrosis ruins a single gene, but that one change in the genetic makeup has a multitude of phenotypic consequences. Anyone who fixated on lung function at the expense of other symptoms could easily lose sight of the fact that the cystic fibrosis gene does not control breathing, but the movement of salt and water.

IN 1991, A DUTCH woman approached researchers in the Department of Human Genetics at Nijmegen University Hospital and asked them if they could explain what was wrong with her family.[13] In each of the last four gener-ations, a small number of men had suffered from a peculiar form of mental re-tardation that altered behavior as well as learning ability.[13–15] Unable to puzzle out an elementary school curriculum or manage a full-time job, affected men usually kept to themselves. But if they became angry or frightened, they could erupt in aggressive outbursts lasting as long as three days. During these out-bursts, they threatened, verbally abused, and occasionally even assaulted oth-ers. They set fires. They exposed themselves. One raped his sister. At least one attempted suicide. They had trouble sleeping and experienced the super-nightmares physicians call "night terrors."

The woman noted that one of her granduncles had constructed a rudimentary pedigree thirty years before, after painstakingly interviewing all family members alive at that time, and discovered a fascinating pattern. He wrote, "I concluded that mental deficiency was hereditary in this kindred, and that it was transmitted by females. However, only males were affected."[13]

The diligent uncle was a gene mapper's dream. Not only had he amassed pedigree information it might have taken researchers years to piece together, he'd also provided an important clue suggesting where to begin looking for a guilty gene. Girls inherit two X chromosomes. Boys, on the other hand, inherit one X chromosome—from their mothers—and one Y chromosome. Genes on the X chromosome do not have a counterpart on the Y chromosome, meaning that a boy who draws a disease gene on that chromosome (from an unaffected, heterozygous mother) is in the same unenviable position as a child who inherits a pair of recessive cystic fibrosis genes. If the syndrome described by the worried woman was, in fact, sex linked—as the uncle's pedigree data suggested—researchers didn't have to rummage through the entire genome but could focus on the X chromosome.

Using previously identified fragments of X-chromosome DNA as guideposts, a team headed by geneticist Han G. Brunner charted inheritance patterns in twenty-four family members (including eight affected males) and uncovered a heritable defect at the chromosome address Xp11.23-11.4, the gene for the enzyme monoamine oxidase type A (MAOA) that breaks down norepinephrine, dopamine, and serotonin at the conclusion of neurotransmission.[15] In collaboration with Xandra Breakefield at Harvard University Medical School, the group pinpointed the site of the mutation: a short run of DNA right in the middle of the protein, mistakenly changed to an "all done" signal. Born too soon, the half-finished enzyme was essentially useless.[14]

Outbursts of aggressive behavior. A clearly defined error in a gene. How easy it would be to glide to the conclusion (as some reports in the popular press did) that the MAOA gene is a "mean gene" and MAOA deficiency, "familial aggression disorder."

This is just the sort of trouble you can get into when you use entry-level skills to tackle advanced genetics. The MAOA enzyme, a protein as widely distributed as the salt pump that optimizes sweat and mucus composition, is a protein entrusted with the care of three critical neurotransmitter systems, in a brain where a single change at the synapse can transform the activity of an entire pathway. As a result, the genetic disorder described by Brunner, Breakefield, and their colleagues (now called "Brunner's syndrome")[16] is characterized not by one symptom—aggression—but by a constellation of symptoms, including, in addition to episodic aggressive outbursts, cognitive deficits, sleep problems, poor impulse control, and repetitive movements (hand wringing, plucking at clothes or skin). Aggression is only one of several traits caught in a wide-ranging net. Learning, problem solving, motor function, and even survival may also be affected.[17]

Defects in the MAOA gene don't have to lead directly to violence to provide useful information about the role of the enzyme in the generation and regulation of aggressive behavior. Brunner believes that the true significance of single-gene disorders like MAOA deficiency lies in what they can tell us about process rather than program; although "genetic studies cannot explain *why* impulsive aggression occurs, they may still help to improve our understanding of *how* aggressive behavior happens."[13]

All in the Family

The "one gene, one disease" model—medicine's spin on the principles of inheritance laid out by Mendel—reduces genetic analysis to its simplest level, a yes-or-no decision with either-or consequences. It's the clinical equivalent of white and tawny lions or green and yellow peas—and it's what "heritable" means to most people. Inherit a "good" gene from your parents, and you luck out. Draw a "bad" gene, and you're sentenced for life.

But heredity just isn't that simple. Many of the traits that interest animal breeders, clinical researchers, and behavioral biologists don't act quite the way Mendel predicted. Geneticists call these dissident traits "complex traits" because they turn what ought to be a simple relationship between a gene and an outcome into an elaborate exercise, by violating the basic assumption of Mendelian-pattern inheritance: a one-to-one correspondence between phenotype and genotype.[4,10,18] For example, dominant genes don't have to be tyrants. "Remember that most of Mendel's sweetpeas turned out pink," quips Elise Hancock, senior editor of the *Johns Hopkins Magazine*.[19] Gene expression isn't necessarily an all-or-nothing phenomenon either, an ambiguity that invites some collie breeders to gamble with a merle-to-merle cross in the hope that the resulting litter will contain more promising puppies than dead ones. A phenotype linked to a particular gene may turn up in an individual who doesn't have that gene. Two individuals with identical phenotypes may turn out to have very different genotypes. And few genes limit themselves to just two options. The gene that determines human blood group antigens comes in three varieties (A, B, and O), or *alleles;* the mouse coat-color gene A comes in at least four; the genes that determine tissue compatibility, in dozens.

Many complex traits are also quantitative traits, characteristics that can be measured along a continuum (e.g., height), rather than sorted into categories (e.g., blood group).[4,10,18] Quantitative traits further complicate genetic analysis. Traits determined by committee, they are the result of a group effort on the part of several genes. And they are also traits that exceed the genome, the physical and behavioral characteristics that most clearly illustrate the collaboration between genes and the environment.

SOME FAMILIES PASS ON red hair, a penchant for music, a distinctive facial feature. Other families seem born to sorrow and trouble, breeding generation after generation of abusive, violent, disruptive, disturbed individuals. But just because bad behavior runs in families doesn't necessarily mean it's hereditary. Troubled parents inflict troubled homes as well as their DNA on their unsuspecting children. How can an outside observer tell if the behavior of the younger generation—their behavioral phenotype—is the fault of their environment, their genes, or some combination of the two?

To compare the impact of genetic and environmental factors on a complex, quantitative trait like aggression, you have to find a way to disentangle the two, so that the contribution of each can be estimated independently. You could, for example, minimize environmental input by rearing genetically unrelated individuals under similar conditions. Or you could compare the expression of the trait in genetically similar individuals, reared in different environments, to factor out genetic influences.

Geneticists call the proportion of the overall variation in phenotype that can be assigned to genetic differences the *heritability* of the trait in question.[4,10] The closer this value is to 1, the greater is the contribution of genetic factors. In contrast, as heritability approaches 0, the voice of the genome grows weaker, and environmental factors have a greater sway in the differences between individuals.

Estimates of heritability, says Gregory Carey, a geneticist at the University of Colorado, "can verify that the genome is a logical place to look for sources of familial patterns of behavior." But they can also be misinterpreted. Heritability attempts to answer the question, "Why do individuals differ from each other?" It partitions the variability in a trait, not the trait itself. It does not divide behavior into competing states, one controlled by the outside world and the other in thrall to the genes—and it is not proof that behavior is preprogrammed into our DNA. According to one reviewer, "[Heritabilities are] like snapshots of a ballerina. They won't tell you about the ballet."[20]

MENDEL HAD HIS pea plants. Scientists who rediscovered his laws of inheritance in the early years of this century turned to the fruit fly *Drosophila melanogaster*, an animal alternative that is easy to keep, easy to manipulate, and quick to reproduce in large numbers. Modern geneticists, searching for a mammalian equivalent of the fruit fly, have settled on the mouse.

Like flies, mice are inexpensive and breed rapidly. Researchers can shape their genetic constitution almost at will, both by selective breeding and by adding, subtracting, or modifying genes. More than ten thousand available

DNA "marker" sequences provide landmarks for navigating the mouse genome.[21] Best of all, the mammalian genome has changed little on the road from mice to men; as a result, the physiology, biochemistry, and neurobiology of the mouse are no more than a heartbeat away from our own. No wonder Ken Paigen, director of the Jackson Laboratory—a world leader in the propagation of "designer-gene" mice—glorifies these little animals as "the most powerful experimental system we have for mammalian research."[21]

Mice simplify the task of differentiating genetic and environmental influences because it's so easy to control their lives. Rearing and housing conditions can be readily equalized, and if you want genetically equivalent individuals, you can breed them yourself.[22] Human beings, however, present more of a challenge. Observation and statistical analysis must take the place of experimental manipulation and pure inbred strains composed of nearly homozygous individuals.

The family itself is a good place to begin the genetic analysis of a trait that seems to run in families. Brothers and sisters have more of their genes in common than cousins. Children are more closely related to their parents than to an aunt or a great-uncle. If violent behavior had a genetic component, violent people would be expected to have a greater-than-average number of violent mothers, fathers, brothers, and sisters. However, children share more than genes with their parents and siblings. They usually live under the same roof, blessed with the same opportunities or subject to the same stresses. If the parents are poor, the children are poor. If they live in fear, so do the children. If they are lonely and isolated, their children will have few opportunities for social interaction.

Too much togetherness can make it easy to mistake environmental influences for genetic ones. As a result, you can't rely on family studies alone to define the heritability of a human trait. To isolate genes from environment more effectively, you need genetically identical individuals. You need twins.

Approximately one of every 90 pregnancies in the United States results in twins.[23] Two-thirds are dizygotic ("two-egg"), or fraternal, twins,[24] who begin as two individual eggs fertilized by two different sperm cells. Although they are born at the same time and may look similar, fraternal twins have, on average, only half their genes in common—no more than any other pair of siblings. Identical twins, on the other hand, are clones, the result of a single fertilized egg that splits into two separate embryos. Because identical twins come from the same egg (making them "monozygotic"), they not only look alike but have the same genes.

Francis Galton, Darwin's politically incorrect cousin, was the first to propose taking advantage of similarities and differences between monozygotic and dizygotic twins—an "experiment of nature"—to estimate the relative contributions of genes and environment to a given trait. Genetically equivalent, identical twins are to the human geneticist what inbred lines are to the mouse geneticist. In contrast, dizygotic twins may share a home environment, but they don't share all of their genes. By comparing *concordances*—the proportion of identical twins who both have the trait and the proportion of fraternal twins who both have the trait—researchers can estimate heritability.[25]

Studies of the heritability of human traits can also take advantage of a socially sanctioned environmental manipulation—adoption.[25] In this case, the relative contributions of genes and environment are assessed by comparing the phenotype of adopted children with the phenotypes of their biological and adoptive families. If the trait occurs with a higher frequency in biological relatives than adoptive relatives, this suggests that genes play a significant role. If it can be statistically linked to dysfunctional behavior in the adoptive family, on the other hand, the role of the environment overshadows heredity.

Data on the heritability of violent behavior are tucked away inside the results of twin and adoption studies of antisocial behavior, a category that includes adolescent misdeeds, such as truancy or running away; buying, selling, and using drugs; lying; gambling; and general irresponsibility and callousness in addition to physical violence—and studies of criminal behavior, which tally arrest records or self-reports of lawbreaking activity, both violent and nonviolent. The largest and most comprehensive of these studies have come from Scandinavia, a geographic region cherished by geneticists for the homogeneity of its peoples and the compulsive thoroughness of their record keeping. Two studies suggest a modest genetic influence on antisocial behavior. One emphasizes the importance of the environment.

In Denmark, twin and adoption data favor a role for genetic influences. A survey of the criminal records for more than 3,000 twin pairs reported a concordance rate of 0.74 for identical twins, compared to rates of 0.23 to 0.47 among fraternal twins.[26,27] Similarly, a Danish study of 1,145 adopted boys found that being fathered by a known criminal was bad for a boy's social health; nearly twice as many sons of criminal fathers had committed crimes themselves as sons of law-abiding fathers.[26,28] Less than a hundred miles of ocean away, however, Swedish adoption data found nearly identical rates of criminal offending among boys whose biological fathers had criminal records and those whose fathers did not.[26,29]

But when concordance rates for violent crimes were extracted from the Scandinavian data, none of the studies made a very convincing case for an appreciable genetic influence on violence.[26] Correlations between identical Danish twins were still higher than the correlations between fraternal twins, but the difference was no longer statistically significant. Danish adoptees with a criminal father were somewhat more likely to commit violent crimes—but this difference was not significant either. And among Swedish adoptees, fewer of those with a father who had been convicted of a violent crime had a criminal record.

"Overall, the data suggest that genetics don't play a big role when you come to actual physical aggression," concludes Carey. But the heritability estimates for antisocial behavior in general reported by twin and adoption studies caution that we ought to be careful before concluding that genes are irrelevant to aggressive behavior.

The genome may still hold answers to our questions about how aggression occurs—provided we're asking the right questions and looking in the right places. What if we move in a step, below the level of behavior to the level of more global characteristics—temperamental and personality traits, sensitivity to stress, social responsiveness—that determine an individual's overall style of interacting with the world? Do they define the role of heredity more clearly? Can we trace them to single genes or, like the complex behaviors they launch and shape, are they the product of the interaction between multiple genes and an interlocking environment?

The Genetic Tapestry

Border collies herd by intimidation. On the other hand, some bearded collies, as well as the New Zealand herding dog known as the "huntaway," herd by shouting; they bark at the sheep to get them to move.[3] Australian shepherd dogs drive as well as gather.[3] Welsh corgis and Australian kelpies move herd animals by nipping at their heels.[3,30] Rough collies are big enough to push smaller animals around. In other words, no two herding breeds work in exactly the same way. And no two individuals from the same breed have identical herding styles. If you watch Marschak's dogs carefully, you can see three distinct ways a supposedly predetermined trait is actually custom designed.

Meg, a four-year-old with a salt-and-pepper face and a thick, curly coat not unlike a sheep's dense wool, reacts so acutely to the flock and the moment-by-moment fluctuations in her trainer's posture that she can have trouble settling enough to work effectively. Like a young gymnast powering through a tumbling run, her intensity, drive, and eagerness sometimes get in the way of getting the job done. She needs regular training to keep her focused and at her best.

At seven, Spin is a champion in her prime, a fluid, experienced athlete who's always "in the zone." Spin thinks like a sheep. She can anticipate their movements so accurately that she needs only the slightest of corrections to keep her drive on balance.

Tea is a serious young bitch splashed with tan, leggy and bounding with promise. When she crouches, her back humps into an S-curve, her neck extends and drops, and you can almost hear a fierce snarl escape from her panting mouth. Like Spin, Tea is clean and even, not harsh, in her movements. She controls her drive well enough to avoid stampeding the sheep past the small holding pen at one end of the field. A final dash, and she deftly corrals the whole flock—the first of the three to finish the job this morning—then drops expectantly at Marschak's feet, fishing for praise. Within a few months, Tea will be this season's new star on the local herding trial circuit.

"Think of a whole dog as being a circle, and then picture that circle divided into equal parts, each contributing to the whole," advises one professional stockdog trainer.[3] "That is what herding qualities are like—all must be present for a dog to be 'whole,' or really first-rate."

"Eye"—one of the signature characteristics of the border collie—is perhaps the most celebrated of these herding traits.[3] When Meg, Spin, and Tea drop their heads and lock eyes with the sheep, they're showing a lot of "eye." So is the sheltie that never takes her eyes off her handler in the obedience ring, or the rough collie that peers intently at her master's face when he talks. Eye is the source of a herding dog's ability to concentrate intensely on the job at hand, but it's not the only factor that goes into making a solid worker. Assertiveness, or power, is a second. A timid dog that lacks power may not be able to translate eye contact into physical action. In contrast, what a confident, powerful dog lacks in concentration, he may be able to make up in attitude. The quick reaction time that underlies balance is another essential trait, as is the self-control that keeps the dog from resorting to overt aggression. Intelligence, temperament—the dog's overall way of relating to the world—and diligence must be factored in as well, for even a talented dog is not good for much of anything unless it is willing and able to respond to the trainer.

Breeders in search of a champion herding dog strive to draw a perfect circle—breeding for an optimal combination of these component traits, adding a little more eye here, outcrossing to large, strong-willed dog to improve power there. Researchers studying canine behavioral genetics would agree that this strategy makes good biological sense, for they believe that basic traits like eye and temperament have a measurable genetic component. At least one, eye, even appears to follow a classic Mendelian pattern consistent with a single dominant gene.[31]

"Behavior isn't inherited like blood type," stresses Gregory Carey. "There's no gene for joy riding. It's at the level of attitude, temperament, and personality—that's where genes might really be working."

"WHY DO SOME BAD PARENTS have good children, while some good parents have bad children?" It was this question that led child development expert Stella Chess and colleague Alexander Thomas to initiate the first in-depth studies of human temperamental traits and their impact on the relationship between parent and child and between the child and the world at large. Chess and Thomas identified nine core elements that fashioned a child's temperament—his or her typical "style of behavior" (Table 8.1).[32] Temperament, as defined by these features, is a circle of qualities related to responsiveness, adaptability, activity, and emotional intensity, qualities with a strong physio-

Table 8.1. Temperamental Categories of Chess and Thomas

Activity level	Motor behavior, expressed as the relative proportion of active and inactive periods
Rhythmicity/regularity	The predictability or unpredictability of feeding, sleeping, or other behaviors
Approach/withdrawal	The initial response to a novel stimulus
Adaptability	Responses to new or altered situations
Threshold of responsiveness	The intensity level of stimulation needed to provoke a response
Intensity of reaction	The "energy level" of responding
Quality of mood	The amount of pleasant behavior versus the amount of unpleasant or unfriendly behavior
Distractability	The effectiveness of extraneous stimulation at interrupting ongoing behavior
Attention span/persistence	The length of time an activity is pursued and the degree to which it is maintained in the face of interference

Adapted from S. Chess, Temperament theory and clinical practice: The functional significance for psychiatrists and allied professions, paper presented at the American Psychiatric Association annual meeting, May 4–9, 1996.

logical component that suggest close ties to neural pathways mediating safety, reward, and mood.

The core qualities could be grouped into three constellations that described three very different types of individuals. Children with an "easy temperament" pattern were flexible and predictable, easy-going and gregarious. Those with a "slow-to-warm-up" temperament were the children often labeled as "shy"—quiet children who made friends slowly and preferred staying at home to new adventures. And children with a "difficult temperament" were the challenging ones: they ate and slept irregularly, cried loudly and often, threw tantrums, resisted direction.

Chess and others proposed that temperament in infancy and early childhood is the foundation of the more complex interaction style known as personality, as the core traits are refined and expanded to yield an intricate mixture of sensitivity and ability, talent, preferences, and motivation. Examined over time, temperament and personality traits were consistent as well as persistent, leading, in some cases, to enduring patterns of behavior, such as those described by Harvard psychologist Jerome Kagan. Kagan observed two complementary temperamental profiles that could be clearly differentiated in infants as young as four months of age and that foreshadowed behavior five years later.[33,34] Slightly more than a third of the infants Kagan tested had a "sanguine" temperament, characterized by low reactivity to novel stimuli, fearlessness, sociability, and extroversion, as well as a lower resting heart rate—suggestive of a sympathetic nervous system that responds in a leisurely fashion. Children who had these low-reactive temperamental features as infants tended to be bold, outgoing toddlers; at age five, 20 percent could still be described as uninhibited. In contrast, about 20 percent of Kagan's babies were easily overstimulated, cried when they encountered an unfamiliar object or heard a strange voice, and had high resting heart rates. Two-thirds of these high-reactive infants were shy and reserved at age two, and about 20 percent of this group remained fearful, inhibited children when they entered kindergarten.

Are uninhibited five-year-olds starting material for antisocial teenagers? Psychologist Terri Moffitt, of the University of Wisconsin, has followed the relationship between behavior and temperamental differences beyond childhood into adolescence and early adulthood. Moffitt and her colleagues found that children characterized by a lack of control at age three had a higher incidence of behavior problems, including aggression, at age fifteen than three-year-olds with a more inhibited or sluggish temperament.[35] "Undercontrolled" children were also more likely to remain belligerent as adults; at age twenty-one, they reported higher levels of conflict and were less well adjusted than their peers.[36]

Inhibited children are more likely than low-reactive, uninhibited children to have inhibited parents, and the reactivity of identical twins is more similar than that of fraternal twins, suggesting that temperamental and personality traits like inhibition may have a genetic component.[34] And in fact, over the past two decades, large twin studies *have* found that identical twins—whether they're reared together or separated and reared by different families—are more similar in temperament and personality than fraternal twins. According to these studies, 40 to 60 percent of the variation in temperamental and personality traits can be attributed to genetic factors.[37,38] And studies of specific traits that could contribute to the development of violent behavior, such as reactivity, impulsiveness, or "novelty seeking"—the appetite for new and exciting experiences—also find evidence of a genetic influence.[34,38–40]

Twin and adoption studies don't provide convincing evidence that violent behavior is a heritable trait. But when checklists or pencil-and-paper measures of aggressive personality traits are substituted for arrest rates or criminal convictions, genetic influences on aggression begin to emerge more clearly. Identical twins as young as three years of age were rated as more similar than fraternal twins by parents answering questionnaires about their children's impulsivity, reactivity, responses to novel or highly pleasurable experiences, and tendency to become angry.[41] In studies of adolescents or adults, personality features related to aggression, measured by standardized psychological tests, were more highly correlated in identical twins than in fraternal twins.[25,26] Adoption data also document a small but positive correlation between the responses of adopted children on such tests and the scores of their biological, but not adoptive, relatives.[26]

Knowing that the traits that make up complex behaviors like herding and aggression respond to genetic influences is only a first step, however. Temperamental features like novelty seeking and adaptability are still complex, quantitative traits able to assume a wide range of values. "They're like height," says Carey. "You can't separate them into just two or three categories or say that certain people have it and others don't—everyone has some." And that means that personality traits are likely to be the result of a collaborative effort among multiple genes. Each contributes a small piece to the behavioral whole—some directly, others by regulating the expression of their partner genes.

George Uhl, chief of the Molecular Neurobiology Branch at the National Institute on Drug Abuse's Addiction Research Center, says, "If anyone's still thinking a behavior can be linked to one gene, well, sorry, guys, it's more complicated." Finding one gene—especially when you're not quite sure what you're looking for—is difficult enough. Finding several, and determining how they interact, represents a real challenge. When molecular biology was in its infancy, before the mid-1970s, the genetic dissection of complex traits was more dream than reality. Then the discoveries that led to gene cloning and new ways of mapping gene location revolutionized genetic analysis. Today, researchers using sophisticated molecular genetic techniques have the tools to begin unraveling the tangle of interrelated genes that collaborate in the making of the complex traits most relevant to social behavior.

The time-honored way to locate single genes is the matching exercise known as linkage analysis,[18] a high-tech version of the way many of us locate those household objects that have a tendency to "go walking." For example, if I want to find my scissors, I know better than to look in my desk drawer. Instead, I look for my oldest daughter, who has inevitably borrowed them for one of her endless little projects. Find her, and the scissors are sure to be close by. To find a tiny needle of a gene buried in the haystack of human DNA, scientists use the same principle of guilt by association. By comparing the presence of a known gene or a scrap of DNA known as a "restriction fragment length polymorphism" (RFLP; insiders call it a "rifflip") and the occurrence of the trait in the pedigree of affected families, the molecular geneticist can calculate a "lod score," an estimate of the likelihood that the marker sequence contains or holds hands with the sought-for gene. This was the procedure Han Brunner and his colleagues used to link their behavioral syndrome and the MAOA gene on the X chromosome.

But when multiple genes contribute to the trait of interest, classic linkage analysis is no longer powerful enough to locate the individual genes, particularly those that make only small contributions.[18,42] Hunting down the many genes that influence a quantitative trait—the *quantitative trait loci*—demands a new strategy that's up to the challenge. If you're a good statistician and have access to a large number of subjects and plenty of DNA markers, you can go on a genome-wide search for links between genes and traits known as QTL analysis.[18,22,42,43] A mouse geneticist, for example, starts with two inbred strains of mice whose phenotypes lie at opposite ends of the spectrum of possible values for the trait. A series of clever matings produces heterozygotes with a range of phenotypes corresponding to particular combinations of genes from each of the two strains. Next, the researcher scans the genome for marker sequences known to differ between the original inbred lines. Mice with the same version of each marker are grouped together; if the value of the trait differs significantly between groups, that marker defines the approximate location of a QTL. Additional rounds of selective breeding and mapping help to pinpoint the relevant gene contained in each identified marker sequence. In the case of aggressive behavior, this combination of extended linkage analysis and fancy inbreeding has identified two sites on the Y chromosome (and suggested the presence of modifier genes on others) associated with offensive aggression in male mice.[44–46]

QTL analysis is encyclopedic, but it can pick up a lot of chaff along with the wheat. Multiple comparisons and too much number crunching increase the chance of making a spurious connection—unless the researcher pumps up the number of subjects to compensate statistically for the risk. With mice, that's no problem. With humans, who have smaller families at a slower pace, it means that QTL analysis is an elegant idea that just isn't practical.[43,47]

Fortunately there's an alternative. If you know something about the neurochemistry or neurophysiology of a trait but don't have hundreds of subjects or detailed pedigrees, you can begin your search for relevant genes with an educated guess. After choosing a plausible candidate gene, you can compare the frequency with which various alleles for that gene appear in individuals with or without the trait (or those with high levels and those with low levels).[18,42,43] When there is a statistically significant correlation between one allele and the trait in question, you've found yourself a QTL.

If you're interested in a trait associated with aggressive behavior—novelty seeking—constituents of the dopamine system might be a logical place to start. Cocaine and amphetamine, drugs that target dopamine pathways, increase risk taking and drive users to seek out their thrilling and highly pleasurable effects.

On the other hand, people with Parkinson's disease, who have lost a significant proportion of their dopamine-containing neurons, also lose their taste for new experiences; their scores on questionnaire assessments of novelty seeking (but not other personality traits) are lower than those of healthy people.[48]

A receptor might be a good choice for a candidate gene. But which of the five known dopamine receptors represents the best choice? Two research teams, one located in Israel and the other affiliated with the National Institutes of Health in the United States, settled on the D4 dopamine receptor, which just happens to have a convenient distinguishing feature: a loop of DNA containing a short sequence that repeats itself several times—two in some individuals, four in most, and seven in still others. Both groups reported in 1996 that individuals with the seven-repeat allele had higher novelty-seeking scores than those with shorter forms of the variable region.[39,40]

Herding is more than eye, however, and aggression is more than novelty seeking. The NIH group estimated that variations in the dopamine D4 receptor could account for at best 10 percent of the variation in novelty-seeking behavior.[40] To account for all of the variation explained by genetics, they estimate, might require as many as ten genes. Which are the most important? Which play lesser roles, or fine-tune the others?

And complex traits are, by definition, more than genetics. Temperament also listens to the environment, and the mature personality, like social behavior, is ultimately the result of a developmental process. Stella Chess notes that temperamental features of the child define the first intersection between the newborn nervous system and the parents who are the infant's environment. Their responses (shaped by *their* temperaments and personalities) interact in turn with the child to refine his or her view of the world, to update the chemistry that will underlie future responses. These interactions can set in motion an upward spiral—if parent and child can find a way to reconcile their individual styles—or initiate a vicious circle that culminates in destructive behavior.

Chess emphasizes that the important variable in this process is not the degree of similarity between parent and child, but the goodness of fit between the two. Of course, it's easy to see how a child with a difficult temperament and a parent who's a slow-to-warm-up pushover or an undercontrolled sociopath are a poor match, how the worst in the child brings out the worst in the parent and vice versa. But interactions between temperament and environment can also take an unexpected turn. In the wrong circumstances, even a desirable trait can turn out to be a problem.

NOT ALL BORDER collies are as lucky as Turn. Some must adapt to an environment that turns all of their farm virtues into vices: the family home. "Border collies are smart and beautiful," says Eve Marschak, "but they don't make

great pets." Like many other border collie breeders, she's reluctant to sell puppies to people who have no constructive work for them to do, explaining that
the average suburban family with two or three children, a small backyard, and
little free time is as poorly suited to raising a spirited dog as a chaotic family on
the edge of disintegration is to raising a spirited child.

A noisy toddler and an energetic six-year-old are painfully overstimulating
to an animal with supernatural eyesight and hair-trigger reflexes. Joggers, cars,
or the neighborhood play group have to do when sheep aren't available. A dog
that can cover a hundred miles in a day's work isn't likely to find a ten-minute
game of fetch much of a challenge; according to the Border Collie Association
of America even "a one- or two-mile run is barely a warm-up for these dogs."
And while owners may think they want a smart dog, they may be less than
thrilled when that dog decides to match wits with them in puppy kindergarten. Bored and irritated, the border collie dissatisfied with its role as a family pet may snap, harass passers-by, chase birds, chew through doors, smash
plate glass windows to get at shadows or dripping water, plow up the garden, or
shred the curtains. Is it any wonder so many make their final home in an animal shelter?

Traits like novelty seeking can also be an asset or a liability, depending on
the circumstances. But an overemphasis on their association with aggression
stresses vice at the expense of virtue. More than merely challenging, they may
be dangerous, hovering at the surface of the personality where they are easily
hooked by outside influences fishing for trouble. Strong rules and a short leash
offer the only hope of keeping these suspicious traits separated from provocative events.

But phenotypes have no intrinsic moral value—and making the most of
volatile traits is more complicated than enforcing rules. The ancestral border
collie's perceptual acuity, intelligence, and hunting skills were effective or ineffective, rather than right or wrong. On a sheep farm, food is plentiful, but intelligence and quick reflexes still make for a good fit between a dog with an
energetic personality and its environment. But in the suburbs, a box of squeaky
toys and a few obedience lessons are no substitute for stimulation. Out of step
with its surroundings, the herding phenotype can turn aggressive. It's not the
traits that are bad or the environment that's antisocial; it's the way they misconnect, ensuring that behavior will be inappropriate.

What's important is the fit between temperament and environment. As
Chess notes, "There are risk factors for all temperamental qualities. Even an
'easy' child can get lost if they're overlooked." A "prosocial" environment is not
one that tries to contain temperament but one that makes the most of it.

Some environments are not fit for anyone. Some temperaments present
special challenges. But even the best environments can be suboptimal, and any

phenotype can be ruined when innate features and the environment operate at cross purposes.

On the Front Lines

Molecular biology can fish out interesting genes, but immersion in the outside world is what makes them come alive. By themselves, genes are potential. Animated by the environment, they help to form individuals. As Gregory Carey points out, "Even identical twins are never identical in their behavior—only similar."

Fatalism is not a feature of modern genetics. Even "evolutionary psychologists"—second-generation sociobiologists who claim that behaviors as diverse as language and infanticide are the genetic legacy of encounters between our Stone Age ancestors and the forces of natural selection—grudgingly acknowledge that the brain they believe to be largely hard-wired must still pay some attention to environmental influences.

Context and experience have many supporters. Unfortunately, when some advocates try to explain how genes and environment might fit into the same nervous system, their thinking can grow a little fuzzy. They either reduce the interaction to a magic formula—40 percent genetics plus 60 percent experience—or inflate it into a spacey holism, the kind of "everything interacts with everything" assertions that lead frustrated representatives of the evolutionary psychology camp, like cognitive psychologist and author Steven Pinker, to write off most "interactionist" explanations as "ideas that are true but useless."[49]

University of North Carolina psychobiologist Robert Cairns maintains that it's possible to discuss behavior, genes, and environment in the same sentence without resorting to magical thinking. Cairns would agree that merely stirring genetic and environmental influences together in the same explanatory pot encourages the undiscriminating to "conjure up all kinds of mysticisms and make them into whatever happy myth they already believe anyway." And he's equally critical of strategies that partition behavior into a genetic sector and an environmental sector, arguing that genes and environment "do not affect behavior in an additive manner, but are intimately fused within the developing organism."[50] But unlike less rigorous advocates, Cairns doesn't believe that behavior is an interactive process just because he finds this idea philosophically appealing; he has experimental evidence to back up his claim. By bringing behavior in from the field and scrutinizing it in the bright light of the laboratory, he has been able to extract living examples of gene-environment interactions from the intellectual silt and forge them into a hard science.

Cairns started studying mice because he wanted to help kids. "I was drawn to psychology because it seemed to offer a balance between the real world and the scientific," he explains. "That interest led me to the study of delinquency, with the idea of identifying kids who could potentially be helped."

Cairns thought he'd found the perfect way to reconcile a fascination with science and a longing for relevance. But his research quickly hit a theoretical roadblock when he discovered that his data didn't fit the accepted notion that aggression is a learned behavior. "The research showed that the popular explanations offered by social learning theory or psychoanalysis didn't hold much water. In effect, they were post hoc rationalizations that had a lot to do with what general society already believed were the causes of aggressive behavior. So I began working with animals to see if there really was any truth to these assertions about the importance of experience. And that's when I learned there were *real* phenomena you could track in a systematic, logical way."

Switching from human subjects to an animal model meant that Cairns could track those social phenomena over the entire life span, without signing up for seven decades of follow-up. He could manipulate experience instead of recording it, selecting where his animals lived, how they spent their time, when they socialized and with whom. And through selective breeding techniques, he could define just how far heredity could push behavior.

Could aggression be bred into mice? Cairns and colleagues screened ordinary albino laboratory mice to identify the most and least aggressive members of a litter, based on their reaction to a ten-minute encounter with an unfamiliar mouse. Males who readily attacked the intruder were bred to sisters of similarly aggressive males from other litters.[51,52] Aggressive males and their female littermates were again grouped and mated, while the sissies were removed from the breeding population. The cycle was repeated, generation after generation, retaining the aggressive progeny and weeding out the animals that showed no interest in fighting.

Working in the opposite direction, the group crossed males from the founder generation that had not attacked an intruder with female littermates of other mild-mannered males and selected the least aggressive offspring for the next breeding cycle. In each succeeding generation, the bullies were removed from this breeding population, leaving behind only those mice that avoided conflict at all costs.

Breeding nasty to nasty and nice to nice separated aggressors and pacifists with astonishing rapidity.[50-53] Second-generation offspring of the initial crosses between aggressive and docile mice already differed substantially in their willingness to attack. By the fourth generation, well-defined, true-breeding lines—one highly aggressive and one docile—were firmly established (Figure 8.2).[54]

The difference between the two lines couldn't be explained by any obvious differences in length or weight. You couldn't tell them apart by appearance. Both lines had the poor eyesight typical of albinos; both seemed to have their

Figure 8.2

A high-aggressive mouse and a low-aggressive mouse encounter one another for the first time after 42 days of isolation. Following a five-minute introduction to the test cage, a barrier separating the mice is raised. The high-aggressive mouse (A, animal on the left) immediately threatens his low-aggressive partner, who crouches in defense. Undeterred, the high-aggressive animal attacks (B). The defeated low-aggressive mouse leaps into the air to escape his attacker (C).

Photographs courtesy of Dr. Robert B. Cairns, University of North Carolina.

other senses intact. Nonsocial behaviors like feeding and grooming didn't seem to differ either. The only outwardly observable feature that distinguished high-aggression and low-aggression lines was the way they responded to social contact.[50,53] Low-aggression mice froze the minute an unfamiliar opponent approached or sniffed them. In contrast, high-aggression animals refused to take anything lying down. Contact between these mice and their test partners rapidly escalated into conflict.

The ease with which behavior could be manipulated by a selective breeding program demonstrated that Cairns's reservations about social learning theories of aggression were well founded. Animals from the high-aggression line needed no instruction or role models to teach them how to attack; they did that naturally, on the first encounter with an opponent. And mice from the low-aggression line cowered and froze when they met up with a stranger for the first time, before they'd ever made contact. Contrary to what social science research believed about aggressive behavior, offensive aggression in mice appeared to be strongly influenced by heredity; in fact, the genetic component appeared to have the majority voice in the determination of the phenotype—that is, until Cairns watched behavior that should have been locked in by the genes shape shift in response to the passage of time and the changing demands of the environment.

"IT IS AN ILLUSION," Cairns writes, "to view aggressive behavior as a static phenotype."[55] He views behavior as a developmental process, and just like other developmental sequences—the migration of neurons, the formation of synapses—timing is everything in the development of behavior. Before puberty, neither low-aggressive nor high-aggressive mice attacked an opponent at all, regardless of their breeding. The spectacular line differences didn't emerge until the animals reached sexual maturity.[51] Later in adulthood, the two lines began to converge once more. Genetic influences do not simply step in and take charge at birth, but wax and wane in their power over behavior across the life span.

Data from twin and adoption studies suggest that the relative importance of genetic factors in human behavior varies over time as well. Early in life, antisocial and criminal behavior, including violent behavior, seems more responsive to environmental factors than to genes. For example, Michael Lyons of the Harvard Institute of Psychiatric Epidemiology and Genetics examined self-reported criminal behavior among 3,226 twin pairs of Vietnam veterans (55 percent of whom were identical twins) and found that the common environment—features common to both twins, such as family structure, parenting style, neighborhood composition, and educational background—was significantly more important in determining the likelihood of delinquent behavior before age fifteen.[56,57] But after age fifteen, identical twins were more likely than fraternal twin pairs to have been arrested, to have been arrested more than once, or to engage in setting fires, destroying property, prostitution, gambling, selling drugs and stolen goods, or other criminal behaviors.

As they age, people and animals accumulate experiences as well as pass developmental checkpoints. High-aggressive mice were provoked by meeting an unfamiliar mouse for the first time, and low-aggressive mice were immobilized, but repeated exposure to the stranger normalized behavior in both lines.[50] The outcome of the initial encounter with a stranger resets biochemical parameters that shape the response to a subsequent encounter; that interaction continues to push behavior toward an increasingly better fit with the social environment. Experience gradually mellowed aggressive mice and bolstered the timid; by the fourth encounter, "high-aggressive" mice and "low-aggressive" mice had become equally aggressive mice.

Even short periods of experience proved enough to override genetic background. When Cairns placed a high-aggressive male mouse, a low-aggressive male, and a female together in the same cage, the high-aggressive animal, to no one's surprise, invariably attacked first.[50] But two hours later, more than 40 percent of the low-aggressive mice had learned to fight back—and they did it effectively enough to take charge of the relationship. Their testosterone levels rose in the characteristic fashion of dominant males, their cortisol levels dipped, and their genes no longer mattered.

Colorado's Carey suggests that for human males, experience may also be the driving force behind age-related variations in the relative importance of genetic and environmental factors, an interaction that he calls the "smorgasbord effect." "When people make their first trip up to a smorgasbord, they want to sample all of the tiny wonderful things they see. When they go back for seconds, they pick just the ones they liked. Teens are the same way. During adolescence, they're surrounded by lots of choices—in friends, social activities, the neighborhood—and they explore. But when they become adults, they make choices, and they lock into the behavior that seems to suit them best."

The effect of experience wasn't limited to conflict; in fact, Cairns learned that life events could temper heredity before his mice ever met their first opponent. All he had to do to cancel the result of generations of selective breeding was to make a simple change in living arrangements. Male mice isolated from their peers at weaning and housed in individual cages lived up to their genetic reputation when tested at puberty; animals from the high-aggression line could be counted on to fight, and you could bet a week's salary that those from the low-aggression line would give up and freeze.[50] But when high-aggressive male weanlings were group housed with other males, they grew up to be no more aggressive than ordinary mice.[58] Similarly, communal living transformed cowardly, low-aggression males into confident, assertive animals that no longer shied away from the mere suggestion of social contact. The environment—not a rigid genetic program—was the critical factor that decided whether the mice lived up to their potential or gravitated back to the original behavioral baseline. What's more, subsequent changes in lifestyle could reverse the effects of isolation at weaning. Isolated mice returned to a group living situation after three weeks of solitary confinement lost their taste for conflict; paired with a stranger two weeks later, they attacked significantly less often than their counterparts who had continued to live alone.[50] The rapid shift in behavior, Cairns concluded, "constituted a striking demonstration of the plasticity of the neurobiological system . . . and confirmed the view that its functional organization is eminently open to experiential input."[50]

SOCIAL BEHAVIOR is neither static nor solitary. According to Cairns, the second great illusion about aggressive behavior is the belief that it is "the outcome of the internal processes (genetic or otherwise) of a single organism."[55] The social context in which encounters occur shapes and guides the behavior of his selectively bred mice as clearly as experience. Attacks by even the fiercest mouse are not random, but are customized to meet the specific challenges posed by the behavior of the opponent.

High-aggressive mice are socially sensitive mice. From the moment they're paired up with an unfamiliar mouse, they're on the lookout for the slightest hint of opposition. Edgy and overreactive, even a casual sniff from their partner and they're rattling their tails and raising their hackles in preparation for an all-out assault. But the opponent is not doomed; he actually controls his own destiny. Cairns notes, "What determines whether or not the high-aggressive male will go on to attack is the response of the other to his heightened reactivity. You get very different responses if the opponent freezes than if he continues to move."

The less reactive the opponent, the less likely it is that the attack will escalate. When Cairns ensured that the partner mouse was less provocative—by giving him a moderate dose of a sedating drug—the aggressor backed down.[59] The higher the dose, the more the partner "became like a stone," the less threatening he appeared, and the less often his irritable opponent attacked. Selective breeding may alter the sensitivity to threat. But the perception of threat is also a function of the behavior of the partner, a flexible reaction, not a programmed response.

Genes, like neurotransmitters, receptors, and hormones, are only part of a process that also includes context, timing, and history. "It's not a gene or a biochemical you have behaving," says Cairns. "It's an integrated organism behaving within a real world."

ROBERT CAIRNS is a different kind of evolutionary psychologist. In contrast to those who look to the distant past for evidence of the impact of natural selection on behavior, he has focused on systematic changes in behavioral responses that can occur over just a few generations, for evidence of the impact of behavior on natural selection. Such near-term adjustments in the relationship between organisms and their environment—a process that can be called *microevolution,* to distinguish it from long-term evolutionary processes that span eons—"speak to the *dynamics* of evolutionary change, as opposed to its static properties."

Cairns's results have revealed that aggression is a behavioral trait open to change, not only across the life span but between generations. The rapid divergence of high- and low-aggressive lines suggests, in fact, that social behavior has a special sensitivity to selection pressure. But that does not mean that behavior is fixed by the genes. The powerful responses to housing conditions, experience, and context demonstrate that the genotype is also sensitive to the environment. Behavioral outcomes that override genetic background—such as the domestication of high-aggressive mice by a few weeks of communal living, or the increase in the confidence of timid mice as a result of experience—suggest that social responses may be "more appropriately described as *accommodations* to changes in stimulative conditions than as determined by those same conditions," notes Cairns, who continues, "changes in behavioral patterns did not occur because, but *in spite of,* the genetic, neurobiological, and environmental controls."[50]

Social behavior, Cairns argues, is neither born nor made, but developed. A dynamic reflection of the interaction between a responsive genome and an unpredictable environment, it is anything but stationary. In fact, it is among the most flexible of traits. And because it possesses that flexibility, it serves a critical evolutionary function.

In the war for survival, behavior is the cavalry. It reconnoiters "the space between the organism and the environment"[50] and moves quickly to respond to the challenges it discovers along this front. On the interior of the organism, the success or failure of behavioral accommodations in reducing threat and promoting well-being is recorded in the language of neurochemistry and hormones; these changes bias future reactions to similar environmental challenges—reactions likely to enhance the organism's survival. The first step in repositioning the organism to better fit its circumstances, behavioral accommodation is, by virtue of being "relatively fail-safe and reversible," the "leading edge of adaptation," according to Cairns.[50]

The developmental flexibility of social behavior extends the capacity to adapt beyond the genes. Genetic responses to persistent selection pressure underlie long-term evolutionary changes in brain structure and physiology. But rapid behavioral responses also allow the organism to adapt to temporary selection pressures imposed by today's environment. Unlike the adaptation of natural selection, behavioral adaptation is light on its feet, able to wheel back and regroup to a previous position or change again, in response to new challenges from the environment. If the environmental pressures that called for change vanish in the next generation, it's quick and easy to go back to the old schedule. The organism isn't locked into a now-outmoded response pattern, and the instructions for the original pattern haven't been lost.

PLASTICITY IS A FAR better weapon against an unpredictable environment than determinism. A brain tied to a program is at the mercy of the outside world; a brain responsive to the environment can work around it. Cairns stresses that "any behavior that would become stereotyped or frozen, whether by genes, neurochemistry, or early experience, would lose its most vital function of promoting flexible and reversible adaptations."[50] By realigning the nervous system and the outside world, behavior catalyzes an adaptive allostasis that allows the organism to remain in balance with the unique demands of this time, this place. Behavior remains flexible because the present is as important as the past, and today is the foundation of tomorrow.

Of Mice and Miracles

For several months during the winter of 1995–1996, Ted Dawson's desk overflowed with mail. Dawson wasn't this popular because he'd just released a hit record, directed a blockbuster movie, or written a best seller. He's just a scientist, an affable, soft-spoken fellow more comfortable in his laboratory at the Johns Hopkins University School of Medicine than in the limelight. But on November 23, 1995, Dawson and four colleagues—including University Distinguished Service Professor and opiate receptor pioneer Solomon Snyder—made headlines when they published a paper in the prestigious scientific journal *Nature* describing abnormal, lethal aggression in mice lacking the gene for an enzyme that makes the chemical messenger nitric oxide.[60] That's when the mail started pouring in.

A videotape made by another member of the research team, animal behaviorist Randy Nelson, shows four of the mice busily puffing about in the sawdust. Suddenly one mouse's tail begins to twitch. He leaps onto another mouse's back and sinks his teeth in, like a tiny, furry velociraptor. But his victim is no ordinary opponent. He doesn't surrender; he fights back, rearing and twisting around his opponent in search of an undefended spot to bite. Soon the others join in, and the conflict degenerates into a free-for-all. Wrestling, tumbling, thrashing, the combatants attack each other again and again; unless they are separated, they may fight to the death.

The last paragraph of the *Nature* paper contains a line, added in the final revision, that reads, "Nitric oxide may be a major mediator of sexual and aggressive behavior, relevant for studies of their biological determination in humans as well as mice." It was not meant to be a claim that the nitric oxide synthase gene was the "cause" of violence. In an interview for the *Johns Hopkins Magazine,* Snyder emphasized, "It's very important not to get the notion that nitric oxide is responsible for much or all of human violent behavior. There are many, many studies indicating that other factors—like poverty, social anger, the use of cocaine—have a lot to do with violence and crime."[61]

But once again, the media trumpeted the discovery of a "crime gene." Dawson and the others found themselves in the middle of the same storm that had engulfed Fred Goodwin, the Maryland crime and genetics conference, and Brunner and Breakefield's work on MAOA deficiency.

Dawson gestures at the pile of letters. They range from irate to crazy to menacing. "I have things here you wouldn't believe," he marvels. "Conspiracy theories. One guy wanted to know if it 'wasn't more than coincidence that we made these mice now, at a time of economic cutbacks for social programs.' We had the other extreme, too, like the writer who wanted to know whether the 'military was aware of this and was it technology that could be put to good use.' "

He has some advice for people like me, who won't be dissuaded from insisting in public that behavior is a biological process, a process that begins and ends in the brain. "I think you can help to educate the public better," he says, then warns, "but if you take a more biological than behavioral approach, you will definitely get a lot of letters."

The irony is that Dawson and his colleagues had no intention of creating murderous mice. They were pursuing the chemical murderer nitric oxide, wanted for conspiracy in the third most common cause of death in the United States: stroke.[62]

A colorless, water-soluble gas, nitric oxide (NO, for short) is a recent addition to the fraternity of neurotransmitters.[63] Under normal circumstances, NO earns a respectable living as a messenger, particularly in the network of regions subserving emotion.[63,64] But when local hemorrhage or occlusion of a cerebral blood vessel disrupts neurotransmission, NO floods out of the injured nerve cells. Uncontained NO is as dangerous to neurons as the glutamate liberated by excessive levels of cortisol.[65] Researchers like Dawson and Snyder believe that the swath of destruction created by the overflow of the gaseous transmitter may be responsible for the devastating deficits in speech, movement, and memory experienced by many stroke patients. Drugs that could block its lethal actions could mean the difference between life and death or disability for half a million people every year.

Trapping infinitesimal amounts of an invisible gas with a half-life of only five seconds is a bit like trying to estimate radon levels in your basement from the thimbleful of air you can capture when you clap your hands. Nitric oxide synthase, the enzyme that manufactures NO, stays put longer and is more amenable to experimental probing. In 1993, a former graduate student of Snyder put the team in touch with Mark Fishman, of the Massachusetts General

Hospital in Boston, who suggested using state-of-the-art genetic technology to selectively clip the gene for NO synthase from mouse chromosomes. The Hopkins researchers were enthusiastic. They believed that such "knockout" mice would be an ideal model to clarify further the role of NO in post-stroke damage—provided that animals lacking a crucial element in an important neurotransmitter system were still capable of surviving. "Frankly, we thought at first that it might be a lethal mutation," says Dawson. "So the most surprising thing was that the animals, at least when we first got them, were remarkably normal." The mutant mice not only lived past infancy, they could see, hear, smell, walk, and balance as well as their genetically undisturbed, or "wild-type," compatriots; in some tests, they actually outperformed normal mice.[66] They were confident in a strange environment, resistant to several well-known neurotoxins.

At puberty, male knockouts were isolated for several weeks for breeding purposes and returned to a group cage. Then, inexplicably, they began to die. "We knew they were fighting," explained Valina Dawson, Ted Dawson's wife and coworker. "Their ears were torn and their skin was broken."[61] An observant lab technician who arrived early one morning finally caught the animals in the act, proving that the mice were killing each other instead of succumbing to a mysterious cardiovascular anomaly.

Whatever was an enzyme associated with stroke doing mixed up with behavior? The Hopkins researchers hypothesize that in normal mice, NO may act as a kind of neuronal braking system. Cutting out the NO gene is tantamount to brake failure; as a result, says Nelson, the genetically engineered mice behave "as if they don't understand the rules of social conduct."[61]

Knockout mutants are the mirror image of single-gene defects. Instead of starting with a phenotype and looking for the gene, you start with the gene and look for its fingerprints in the phenotype. Because the experimenter, rather than nature, controls gene expression, "reverse genetics" favors the testing of hypotheses rather than a search for causes. It still cannot tell us why aggression occurs, but it can tell us a great deal about how it happens.

Pondering the intricacy of the prefrontal cortex, Patricia Goldman-Rakic observes, "Probably every neurotransmitter we know of is present in the prefrontal area. The job of the neuroscientist is to figure out the role of each." That is the task facing behavioral geneticists as well. We already know that many neurotransmitters and hormones, dozens of traits, and a multitude of genes contribute to aggression. The challenge is to decipher the role each plays in the etiology of aggressive behavior—in varying environments and contexts, across the life span. Knockout mutants may be one tool to help get the job done.

Karen Overall, a researcher who also directs the Behavior Clinic at the University of Pennsylvania School of Veterinary Medicine, says, "The most crucial contribution of behavioral genetics will be as a tool for dissecting the mechanisms of behavior." With the gene out of the way, environmental influences are easier to define. Eliminate it from the moment of conception, and its developmental impact stands out in relief; switch it off later in life, and its day-to-day function ought to show up against the background of a normal nervous system.

For example, Ted Dawson's NO synthase–deficient mice are remarkable not just because they implicate a new neurotransmitter system in aggression, but because they reveal the sensitivity of this system to a well-documented environmental influence: isolation. Aggressive behavior lies dormant in group-housed knockout males until isolation awakens it. The intersection between genetic deficit and environmental insult confirms that living alone isn't a healthy influence on male social behavior—and that nitric oxide has something to do with it.

Another knockout mutant mouse has started to clarify the details of serotonin involvement in aggression. Columbia University neurobiologist Rene Hen deleted the gene coding for one of the seventeen known serotonin receptors, the 5-HT_{1B} receptor, and found that as adults, *both* male and female knockouts were more aggressive.[67–70] Adult males without the 5HT_{1B} gene, Hen reported, not only attacked more fiercely and more frequently than normal males with an intact receptor gene; they attacked differently. While wild-type mice followed the rules of mouse encounters, circling and sniffing their opponents before attacking, 5-HT_{1B} knockout mutants often skipped the opening formalities and jumped right into fighting, sometimes attacking in less than ten seconds.

These knockout mice tell us something about context and timing, and the role of one component of the serotonin system in these processes. Once again, male aggression surfaced when the animals were isolated at sexual maturity. And once again, a genetic deficit needed environmental input for activation. Again, social isolation proved to be one of the most critical components of that input.

In addition, Hen suggests that serotonin-receptor-deficient mice may begin to tell us more about the relationship of serotonin, impulsivity, and impulsive aggression. Knockout mice are action oriented. Although they are not quite hyperactive, Hen reports that "they seem to be faster in almost all the behavioral tests we have performed to date: They readily acquire an addiction [to cocaine], they attack faster, they press a lever faster, and they quickly decide where to go"—all cardinal features of impulsivity.[67] By using even more advanced techniques that allow researchers to knock out genes selectively in specific brain regions or switch genes off at different times during development, Hen and his coworkers hope to delineate further the anatomical and developmental substrates of serotonin-mediated behaviors.

Ken Paigen, who's fond of a quote from Walt Whitman's *Leaves of Grass*— "A mouse is miracle enough to stagger sextillions of infidels"—explains that mice can tell us how behavior happens *and* how to intervene in this process. From the skillful application of knockout technology to the study of mice and their genes, we can learn the key points of intersection between genes and environment. If we can also learn how to use this knowledge to prevent or reverse the steady downward spiral toward violence, the mouse—and its crucial contribution to molecular biology—may be "miracle enough" to save countless numbers of lives.

GENETIC TECHNOLOGY is one of the sharpest swords ever loaned to humankind for safekeeping, and critics are right to be anxious. But their concern can now move beyond fears that genetic involvement is synonymous with genetic determinism, to the real issue of how to ensure an amicable relationship between genetic research and society.

"DNA is here to stay," trumpets the cover of a recently published children's book on the hidden wonders of the chromosomes.[71] The very real risks of misinterpreting or misusing genetic information—the loss of health insurance, job discrimination, invasion of privacy, interventions that push the ethical envelope—are not unique to behavioral genetics, and they're not located somewhere in the future. They confront us today. Shouldering our genetic responsibilities will require an educated public, courageous enough to set boundaries instead of cowering behind prohibitions. If we don't want DNA to master us, then we must master it.

Talking about "aggression" and "genetics" in the same sentence offends people who mistake genetic influences for mind control, don't realize that gene expression is an environmentally driven process, suspect that scientists who talk about gene-environment interactions are merely paying lip-service to social factors, or harbor sentimental notions that biology is inherently cruel and social science unfailingly kind. But is a "nurture-only" philosophy really as benevolent as it purports to be?

The environment cannot be disconnected from the body any more than brain function can be understood out of context. And the belief that social factors, divorced from their biological impact, are the "cause" of violence is just as misguided as the belief that violent behavior is written in the genes. Despite its good intentions, it has unwittingly hurt the people it set out to help, saddling those who already bear the brunt of economic dislocation, urban deterioration, and educational decline with the greatest responsibility for the violent behavior of an entire society. As long as violence remains a social problem rather than a human problem, the unscrupulous and the unjust won't need pedigrees or genetic screening to discriminate against groups of people on the basis of their "violence potential." All they need is an address.

Nine

PICKING UP THE PIECES

Crime is down. For six straight years, Justice Department statistics show that the rates of murder, rape, armed robbery, and assault have plummeted, to the lowest levels in over a decade. Big-city mayors, police officials, legislators, even the president have congratulated themselves on the success of their anticrime policies. Nevertheless, in 1996, there were still 19,645 murders, 95,769 rapes, over 1 million cases of aggravated assault, and 537,050 robberies amounting to a total loss of nearly $500 million in stolen property.[1]

A woman is sexually assaulted every forty-five seconds.[2] More than one-third of Americans say they have watched a man batter his wife or a girlfriend.[3] Between 1986 and 1993, the number of abused and neglected children doubled, and the number of those seriously injured increased fourfold.[4] According to FBI reports, workplace violence is now our fastest-growing form of murder; in fact, it's second only to traffic accidents as a cause of death on the job.[5] Angry drivers in the throes of road rage kill fifteen hundred people annually, are single-handedly responsible for a steady rise in traffic fatalities over the last four years, and have prompted congressional hearings and the formation of special antiaggression highway patrols.[6] In Philadelphia alone, aggressive "road warriors" injure someone every fifty-seven minutes and kill two to four times more people than drunk drivers do.[7]

We may have made headway with the problem of crime, but we still have a way to go before we've solved the problem of violence.

"Getting tough" has replaced social reform as the answer to violence. Advocates of longer prison sentences and strict discipline claim that rehabilitation has only led to "revolving-door justice": violent criminals leave prison, commit another crime, return to prison, then repeat the cycle, over and over again, at the public's expense. If we can't seem to change these individuals, they argue, we should give up. Get them off the streets, keep them off as long as possible, and make sure their prison experience is a miserable one. No television, no weightlifting, no college courses, no parole. Teach them a lesson, and make them a public example that will scare potential offenders into putting their guns down.

If the violent really feared punishment, they'd think twice—or would they? We have exiled, imprisoned, burned, maimed, shot, hanged, and electrocuted violent people for centuries, and the killers keep coming. According to Bobbie Huskey, president of the American Correctional Association—the people who run prisons—"our experience shows most offenders don't even think once. They commit crimes on impulse, when they're high or angry, or have no regard for the consequences, and they believe they won't get caught."[8] Punishment will not end violence. The problem keeps coming back; like the sorcerer's apprentice, we march people off to prison, only to find a new generation of violent offenders stepping up to take their place. We discover and prosecute a single act of violence, and a dozen others occur while our attention is diverted.

If we want to make real progress in reducing the level of violence in our society, we have to stop reacting and start thinking. America doesn't need more prisons. We don't lack for resources, programs, people, or character. What we need is a whole new perspective. In place of rhetoric, we need the courage to consider another possibility: that behavior is a dynamic process integrating physiology and experience. If any entity deserves a greater voice in the ongoing debate about violence, it is the brain.

We cannot help but be disappointed by interventions based on an understanding that is only skin-deep. A biological perspective, on the other hand, shows us the whole person, down to the neural core that is the starting point for all behavior. It redefines violence as a developmental process rather than a character defect or an economic outcome, charting the impact of the environment on the nervous system, as well as the influence of the nervous system on the individual's response to the environment. It recognizes victims and perpetrators as more than adversaries, as two inseparable elements of a complex social process.

The lessons of biology are not always easy, and the answers it offers are not always the ones we'd like to hear. But they make sense, because they respect the idiosyncrasies of our physical being. By acknowledging the body instead of pretending it doesn't exist, including rather than excluding what science has to say about violent behavior, we may actually begin to save lives instead of merely putting out fires.

Staying in Line

A world totally devoid of aggression would be biologically unreasonable. Human nature, in the words of ethologist Frans de Waal, is basically good natured; affiliation, not confrontation, guides the overwhelming majority of social interactions. But without the means to protect ourselves and our children, we would be as helpless as plants. Arguments and power plays, as well as displays of affection, are needed to define the social structure, to create the dynamic tension needed to avoid complacency. Provided quarrels do not get out of hand, the result is a more lasting peace.

De Waal emphasizes that "to present all aggression as undesirable, even evil, is like calling all wild plants weeds: it is the perspective of the gardener, not the botanist or ecologist."[9] However, if the protective and organizational dimensions of aggression are to be retained without devolving into destructive violence, it must be contained. Self-control is a more realistic goal than the peaceable kingdom.

Aggression is nature's life insurance. But like your State Farm policy, it's intended to be reserved for life-or-death crises, not cashed out every time you encounter a minor emergency. Controlling aggression requires a nervous system correctly attuned to the demands of the outside world, that recognizes and accepts unambiguous rules, correctly taught. It requires intact executive functions capable of integrating emotional and representational data. It requires a brain unencumbered by disease and uncluttered by interference from recreational drugs.

Violence is the failure to respect the boundary between acceptable and unacceptable aggression. If we want to prevent this breakdown, to have people reserve their strongest responses for true emergencies, we must protect the nervous system from injury, destabilizing levels of stress, drugs, isolation, and victimization. We must strive to create a safe environment where people are not constantly on guard, an environment flexible enough to accommodate some risk taking, structured enough to prevent confusion. And when physiology and environment have conspired to erode control, prompt, decisive action is essential to regaining self-control before it's too late.

Contrary to popular belief, a neurobiological perspective emphasizes the possibility of change. Behavior is developed, not determined. And because social behaviors like aggression lie at the cutting edge of adaptation to the environment, they are among the behavioral elements most open to change. The pathways that determine whether the world is safe and agreeable, differentiate actions that work and are rewarded from those that fail, and build representational memory and assign emotional values to people, places, and things, are built to be malleable, to mold around the unique social features of a specific place and time.

Marian Diamond, a professor at the University of California, Berkeley, knows that adult brains continue to develop. She has seen the brains of rats of all ages respond to the challenge of novelty following a move to a rich and complex environment, with a burst of luxuriant new connections.[10,11] "This is how I explain it to children," she says, holding up her hand, palm out, fingers spread wide. "Imagine my hand as a brain cell. The palm is the cell body, the arm the axon, my fingers, the dendrites." She waves each finger individually, like the tentacles of a sea anemone. "This," she emphasizes, "this part can change."

Strategies to reduce violence have this miraculous plasticity at their disposal. But even the most resilient spring can be stretched far enough to lose some of its elasticity. Despite its agility, the brain is not indestructible. An environment that is relentlessly or catastrophically hostile may overrun stress response mechanisms. Neurons that die can't be replaced. If disease and injury interrupt critical circuits, we don't yet know how to repair the damage.

Change is always possible, but it can be so difficult as to be unattainable if body systems are damaged or deregulated beyond the limits of our knowledge. Ignored, hypertension will batter the cardiovascular system into congestive heart failure. Cigarette smoke will take potshots at the DNA of lung cells until they surrender to cancer. And once the network of mechanisms designed to contain aggression begins to unravel, violent behavior can spiral into an equally intractable problem. Time is of the essence. The longer we delay intervention, the greater the risk that the aggressor will slip out of our grasp. The longer we delay, the more retaliation will seem to be the only answer.

Reconstruction, Not Rehabilitation

Magic was a puller, a horse that leaned so hard on the bit I often felt as if he might tip forward like an unbalanced canoe, sending me rolling over his shoulder. Lucy was just "strong"—a term referring to her attitude, not her weight-bearing ability. She didn't shift to a faster pace; she launched. When she jumped, she landed charging.

A horse that's ahead of itself and a novice rider are working at cross purposes. Unless the rider gains the upper hand, the horse's willfulness can lead to a confidence-shattering fall. But restoring balance requires more than mastery. Both horse and rider contribute to the problem behavior, and both must change if there is to be a lasting improvement in their relationship. The animal must learn to see the rider in a new light, as a leader rather than a learner. And the rider, by improving her equitation skills, must create a new atmosphere of confident authority.

Because the horse's understanding is informed by the physical contact between mouth and bit, switching bits can encourage a more congenial perspective. In Lucy's case, that meant switching to a stronger bit designed to slide upward when I pulled on the reins, raising her head to shift her center of balance and counter her urge to lunge. For Magic, the key was lightening up rather than bearing down. His forging ebbed with a change to a less demanding bit that he agreed to accept rather than fight.

With less to fear, hands can come to their senses. The rider's center of gravity shifts along with the horse's. Backs and legs grow useful. Secure, calmer, and more articulate, the rider can give directions instead of pleading for compliance. Confronted with authority rather than anxiety, the horse's opposition gives way to respect.

When the horse's mouth and the rider's actions are in alignment, problem behavior disappears. Transforming Magic's and Lucy's responses could be achieved only by equilibrating comprehension and experience, pairing a more suitable bit with a more agreeable riding environment, in the form of a more educated rider. Changing one without changing the other would have left half the problem unsolved. But combining physical and environmental interventions restored a sense of balance, a space quiet enough for new learning to take place.

REHABILITATION IS SUPPOSED to fix the problems that lead to violent behavior. It can't possibly work, however, if it tries to fix only half of the problem. Job skills are useless if jobs are nonexistent—or if the worker is still too hotheaded to get along with coworkers, too fascinated by alcohol to get to work on time. Counseling isn't a bad idea. But the antisocial don't need a listening ear; they need step-by-step instruction in the mechanics of acceptable social behavior and the incentive to practice these skills. College courses don't cover risk assessment, addiction, coping with mental illness.

Violent behavior is open to change, but only when *both* the external and internal conditions that have led behavior over the boundary of acceptable force have changed as well. To break the vicious circle between environmental cues, negative perception, and maladaptive behavior, the brain must develop a different attitude toward the outside world, and the world itself must be different. Changing only one side of the equation allows the other to continue to push behavior in the wrong direction. Reconstruction—changing both—establishes a new equilibrium.

Lasting changes in behavior require attention to both physical and environmental elements. Interventions that ignore the neural origins of behavior leave behind a nervous system that's still out of step with the environment, vulnerable to the stress of life or unresponsive to the push and pull of emotion.

And the good of interventions that target the brain but ignore environmental forces will soon be undone by the same insults that fueled the destructive dynamic in the first place—especially if the individual has not learned more effective and socially acceptable ways of protecting himself in potentially threatening circumstances.

The key to tempering violent behavior is adjusting the calculation of threat so that the intensity of the response matches the true demand of the situation, without overshooting or undershooting. But how do you rein in a brain that consistently misses the mark? You could do what the brain itself does: take advantage of the unique ability of the frontal cortex to reconsider emotional significance in the light of additional information, to postpone responding until reason, analysis, and insight have had an opportunity to conduct a reality check.

Work by New York University's Joseph LeDoux demonstrates how cortical input can influence risk assessment. His fear-conditioned rats freeze at the sound of a tone that is paired with an unpleasant shock. When LeDoux and graduate student Maria Morgan removed the medial sector of the prefrontal cortex, fear responses persisted even if the current to the test cage was turned off and the anticipated shock never occurred again.[12] In contrast, when rats with an intact prefrontal cortex hear the tone repeatedly without the shock, they gradually learn a new association: this noise, which warned of danger in the past, is no longer a dependable prelude to shock. An intense fear response to every tone is no longer necessary, and freezing steadily declines. The emotional memory persists, but the cortex can override it. Without the prefrontal cortex, the perceived threat lingers, despite little basis in reality; as LeDoux explains, "The association between event and outcome has become irrational." With help from the prefrontal cortex, however, the rat can trade up to more reasonable responses.

Cognitive behavioral therapy (CBT), a short-term, goal-directed method of psychotherapy, skips the insight that's the focus of many traditional talk therapies and proceeds directly to dismantling the link between environmental cues and destructive responses, recruiting the analytical talent of the frontal cortex to rebuild the brain's relationship with the world.[13,14] In CBT, the therapist does more than just listen. He or she plays an active role in the process, helping the patient identify the external events that provoke overly aggressive responses, question the assumptions that lead to those responses, formulate alternative reactions, and evaluate the results. CBT teaches patients to recognize their violent behavior for what it is—an automatic, inappropriate reaction to unsubstantiated allegations of danger—and develop a more objective approach. Through a series of work sessions with the therapist and individually designed homework assignments, the patient learns coping and problem-solving skills that enable him to resist, ignore, or avoid provocation.

CBT improves reality checking. By asking the violent person to *think*—to seek the opinion of the frontal cortex—the technique effectively walks him back down the spiral of negative interactions that established aggression to a new reality, one that's based on a realistic assessment of risk. When the estimate of risk goes down, so does the need to overreact in self-defense. When there's time to think about the consequences, immediate gratification loses some of its allure.

CBT is structured enough to stay focused on the problem behavior of aggression, flexible enough to accommodate a range of personalities, and practical enough that it can be adapted for use in a community or prison setting, not just by a handful of experts. In addition, brain imaging shows that CBT is a "talk" therapy with physiological results. PET scans of patients who responded to CBT for the treatment of obsessive-compulsive disorder demonstrate changes in brain activity in regions controlling voluntary movement—cortically driven responses—after just ten weeks of CBT.[15]

But what if violent responses surge past control mechanisms without paying attention? What if the violent individual can't hear the voice of the therapist because the alarm raised by the brain's rapid response system keeps drowning it out? What if he feels so threatened he can't think of alternate responses, if the circle from environment to aggression and back again is too deeply ingrained for learning to take place?

When destructive interactions between brain and environment have raised a formidable barrier to communication, translating the message into the chemical language of neurons is a way to talk so the brain will listen. The selective and thoughtful use of medications that normalize stress responses, delay impulsive reactions, block the craving for drugs, or suppress nonlethal paraphilias can break the link between stimulus and aggressive response as quickly and decisively as raising a horse's head can check a reckless urge to run. And when the call to action is less strident, aggressors can finally benefit from interventions like CBT, which teach the mental skills needed to effect long-lasting changes in behavior.

Psychopharmacology—whether it's for depression or aggression, psychosis or substance abuse, even terminal cancer pain—has a curious power to unnerve our caffeine-consuming, nicotine patch–sporting, Viagra-popping society. For a country with a $240 billion drug problem,[16] you'd think we'd be less emotional. After all, we have been talking to the brain in its own language for thousands of years. Egyptian medical records dating from 1500 B.C. contain hundreds of prescriptions, including preparations for treating neurological and psychiatric disorders.[17,18] Ancient Hindu texts prescribe extracts of the plant *Rauwolfia serpentina*—the source of reserpine, a compound that played a critical role in the development of modern antidepressants—for the treatment of anxiety.[17–19] Medieval physicians and folk healers expanded the use of medicinal plants for emotional complaints, and the sixteenth-century writings of Paracelsus contain specific recommendations for the pharmacological treatment of a range of psychiatric conditions.[19]

Our queasiness about behavioral drug therapy is rooted in a profound misunderstanding about what the drugs do and don't do, a suspicion that some who prescribe these drugs don't always know what they're doing, and a familiar helplessness when it comes to setting responsible guidelines for their use. Some people, for example, confuse stalling automatic responses with knocking people out. To these individuals, psychopharmacology is no more than a "chemical lobotomy." What they don't realize is that except in extreme circumstances, sedative drugs are among the worst choices for bringing aggression back in line. Some, in fact, actually make a bad problem worse.

Well-known tranquilizers like Valium and Xanax can increase or even provoke violent behavior.[20,21] One case report, for example, described a twenty-five-year-old man who was given Valium to control his angry outbursts—and became even more hostile, assaulting his wife and breaking her jaw.[22] A clinical trial of the Valium-like drug Xanax precipitated violent episodes so consistently that the trial had to be discontinued.[23]

Effective pharmacotherapy for hostile, impulsive, and drug- or sex-related aggression ought to *increase* the aggressive individual's ability to function rather than decrease it. As a result, the drugs of choice are not sedatives but newer antidepressants like Prozac; drugs that treat episodic or paroxysmal conditions, such as anticonvulsants and the antimanic drug lithium; drugs that block hormone action; and medications intended to treat addiction and block drug craving—agents that interpose a suggestion to "stop, look, and listen" between perception and response. The ensuing pause—impulsive aggression expert Emil Coccaro calls it "reflective delay"—gives slower analytical mechanisms a chance to learn new responses and a fair chance to use them.

Medication cannot "cure" violence any more than it can cure heart disease, arthritis, or migraine. But it can limit flare-ups, relieve pain, prevent relapses, and save lives. Used carefully, after proper evaluation, and in conjunction with cognitive therapy or other psychotherapeutic techniques that teach coping and social skills, it does not control people but frees them from behavior that imperils their lives, destroys their relationships with others, and compromises their ability to stay within socially sanctioned boundaries.

We can set guidelines for behavioral drug therapy, just as we can for behavioral genetics, if we are ready to become informed and involved. We can reject coercion, knowing that people will not be helped by drugs they do not want to take. We can insist that research belongs on the outside of prison walls where participants are true volunteers, with the freedom to give truly informed consent and the right to withdraw at any time—and where we have no shortage of violent subjects. But the need for caution does not mean that we cannot design correctional programs that include pharmacotherapy as sentencing options, just as we already recommend drug treatment programs or boot camp.

Thirty-two states have death penalty laws that authorize execution by lethal injection.[24] If we believe it is acceptable to kill violent individuals with drugs, can we also find it acceptable to save them?

DAMPING DOWN risk assessment without reducing threat is likely to effect a short-lived improvement in behavior. Neither therapy nor pharmacology can maintain socially acceptable behavior under the pressure of an environment that continues to call for aggression.

Dangerous, abusive environments can reinitiate the escalating interchange that led to violence in the first place. Chaotic environments perpetuate confusion. Sterile environments invite people to experiment with inappropriate diversions; fragmented, desolate environments burn away the social bonds that give rules their meaning.

The environment matters because "strong emotions make strong memories."[25] Emotional responses can be overridden, but they are not forgotten. Sunk deep in the well of memory, they can still be retrieved if the environment signals a desperate need. Conversely, the environment also matters because it can provide an external mechanism for buffering stress and correcting misconceptions.

Social interventions, however, have almost as bad a reputation in some circles as pharmacotherapy has in others. Here the issue is money. Intervention is a code word for a tax-escalating free-for-all, a conglomeration of ill-conceived, scattershot, ineffective initiatives with a tendency to multiply faster than a pair of rabbits.

But are efforts to improve the social environment really more expensive than prisons? Longer sentences and mandatory life imprisonment ensure that we'll be feeding and clothing the multitudes flowing into the prison system into their dotage. Repairing communities, ensuring the welfare and safety of children, sheltering battered women, supervising recovering substance abusers, and actually caring for the mentally ill does cost money. On the other hand, the alternative—building, staffing, maintaining, and populating more prisons—is going to cost a fortune.

The good news is that we don't need to solve all of society's ills to make a meaningful impact on violence. Violent behavior responds best to programs that do three things: preserve safety, add structure, and promote attachment to others. Interventions that work toward these goals will reduce violence and pay for themselves in the long run.

Biological intervention means environmental intervention. Well-designed, carefully targeted intervention programs don't have to take over government budgets. They don't have to cost billions of dollars, make social work a growth industry, give people a free ride—or a free lunch. They do require giving up on the idea of vengeance. It's up to us. We can change the environment, or we can allow it to bankrupt us.

Why Prisons Fail

Prison wardens know that harsher punishments for violent offenders don't work. Asked their opinions of longer prison terms, "three-strikes" laws, and proposals to strip prisons of such "frills" as television and coffee makers, prison officials attending a 1996 convention agreed that overcrowded, get-tough prisons "have less of everything except violence."[8] A survey of 641 wardens, carried out at about the same time by researchers at Sam Houston University in Huntsville, Texas, concurred; according to the majority of those surveyed, get-tough policies are counterproductive.[26]

If those who work in no-frills prisons have their doubts, it's safe to predict that outsiders championing this approach as the answer to violence are headed for a disappointment. But even prisons built with the best of intentions have not quite worked out as planned. Without an appreciation of the neurobiological foundations of violence, it's easy to enact policies that do just the opposite of what was intended.

THE IMPOSING MEDIEVAL facade of Philadelphia's Eastern State Penitentiary, opened in 1829, was meant to terrify would-be wrongdoers.[27,28] The prisoners themselves, however, never saw this side of the building; they were brought to the prison with their faces covered, and they entered through a rear door.

The entire sentence was served in total isolation. Prisoners were permitted to talk to no one except their warden and the minister who walked the corridors every Sunday morning. A black cloth was hung down the center of each cell block to prevent inmates from catching a glimpse of those housed on the opposite side as they listened to his sermon.

For as long as fourteen years, inmates endured their silent punishment in rooms smaller than the average suburban bathroom (Figure 9.1). A narrow skylight was the only escape from darkness. The hot air blown into the cell in the winter to keep it bearably warm could lead to death from carbon monoxide poisoning. The cell had no shower. Toilets were flushed every two weeks.

Dozens—perhaps hundreds—who entered this "monastery for criminals" left insane.

Eastern State, conceived and implemented by some of nineteenth-century Philadelphia's greatest and kindest intellects, was actually an advance over the squalid Walnut Street Jail it replaced. The reformers called themselves the Philadelphia Society for the Alleviation of the Miseries of Public Prisons, and they had a better idea than penning dozens of men, women, and children together in a single filthy cell. Their new prison would house offenders in solitary confinement, an environment that would force prisoners to be penitent—hence the name *penitentiary.* At a cost of nearly $700,000, Eastern State was the most expensive building that had ever been constructed in the United States. It was one of the first to have central heating (the carbon monoxide problem was corrected with the installation of steam heat in 1889), running water, and electricity. The prison had indoor plumbing (even if it wasn't efficient) at a time when residents of the White House were still using outhouses. More than three hundred prisons worldwide were modeled after Eastern State and the "Pennsylvania system" for reforming behavior.

Eastern State was the most humane and intelligently designed prison the world had ever known, but that did not keep it from being a colossal failure. The principles of silence, humility, and simplicity that worked so well in the Friends meetinghouse did not translate well when forced on individuals who had no calling to a monastic life. Prisoners did not become humble and remorseful; instead, they ended their sentences in a worse state than when they started.

The Pennsylvania system was an improvement in many ways. Procedures such as separating male and female inmates; encouraging daily exercise; providing heat, running water, clothing, and sufficient food; teaching job skills— radical ideas in their day—corrected many of the most egregious of human rights violations rampant in nineteenth-century prisons. The problem was that society members were moral philosophers, not biologists. The correctional program at Eastern State failed because it ignored a fundamental principle of behavioral biology: social animals don't thrive in isolation.

Familiarity may breed contempt, but isolation breeds violence. The most reliable way to ratchet up aggression in animals, especially male animals, is to isolate them. Mice, for example, grow steadily more aggressive the longer they've been isolated.[29] Luigi Valzelli, who described this socially destructive effect of solitude, observed that "the best age for obtaining aggressiveness by isolation is . . . mature youth"—the prime age for incarceration. And the worst fighting, Valzelli noted, occurred when three or four previously isolated young adults were grouped together in a small space.

As Robert Cairns has shown, isolation is more powerful than genes. It dictates the expression of aggressive behavior even in animals selectively bred for their social reactivity. Mice from high-aggression lines must be isolated at weaning for their breeding to show; group housed, they're no fiercer than animals from low-aggression lines. Aggression emerges in knockout mice missing the nitric oxide synthase gene or serotonin receptor genes when they're isolated for breeding.

Figure 9.1

A monastery for criminals: Eastern State Penitentiary, Philadelphia. Guard's-eye view of Cell Block 10.

Photograph by David Deutsch.

Figure 9.1 (continued)

Inside a typical cell. The door at the rear leads to a walled exercise area. The prisoner lived and worked in this space in silence for as long as fourteen years.

Photograph by David Deutsch.

Living alone is stressful, but it may not be the only reason that isolation leads to belligerence. Cairns believes that isolation also encourages delusions of grandeur. "Without cagemates, an isolated mouse has no reason *not* to believe he's king of his domain," says Cairns. He also points out that isolation increases reactivity to the behavior of others. Isolated animals are more readily startled, more easily provoked by movement on the part of the partner. Living in a group, however, reduces the novelty of "the other." And group-housed animals learn quickly that excessively aggressive moves on their part aren't tolerated by their cagemates. Over time, the group social structure sets boundaries that keep aggression in check.

Isolation, an environmental maneuver, has demonstrable physical consequences, upsetting the balance of neurochemical pathways crucial to the control of emotional and stressful responses. Valzelli and others demonstrated deficits in serotonin function during periods of prolonged isolation that paralleled increases in isolation-induced aggression.[30-32] In contrast, strains of mice that were indifferent to the aggression-promoting effect of isolation also failed to show an association between housing conditions and serotonin activity.[30]

Despite the failure of isolation as a correctional technique at Eastern State and copycat prisons modeled on the Pennsylvania system, as well as neurobiological evidence of isolation's powerful aggression-enhancing effects, we remain stubbornly dedicated to the idea that solitary confinement will convince violent individuals to behave. Ironically, today's high-security prisons—known as "control units" or "supermax" prisons—isolate the worst of the worst, inmates who have proven too violent to be held at other prisons.[28,33] Here, they are confined to their cells for as long as twenty-three hours a day. They eat and exercise in isolation. They may be handcuffed and shackled when they are out of their cells to provide an extra level of control. The guards carry serious weapons. Unlike their Eastern State counterparts, expected to spend their solitude performing constructive work, supermax inmates have little to do except seethe.

"If you ain't wrapped too tight, 23-hour lockdown can be enough to make you explode," says a priest assigned to a Colorado control unit prison. Most individuals who spend time isolated in supermax facilities ultimately return to a general prison population—or the community. When we consider the biological implications of their time in solitary confinement, it should be no mystery why they often explode once they're out.

Isolation is only one example of how prison organization works against human nature. Primates also react aggressively to the stress of disruptions in the social structure—and prisons operate in a continual state of social upheaval. New prisoners arrive. Contentious prisoners are removed to solitary confinement, then returned. Other inmates are transferred; shifted; and regrouped to control them, assign them to new programs, or simply make room for the flood of newcomers. Cells intended to hold ten prisoners now hold twenty. This crowding and shuffling of violent individuals who don't know each other and have no escape blatantly ignores the biological need for social continuity. As a result, it invites aggressive attempts to define and redefine social relationships and status.

Our cherished idea that pain deters violence flies in the face of what behavioral biologists know about the relationship between painful experiences and future aggression. In the case of violent behavior, this reasoning is counterproductive. "Punishment is a less desirable method of control, since pain itself is a stimulus to fight," writes pioneer ethologist J. P. Scott.[34] Shock-induced aggression, for all its inadequacies as a model of real-world aggressive behavior, demonstrates conclusively that confinement and pain are highly effective at provoking abnormal aggression. Punitive sentences and strict prison environments may satisfy our desire to "do something," but in the long run, they are unlikely to reduce—in fact, they almost certainly magnify—the risk of violence.

Prisons fail because they rely on procedures that make aggression worse. If we are to continue to depend on incarceration as our first-line defense against violent crime, we must stop using the time spent in prison to perpetuate violence. As Dorothy Otnow Lewis, a pioneer in the study of the relationship between biology and human violence, summed up the problem in an on-line interview: "Our correctional system reproduces all of the ingredients known to promote violence: isolation, discomfort, pain, exposure to other violent individuals, and general insecurity. In our prisons we have created a laboratory that predictably reproduces and reinforces aggression. Perhaps with a bit of ingenuity we could do the opposite."[35]

Prisons also fail because the people who run them don't understand that *aggression* is a plural noun. As a result, inmates are separated on the basis of age, sex, or estimates of how likely they are to misbehave in prison, not according to the nature of their aggressive behavior. The hostile and the antisocial may be cellmates. Inmates addicted to crack, methamphetamine, heroin—individuals with a nervous system primed to overreact by drugs—are beaten, threatened, sodomized, terrorized, a perfect formula for eliciting an angry retaliation. Predators have an endless source of victims; the chronically mentally ill face an endless series of abusers.

The answer is not more prisons but more detail. Correctional policies can no longer afford to ignore the biological complexity of aggressive behavior. Human violence comes in a range of shapes and sizes. To be effective, we need programs that acknowledge and address these differences.

But the most important reason prisons fail is that they take over far too late in the game. By the time violent individuals score ten to fifteen years, they've experienced decades of destructive interchanges between the brain and the environment. If we were to do only one thing differently, it ought to be interrupting the cycle of violence as early as possible. When more people are pulled out of the system before they are ruined beyond repair, prisons will become what they ought to be: the option of last resort.

An Ounce of Prevention

The evolution of violence is a developmental process. But not all components of the many intersecting systems that contribute to aggression—neural pathways and connections, transmitters and receptors, hormones, stress responses—develop at the same rate. At each stage in development, certain elements mature and lose a measure of flexibility while others remain more plastic—as well as more vulnerable. These systems still demand protection from toxic influences but also offer unique opportunities to influence behavior.

The first two years of life represent the most critical period in human neural development. These earliest years are crucial because they are the genesis of the nervous system, the road-building time when neurons are born and axons navigate the chartless expanse of the embryonic brain to found synapses. They are a time of introductions, when the brain first meets the surrounding environment. And they are a time of learning—learning to speak and recognize familiar faces, learning whether the world is safe and dependable, learning how to use behavior to interact with others.

The newborn immune system uses early experiences to elaborate a repertoire of antibodies that will determine future vulnerability to infectious disease. Similarly, brain and body rely on the experiences of infancy and early childhood to "educate" a naive stress response system. After the earth-shattering experience of birth, stress responses are shut down, awaiting further instruction from the environment.[36–39] Basal cortisol levels drop, and less cortisol is released in response to stress. Cortisol secretion will remain submaximal over the next decade, as stress response mechanisms gradually mature (Figure 9.2).

Just as the infant with an immune system in progress is susceptible to all sorts of infections, this *stress hyporesponsive period* represents a window of vulnerability, in which the developing nervous system is easy prey for challenges it might successfully resist later in life. Exposure to virulent stressors during this period may be as detrimental to the future sensitivity of the stress response as exposure to human immunodeficiency virus can be to the integrity of the immune system.[36,40]

If the brain is to perform optimally in adult life, it must be protected during development from factors that impair growth, damage neurons, or interfere with the formation of synaptic connections. And if the HPA axis is to

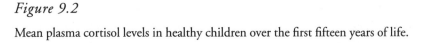

Figure 9.2

Mean plasma cortisol levels in healthy children over the first fifteen years of life.

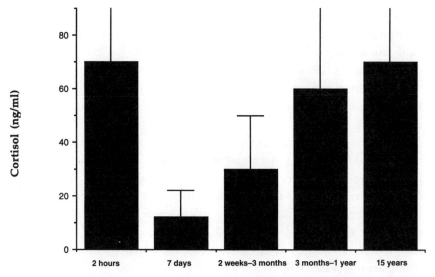

Reprinted with permission from G. W. Sippel, H. G. Dorr, F. Bidlingmaier, et al., Plasma levels of aldosterone, corticosterone, 11-deoxycorticosterone, progesterone, 17-hydroxyprogesterone, cortisol, and cortisone during infancy and childhood, *Pediatric Research 14* (1980): 39–46.

respond effectively to the stress of life, without either collapsing or overheating, stress response mechanisms must be protected from challenges they are not yet prepared to handle.

FOR NINE MONTHS, the fetus's relationship with his or her mother is the only one that matters. Everything the mother eats, drinks, smokes, or injects has access to the developing brain. When she's under stress, the fetus feels the pressure as well. If her blood sugar, body temperature, blood pressure venture outside normal limits, the fetus's physiology will also be challenged.

Prevention, which is really the earliest of interventions, starts here, with prenatal care that emphasizes emotional as well as physical health. The "body" side of an ideal prenatal program would include such commonsense goals as monitoring for pregnancy complications and ensuring adequate nutrition, while the "brain" side would focus on preventing brain injury and protecting the stress response. Screening for evidence of substance abuse and domestic violence would be considered as vital as detecting gestational diabetes and monitoring blood pressure. Mothers would learn not only how to feed, bathe, and carry their babies, but also how to understand and appreciate temperamental differences.

Of all the insults that can be heaped on the developing nervous system, none has drawn more public attention than prenatal substance abuse. Drug and alcohol abuse by pregnant women is not a trivial problem, and it's not limited to inner-city neighborhoods. According to the Centers for Disease Control in Atlanta, the number of women who admit to drinking at least a moderate amount of alcohol while they're pregnant has gone up fourfold since 1991.[41] A 1994–1995 survey found that 20.4 percent of pregnant women sampled had smoked cigarettes at some point during their pregnancy, 8.8 percent drank alcohol, and more than 5 percent admitted using illegal drugs.[42]

New research on the long-term effects of cocaine exposure in utero shows that prenatal drug abuse is not a one-way ticket to a juvenile detention center. But the toxic effects of alcohol or the subtle structural derangements caused by cocaine can force exposed children to start life with one more obstacle to overcome, in addition to coping with a less-than-functional parent. Even a single mother who's sober can do an acceptable job of raising a child. But a woman compromised by alcohol or drugs—whether she is married or unmarried—is poor parenting material.

Detecting and treating maternal substance abuse are pivotal to any prevention program that hopes to reduce future behavior problems. But spotting pregnant women with a drug or alcohol problem isn't simple. Abusers who live in "problem-free" suburban neighborhoods or don't look like destitute crack addicts are readily overlooked by doctors and not eager to enlighten them. "Don't ask, don't tell" is the prevailing philosophy. And that's not likely to change as long as women who admit to a problem are more likely to get into trouble than to get help.

If you took a few more deductions last year than you were entitled to, would you invite the IRS to audit your tax return? Would you want your auto insurance carrier to know how fast you really drive? How about those office supplies that followed you home? Why not schedule lunch with your boss and confess? If you're reluctant to admit to small indiscretions with less-than-life-threatening consequences, you might understand why a pregnant woman who's also a substance abuser would want to keep her problem under wraps: she stands to lose her other children, go to jail, even be charged with child abuse.

The prevailing attitude toward substance-abusing mothers is ugly. Public opinion and political expediency favor prosecution, not treatment. This global, short-sighted vindictiveness is another example of how we waste energy and resources on punishment rather than saving lives. The risk to the infant increases every day the mother takes drugs—and the more she consumes in each binge, the greater the risk. "If you're out there trying to do early intervention for the child, but don't give treatment to the mother, you're going to lose," says

University of Illinois researcher Ira Chasnoff. We stand the best chance of protecting babies by treating mothers, as early in their pregnancies as possible. Women who fear prosecution won't seek treatment, won't stop using drugs, won't accept even the most rudimentary medical care. More children will be hurt as a result. The final cost-benefit analysis may be difficult to swallow, but it's simple: less exposure = less damage = less violence in the future.

IT TOOK CONNIE BASTEK-KARASOW six years to build her dream house. Libertae Family House, a sixty-room complex located about forty-five minutes north of Philadelphia, was constructed to provide shelter for up to sixteen women and their children. It welcomed its first residents—a three-month-old baby and his mother, a recovering drug abuser—in March 1997, making it one of a handful of programs that keep mothers and children together during addiction treatment.

Libertae (a Latin word meaning "freed women"), which has treated women with drug and alcohol problems since 1973, was already special. The program emphasizes financial and social responsibility as well as sobriety, moving 95 percent of its clients from welfare to paid employment in the first six weeks of their stay. Women are expected to work with a counselor to construct a budget that includes payments to creditors, outstanding fines, and a contribution toward their treatment costs. Compared to the national average of 55 to 60 percent,[43] Bastek-Karasow estimates that up to 80 percent of Libertae clients are still sober one year after discharge.

Women who enter treatment for substance abuse are nearly always separated from their children—even if those children are newborn babies. Bastek-Karasow, director of Libertae, was determined to change that. Residents in the new Family House can keep their children with them for their entire stay. Counselors take advantage of the opportunity of having mothers and children under the same roof to teach the women how to be better parents, as well as sober, healthy parents.

The residential program at Libertae Family House is not only successful, according to Bastek-Karasow; it's cost-effective. Children are kept out of foster care, at a savings of more than $10,000 per child per year; overall, treatment averages a seven-dollar return for every dollar expended. And by preserving the first and most important of social relationships, programs like this can prevent the disastrous neurochemical and hormonal consequences of maternal separation.

DRUG ADDICTION ISN'T the only problem pregnant women are reluctant to talk about. Pregnancy is no deterrent to an abusive husband or boyfriend. Courtney Esposito, a counselor who teaches physicians and other health care professionals how to recognize domestic abuse, told *American Medical News,*

"Abusers go for what represents the baby. If a pregnant woman presents with injuries to the breasts, genitalia, and abdomen, it is very likely the result of a beating."[44]

Abuse during pregnancy sets an ominous pattern for family life. Men who abuse their wives often go on to abuse their children as well.[45] Even if they don't hurt the children directly, abusive men can still do lasting damage. In a study of children referred to the Child Witness to Violence Project at Boston City Hospital, children as young as two who had observed domestic abuse showed signs of PTSD, including sleep disorders, excessive anxiety, emotional detachment, and aggressiveness.[46]

When battered women want out, they have almost as few options as women who are substance abusers; some cities have more shelters for animals than for abused women. At most, the majority of these "safe houses" can offer victims protection for only a few weeks. Women in danger must have a place to go—and help getting there—before it's too late.

NEVER AGAIN WILL THE BRAIN be as open to change as it is during infancy. And never again will the environment be as simple. In essence, for the infant, parents *are* the environment. If their overtures are positive and "make sense" to their baby, he will respond in kind, and the relationship is off to a good start. If they are at odds with his needs, he will be miserable and they will be frustrated—a pattern that's likely to grow more, not less, hostile with time.

Parents start out as idealists, with a mental picture of what a baby "should" be like. The difference between their expectations and the reality of their child may be the beginning of a disastrous misunderstanding. It doesn't matter if the ideal is based on experience; a mother's advice; expert opinion offered by books, doctors, or talk show guests; a case worker's suggestions; or pure fantasy. If it's incompatible with the baby's temperament—the unique, personal style that is the foundation of the personality—it will run head-on into biological reality. Parents faced with such a conflict can either forget what they think they know and adapt their behavior to the child, or insist on trying to force the child to fit their idea of what a baby should be like. Resistance leads to parental frustration, frustration invites coercion, coercion is ultimately met with defiance. The circle has begun turning.

Parents of infants with "difficult" temperaments will have a harder job, but they can learn to cope. Prenatal programs that prepare mothers and fathers to be biological parents in style as well as substance, that define parenting as a problem-solving exercise rather than an effort to shape children to a rigid system of standards, can give them a head start. Parents who are aware that babies differ widely in sensitivity, alertness, intensity, and adaptability are parents who can anticipate the unexpected, rather than parents who are blind-sided by a difficult baby. An idea of how to work with their baby's nature rather than against it leads parents to an approach that is more likely to give them the influence and authority they should have.

Stella Chess, the "grandmother" of temperament research, believes that the "goodness of fit" between parent and child is the key to preventing many childhood behavior problems. "There will never be another first time,"she says. Stanley Turecki, a child psychiatrist who has translated Chess's research into practical parenting advice, stresses that parents "don't need a Ph.D." to achieve a good fit with their children.[47] A short course in the rudiments of temperament doesn't require years of therapy, expensive new programs, or an army of social workers. If we would only agree to guaranteeing the most basic of medical care to all mothers and children, we could institute preventive and educational interventions that have a significant impact on behavior for the rest of the child's life.

Communication Breakdown

In May 1997,the U.S. House of Representatives passed a new bill designed to "crack down" on juvenile crime by rewarding states that agree to prosecute more under-age perpetrators as adults.[48] Prosecuting children as adults satisfies our urgent desire to do something about adolescent aggressors who kill parents, classmates, even total strangers. Conveniently, it also covers up our own failure to pay attention. Teenage murderers do not rise up fully formed on their thirteenth birthday. Their deterioration began years before, in a childhood custom-designed to bring out the worst in them.

It's not as if these troublemakers were invisible. Aggressive, angry children typically command more than their fair share of attention—from families struggling to manage them, schools frustrated by efforts to educate them, angry communities sick of enduring them. In fact, of all the problems common to troubled children—defiance, lying, stealing, truancy, academic failure—none is more likely to translate as a cry for help than aggression.

"Aggression is the number one problem in child psychiatry," contends Carl Feinstein, director of clinical services for child and adolescent psychiatry at Stanford University Medical School and former medical director of the Center for Autism and Related Disorders at Baltimore's prestigious Kennedy Krieger Institute. Other experts who treat childhood behavior problems concur. Tony Rostain says of the children he treats for AHDH at the Philadelphia Child Guidance Center, "probably 75% aren't brought here for their attentional problems, but because of their aggressive behavior." Karen Bierman, a psychologist at Penn State University who heads a research team evaluating a comprehensive intervention program for troubled children, believes that persistent aggression "is one of the most prevalent and intractable mental health problems of childhood and adolescence," and that teenage delinquency, as well as antisocial aggression in adults, "rarely . . . begin without warning signs in early childhood."[49]

How can we be so aware that some children are already too aggressive, and yet so powerless to stop them from evolving into violent adolescents? Are indulgent, preoccupied, self-centered, divorced, or unmarried parents to blame? Television? An extinction of family values? Or have we also overlooked the biological significance of childhood?

EVERYONE WHO HAS TAKEN a psychology course or read a newspaper has heard that "today's child victims are tomorrow's perpetrators." The idea that violence is recycled across generations has been repeated so often that it's taken for granted. But not all abused children grow into juvenile delinquents or antisocial adults. And not all studies that have looked for a link between abuse and subsequent violent behavior have found one.

A recent study, sponsored in part by the National Institute of Justice, provides the best evidence yet that children who are abused and neglected are at greater risk of arrest for a violent crime later in life.[50,51] Children who had documented evidence of abuse or neglect were nearly twice as likely to be arrested as juveniles than children with no history of abuse, one and a half times more likely to be arrested as adults, and significantly more likely to have committed a violent offense. Children who had been physically abused were the most likely to be arrested for violent crimes. But surprisingly, children who had merely been neglected (who constitute 52 percent of the million-plus confirmed cases of child maltreatment each year)[52] were only a percentage point behind.

Hypervigilance, misinterpretation, and an exaggerated response to perceived threats are the behavioral consequences of the strain imposed by the need to compensate for such a heavy load. Even if "abuse excuses" try our patience, it is clear that trauma and abandonment cannot be good for the developing brain.

"Trauma effects are especially important early in development because they encourage adaptation," observes stress researcher Frank Putnam. "And early adaptation resets physiological systems in ways that leave them very different from normal."

THE MOTHER on the other end of the line was trying to strike a balance between sounding calm enough to avoid being judged incompetent and desperate enough to get the talk show host, a local child psychologist, to take her seriously. The problem, she explained, was her two-and-a-half-year-old son, a difficult baby who had matured into a tyrannical toddler.

"He's just unmanageable," she pleaded. "I know two-year-olds have tantrums. But he won't do anything we ask him to without a fight. I've tried asking him politely, saying please and thank you—and he hits me. If I put him in time-out, we end up fighting about him staying in the chair. I've tried being firm, I've tried demanding, I've even bribed him—I admit it. All he does is yell and scream."

"Two is such as exciting age," coos the psychologist ingratiatingly. "Your son can do so many things all by himself now. But he still needs you."

No doubt about it. The exciting little person in question can be heard voicing his need for Mom at stadium-concert volume.

If the psychologist hears, she's not responding. "He's trying to come to terms with his new-found maturity," she prattles on. "You should just spend lots of time playing and having fun. Take him out to places like the mall, so he can try out his new way of looking at the world. He needs to know that someone is celebrating the joy of childhood with him."

Across town, Laurel's parents have talked to three different doctors in two years, trying to understand why their eight-year-old daughter slaps, kicks, and insults her five-year-old brother relentlessly. They've heard from one psychologist that they're "too authoritarian"—but another insisted they were "too indulgent." They're simultaneously "inconsistent" and "overly rigid," "too detached" and "too overinvolved." They've learned how to improve their parenting skills and set up "incentive programs," but Laurel's hostility seems impervious to time-out, sticker charts, and lost privileges. The one thing they haven't learned is how to end the fighting. As for Laurel, the only thing she's learned from any of these experts is that her parents are incompetent. "If we can't control her now," her mother worries, "what will we do when she's a teenager?"

These examples demonstrate how we fail troubled children so consistently: we refuse to see childhood aggression for what it is, when we can no longer overlook it; we don't know what to call it; and when we finally think we've identified it, we still don't know what to do about it. Intervention often focuses on reforming the parents, while sidestepping the question of why the child is lagging in his or her social development or how the relationship between parent and child has gone wrong.

Look behind an aggressive child and you may well find parents who are preoccupied, incompetent, or negligent. Or you may find parents who recognized trouble but prayed that the child would "grow out of it." You may find parents who asked for help and were told that nothing was wrong or that it was their fault, or confused, frustrated, disappointed parents who have tried to follow through on expert advice and failed, even seen the problem grow worse. What you're certain to find is an ignorance of the fact that aggressive adults often start as children who hurt others, children whose social ineptitude already reflects destructive interactions between the brain and an environment that is threatening, unfulfilling, or incomprehensible. Worse, you're likely to find a failure to appreciate that unlike adults, children have nervous systems that are still immature, and their behavior, as a result, is still wide open to change.

Kids are aggressive for the same reason as adults—a vicious circle between brain, behavior, and environment—and they're aggressive in the same ways. Some overreact, to people and situations they perceive as threatening; some underreact, especially to punishment. The spiral may be set in motion by loss, abuse, neglect, fear—or it may have more insidious roots in the temperamental traits that require special management skills.

Children with a slow-to-arouse, fearless temperament, for example, can be easily mismanaged into antisocial tyrants. Because their sluggish sympathetic nervous system assigns lower emotional values to negative events, they seem impervious to punishment. They're more interested in rewards, but the motivational power of candy and gold stars eventually pales in contrast to the thrill of conflict. Intellectually, these children may come to recognize that a system of rules governs social conduct, but the rules themselves hold about as much meaning for them as the tax code.

Child development researchers who study children with autism say that they lack a "theory of mind"—the ability to recognize that other people have their own thoughts, beliefs, and feelings.[53] Uninhibited children on their way to an antisocial personality also fail to recognize that others have feelings. Because the world of emotion is incomprehensible to them, they remain indifferent, cognizant only of their own needs and wants.

Poor emotional understanding results in a social learning deficit. Without emotions to guide them, insensitive children can't seem to figure out how to tailor their behavior to mesh with others, and their negligence often results in inappropriate reactions to emotional situations. For example, if a playmate falls, skins a knee, and starts to cry, a socially aware child will also become distressed or attempt to comfort her companion. Emotionally insensitive children, on the other hand, may act aggressively instead.

Learning is an essential part of changing inappropriate behavior. But unless environmental interventions include the intensive remedial instruction in social skills emotionally insensitive children need to catch up with their more socially sophisticated peers, the aggressive behavior is going to persist. And without relief from the relentless stress of living with the aggression, parents and siblings are going to find it increasingly difficult to cope.

Experts who work with physically disabled children have developed exercises designed to improve mobility, strength, and fine motor control. Similarly, professionals who work with the social, emotional, and cognitive challenges of developmental disorders like autism have come up with strategies to teach social skills to children who need extra help. These techniques include visual aids, such as charts and storyboards, which outline an easy-to-follow script for social interactions; videos illustrating appropriate behavior; modeling; and role playing.[54-56] The goal is to create a consistent, structured environment in which children are surrounded by examples of acceptable social behavior and opportunities to practice nonaggressive solutions to social problems.

Few children currently benefit from such an intensive approach to social education. But new research suggests that "social-skills training," combined with more familiar techniques like parent training and positive reinforcement, could change the way we deal with insensitive, aggressive children, because it addresses the underlying problem directly. Penn State University's Karen Bierman and researchers from four other universities have developed such a program as one element of an ongoing experimental program, designed to evaluate the merit of comprehensive early intervention on aggression and other conduct disorder behaviors.[49] The social skills component of the Fast Track program combines classroom exercises, such as direct instruction, role playing, modeling, and problem-solving exercises, with extracurricular group activities and a "buddy" system that pairs the aggressive child with a classmate for supervised play sessions designed to improve one-on-one, as well as group, social skills. Fast Track shows a lot of promise. In just the first year after its implementation in schools at five independent sites, researchers reported that the program had already shown improvements in the behavior of at-risk children.

FUNCTIONAL DEFICITS can also initiate the process that leads to childhood aggression. Aggressive behavior may be the first sign of a medical disorder, such as vision or hearing problems that confound social learning and make the world seem more threatening than it really is. Or it may herald a neural condition that interferes with executive function or working memory, such as an attention deficit.

A child who can't differentiate what's important and what's not is going to have a hard time learning and remembering which behaviors are acceptable and which provoke an angry response. The details of the conflict get lost in the noise, and the next time the child encounters the same situation, he'll act as if he's seeing it for the first time. Instead of slowing down, he'll speed right by the warning signs that he's about to get it wrong again.

The prefrontal cortex may not know what to make of the angry reaction that meets each mistake, but stress response mechanisms know what to make of it: danger. The child, threatened, shoots back. The parents grow even angrier. In no time, you can have a child who's hostile and aggressive as well as inattentive, locked in battle with an equally hostile family.

"If the environment is threatening," says ADHD expert Tony Rostain, "the child is going to develop coping mechanisms that are maladaptive." He stresses that reducing aggression in these children is a matter of reducing threat, by simultaneously treating the attention deficit and changing the environment—the relationship between parent and child—to develop alternatives to conflict.

Add an attention deficit to a social learning deficit, and all of the tools essential to social learning vanish. Reducing aggression in slow-to-arouse children who also have ADHD means addressing both problems. Treating ADHD without teaching social skills fails to make up for ground lost to the attention deficit. But trying to teach such skills to children who aren't prepared to listen is guaranteed to fail.

Aggressive children with ADHD can't learn alternative behaviors until someone turns down the internal noise level. That's what Ritalin, the drug of choice for treating the ADHD, is designed to do. By optimizing dopamine activity, Ritalin restores the ability to suppress irrelevant information and brings the prefrontal cortex back "on-line."[57] By releasing epinephrine, which slows the firing of neurons in the locus ceruleus, it shushes overzealous noradrenergic input babbling at the cortex.[57] As details emerge from the background, a flat, noisy world finally begins to make sense. In this quiet space, the child is prepared to learn new ways of interacting.

THE PARENTS OF AGGRESSIVE children are not universally abusive, disinterested, preoccupied, overindulgent, or afraid to discipline their children. Parents can also fail because neither they nor the experts they consult recognize the biological underpinnings of a deteriorating relationship, and no one takes effective steps to stop the decline. A problem that often begins as a temperamental mismatch between parent and child, or as fallout from another condition, is exacerbated by ignorance, neglect, bad advice, one-sided treatment. By the time the child reaches adolescence, he or she automatically resorts to aggression at the slightest provocation.

Neither adult prison sentences nor conflict resolution can easily break aggressive patterns of responding overlooked during the formative years of childhood, once they collide with the stresses, social reorganizations, and hormone changes that accompany puberty. Before age twelve, for example, notes ADHD expert Russell Barkley, up to 90 percent of aggressive children with ADHD show improvement with a combination of medication and environmental intervention.[58] After age twelve, that rate drops to 25 to 30 percent.

The frontal cortex is not fully mature until early adulthood. We still have time to intervene in the lives of troubled adolescents, but the window of opportunity, when the brain is at its most plastic, is shrinking rapidly. If we want to complete the process of turning a human being into a habitually violent individual, we can step up to adult-sized punishment, an option likely to increase subsequent violence. If we want less violence rather than more retaliation, we need to use rational interventions based on biology, neurochemical and environmental interventions that may well be our last hope.

Getting Straight

The first lesson of biology is the importance of calling things by their right names. Aggression can give rise to multiple forms of violence, each with its own identity and characteristic behavioral fingerprint. Unless our intervention strategies take this heterogeneity into account, they will be inadequate, incorrect, or even detrimental for a significant proportion of the individuals they wish to change. "One-size-fits-all" solutions simply aren't going to work. Classifying offenders according to their perceived security risk helps prison officials, but misses the point; it's the form, not the intensity, of behavior that matters. An effective antiviolence policy is going to have to account for the varied forms of human aggression; to become, as behavioral biology itself has, more observant, more cognizant of the defining role of personal history, the clues provided by context. When we accept the idea that different people ignore the boundaries for different reasons, we can decide the best way to redefine the limits.

BIOLOGY WON'T MAKE PRISON obsolete. To disconnect chronically violent individuals from the destructive environments that have contributed to their problem, we often need to remove them physically from these environments. To get their attention, we often need drastic action. But not all violent individuals belong in prison.

People with schizophrenia, for example, are ill-served by incarceration. Imprisoning them may keep them off the streets, but it will not reform them, because they cannot understand that the prison experience is intended to deter repeat offenses. Too confused to defend themselves or read the intentions of others, they are easy prey; victimization only reinforces their perception of threat.

Schizophrenia is not simply an insanity defense. It is not a decision, an act, a state of mind, or a fanciful eccentricity. It is a chronic, incapacitating brain disorder that abolishes rational thinking and turns emotional reactions upside down. Schizophrenics, say experts, do not think the way we do, even when they're not actively hallucinating or delusional. They cannot and typically will not ask for help, because unlike the depressed, the anxious, or the traumatized, they don't recognize that anything is wrong. Whether or not they "know" what they're doing when they commit a violent act, the reason why they're doing it is inevitably bizarre. Their behavior may well be premeditated, but it's based on an irrational meditation. Punishing behavior driven by delusions is equally irrational—and it's expensive, cruel, and ineffective as well.

The key to preventing abuse of the insanity defense will be replacing senseless debates about morals and motivations with accurate diagnosis, based on our growing knowledge of the biological foundations of schizophrenia. The key to containing the unpredictable aggression of the minority of mentally ill individuals who are violent is appropriate medical treatment. The violence of paranoid schizophrenia lies beyond punishment. But it can be effectively controlled with antipsychotic medication, ongoing psychotherapeutic support, and adequate supervision.

Treatment reins in delusional thinking—and fewer delusions mean less need to respond violently to an unseen threat. But treatment measures like medication work only as long as the mentally ill individual remains under their protective influence. If he or she stops taking antipsychotic medication, the risk of violence rapidly escalates.[59,60] If medical intervention is not to fail as soon as those remanded to treatment are released, mentally ill individuals who have proven violent need ongoing supervision, in the form of outpatient commitment, a legally appointed guardian, or a long-term residential community.

Advocates who supported deinstitutionalization and the right to refuse treatment have stripped patients and society alike of the right to safety. The mentally ill, who were supposed to be freed, have been abandoned, at the mercy of threatening delusions that prompt tragic attempts at self-defense. It is time to demand that those who have left the mentally ill to their own devices assume responsibility for the welfare of these patients and that they be held accountable for turning our prisons into mental institutions.

DRUG ADDICTS DON'T PROFIT from punishment either. They need sentences that include mandatory treatment for their addiction and correctional facilities that can provide that treatment. The decision to admit illicit drugs into the nervous system is a bad personal choice. Addiction, the unintended and unwanted biological consequence of that choice, is not a moral problem; it's a neural problem. You cannot command an addict to stop using drugs by locking him in a jail cell any more than you can order a paraplegic to get up and walk. Until he's clean, drugs, not rules, direct his behavior. Until he's clean, drugs will encourage him to resort to aggressive methods to satisfy his craving.

Detoxification is not treatment. Insight-based psychotherapy is not treatment. Only strategies that work with the remodeled brain to break the link between environmental cues and craving and prevent relapses will provide a long-lasting solution to substance abuse and the violence it fuels. Twelve-step recovery programs, for example, use structure, repetition, and social support to help the addict resist the pull of drugs. Cognitive therapy has also been adapted for use with substance abusers—so successfully that it's the basis for a new treatment protocol distributed by the National Institute on Drug Abuse.[61] Methadone and a newer, longer-lasting equivalent, LAAM (1-α–acetyl-methadol), step between the addict and illegal drugs, along with the illegal methods used to obtain them.[62] And in the near future, new pharmacological therapies stand to revolutionize the treatment of substance abuse. Antiaddiction medications currently under investigation are aimed at dopamine pathways to block craving at its source, breaking the cue-craving–drug-seeking–drug-taking sequence instead of merely pacifying it.[63] The goal is to capture the addict's attention, so that behavioral interventions have a quiet space in which to teach relapse-prevention skills.

University of Pennsylvania psychiatrist Charles O'Brien points out that while curing addiction is still an elusive goal, as it is for other chronic illnesses, effective control is not only possible but realistic.[43] O'Brien and other experts emphasize that drug treatment is more successful, more reasonable, and more cost-effective than waging war on drugs and drug users. A recent study by the RAND group, for example, calculated that the cost of treatment that reduced cocaine use alone by a mere 1 percent was more than seven times less than the cost of enforcing drug laws.[64] Yet only 2 percent of substance-abusing inmates receive intensive addiction treatment,[65] and drug rehabilitation programs are a favorite target of prison-happy politicians. Prison isn't treatment either. The only action that will have a significant and lasting impact on drug-related violence is a serious effort to reduce substance abuse.

Impulsive aggression is prevalent outside, as well as inside, prisons, a form of violence that makes everyday life miserable for millions of Americans. But the silver lining in this cloud is that the hostile, emotional, stress-disordered aggression of the hothead can be controlled. Like addiction, the critical hurdle is interrupting the cycle between the perception of threat and aggressive responses, to free the brain from survival mode so that the individual can pause before reacting. When the rapid response system doesn't immediately jump to conclusions, working memory has a fair chance to come up with a more reasonable option than conflict.

Reflective delay can be taught; that's the goal of cognitive behavioral therapy. This learning process takes time, however. When intense emotional responses command more than their fair share of attention, learning slows down even more. Time is precious when the hair-trigger response may involve a gun. For some hostile, impulsive aggressors, the only way to buy the time that can save lives is to override hypersensitive reactions until new response patterns are strong enough to take over on their own. Drug therapy can provide that support.

THE TREATMENT of stress response disorders is one of the great success stories of modern psychopharmacology. Prozac was only the beginning. Since 1986, the discovery and introduction of an entire family of new Prozac-like antidepressants—known collectively as selective serotonin reuptake inhibitors (SSRIs)—have allowed physicians to access the overburdened HPA axis via one of its principal controlling elements, the brain serotonin system.

The SSRIs, as their name suggests, disable the recycling mechanism, increasing serotonin concentration in the synapse. But that's not the action that's central to the control of stress responses or to the relief of symptoms. Over the course of four to six weeks, daily treatment with an SSRI also gradually remodels serotonin synapses, altering the number and sensitivity of postsynaptic receptors. It is a long-term adaptation, "normalizing" a system that has wavered far out of balance, not the short-term effect on serotonin reuptake, that's most closely correlated with the onset of a clinical response (see Chapter 5).

When every interaction isn't a life-or-death encounter, the need for excessive responses ebbs. Prozac and other SSRIs brake express responding so that the invalid assumptions that lead to an overestimation of threat can be cleared from the track, then hold it back while new responses are under construction. The medication begins the process of restabilization, giving therapy a real chance to "catch." With time, the individual learns a less overwrought approach to life, and how to deescalate as well as wind up.

Depression, PTSD, and impulsive hostile aggression have a common origin: the strain imposed by chronic activation of fight-or-flight responses. So although each has a characteristic behavioral fingerprint, it's no surprise to find that symptoms of one can crop up in another. For example, a study carried out by investigators at Massachusetts General Hospital and Harvard Medical School found that 43 percent of depressed patients also experienced frequent angry outbursts; some had actually assaulted others.[66] When the researchers treated these people with Prozac, they found that the drug not only relieved symptoms of depression but also significantly reduced the incidence of "anger attacks." Now Emil Coccaro, Richard Kavoussi, and their colleagues at the Allegheny University of the Health Sciences have shown that SSRIs can also dampen emotional hypersensitivity in impulsive aggressive individuals who are not depressed.[67] Among an initial group of forty impulsive aggressive patients in what will ultimately be a two-hundred-patient study, men and women who took Prozac for three months showed a significant and sustained reduction in the number and intensity of hostile outbursts.[68,69] Those who continued to take the drug after the trial period continued to be less aggressive, while others who decided they could do it on their own lapsed back into violent outbursts.

SSRIs aren't the only medications that can check the impulsive responding that's such a barrier to social learning. Lithium, the drug of choice for stabilizing the wild mood swings of manic-depressive (bipolar) disorder, can also motivate reflective delay. No one knows quite how lithium works. It has effects on serotonin and norepinephrine similar to those of pre-Prozac antidepressants, inhibits the cascade of biochemical reactions initiated by second-messenger proteins, and even has antiviral properties.[70] What is known is that it can reduce impulsive aggression. In a landmark study, Michael Sheard, a psychiatrist at the Yale University School of Medicine, compared the behavioral effects of lithium and placebo in violent male offenders with a history of impulsive and antisocial aggression.[71] During the three-month treatment period, the men who received lithium recorded significantly fewer impulsive violent infractions than the placebo-treated group. Forty percent of those who completed the entire course of treatment had no violent outbursts at all compared to 24 percent of those who received the placebo. But if those who responded were taken off the drug, their impulsive aggression returned.

Yet another treatment option are medications more commonly prescribed to treat the paroxysmal electrical activity of seizures. The anticonvulsants carbamazepine (Tegretol) and sodium valproate (Depakote), can stabilize the spontaneous flips in mood that characterize bipolar disorder, and can stabilize impulsive aggression as well.[72–74] Other studies have found positive effects on impulsive aggressive behavior with phenytoin (Dilantin), a third anticonvulsant medication.[75]

FRED BERLIN, of the Sexual Disorders Clinic at Johns Hopkins University, believes that physicians have "three legitimate reasons for prescribing psychoactive drugs: to restore function, so that you have a whole human being; to increase self-control, interrupting the rush to aggression by inserting an extra second or two for cortical mechanisms—the higher function everyone wants people to rely on—to catch up with survival mechanisms; and to diminish suffering." The use of drugs like the SSRIs to intervene in stress response disorders, including impulsive hostile aggression, satisfies all three of these criteria. By normalizing serotonin function, SSRIs also normalize stress responses, reducing destructive behaviors and wear and tear on the nervous system. The resulting shift in serotonin function adds the reflective delay needed for deliberation. Most important, drug treatment can significantly diminish the anguish associated with this kind of violence. It can keep those who live, work with, or encounter hostile individuals from becoming victims. It can keep the impulsively aggressive out of jail, out of divorce court, out of the hospital. And it can save their lives, in more ways than one. Impulsive hostile aggression eats away at the cardiovascular system; no other behavior is more closely associated with an increased risk of coronary artery disease, heart attacks, and death.

"By the time they [impulsive aggressive people] get to me, they have no hope," says Matt Stanford of the University of New Orleans. Pharmacotherapy is not the right choice for all overly aggressive individuals, or even all impulsively aggressive individuals. But for the right individuals, it can restore hope and change lives.

"My first reaction was 'no way, I don't believe in medication for me,'" said one patient, interviewed on the current affairs program *20/20*. But he changed his mind after his doctor—and his beleaguered fiancée—convinced him to give drug therapy a chance anyway. "The old guy, the exploding guy, is gone. Just gone." Another patient, on the road to rebuilding his ruined marriage, had just one lingering question: "Where was this medicine forty years ago?"

Therapy and pharmacology can do only half the job, however. It is useless to detoxify responses if abuse, threat, and community violence still compromise safety. People cannot return to destructive relationships and dangerous communities, or the environment will undo all of the benefit of neurophysiological interventions in short order. One of the greatest merits of community policing, gun control, and neighborhood watches is that by creating an atmosphere of safety, these measures significantly reduce the perception of threat and the need for lethal self-defense.

OUR PRISONS ARE DROWNING in antisocial individuals whose disregard for others has been ignored and mismanaged for so long it has calcified. No other group of violent people needed early intervention more, and no other group is more likely to have fallen through every medical and social service crack.

Punishment isn't any more effective for changing the behavior of emotionally insensitive adults than it is for managing emotionally insensitive children. Remorse is rarely part of their vocabulary. Isolation can't do anything but make their behavior worse. The profound boredom of prison life, totally at odds with their risk-taking nature, either stultifies them or unnerves them to the point of explosiveness. Intimidating other inmates reinforces their view of themselves as masters of their limited universe.

Our disregard for the early warning signs of antisocial behavior comes back to haunt us. We can damp down an overreactive stress response system, but we know little about how to jump-start it. If we miss the opportunity to reach people with social learning difficulties when they are young, we will need to deal with the fact that they have little motivation to change, and their best social skill is manipulating therapists, doctors, and others interested in reforming them.

People with antisocial personality disorder are difficult to engage in therapy, and pharmacology can't touch psychopathy. In Michael Sheard's study of lithium, only impulsive aggression responded to treatment; the drug didn't touch antisocial behavior or the participants' lack of empathy.[71] More recently, Matt Stanford and Ernest Barratt, a colleague at the University of Texas Medical Branch at Galveston, compared the effect of phenytoin on aggression and arousal in impulsive and antisocial subjects. Only the impulsive group responded.[75]

But biology can still pull people back from the brink. To change, the antisocial need strategies tailored to their biological make-up. First, they need a focus on the positive. Resistant to coercion, they are more likely to respond to policies that reward prosocial behavior than those that punish lapses. Their inability to use emotion to organize and guide their behavior calls for a structure that can supply the missing order from the outside. In prison, that means clear rules and a minimum of social upheaval; out of prison, it means a safe, stable environment. Many need treatment for concurrent drug and alcohol abuse; all need intensive training to improve the basic social skills they never learned when they were younger. And the antisocial individual needs a purpose—constructive, challenging work, an opportunity to lead and mentor—a reason for being that matches his fearless personality. Finally, dealing realistically with antisocial aggression means recognizing that change is a long-term proposition; remolding responses embedded in the personality requires ongoing support from the community to keep the individual connected to others and inside social boundaries.

Prevention is the best solution to antisocial violence. If it is allowed to take over the personality long enough, it will exceed the limits of our knowledge. Then the best we can hope for is to chip away at the edges, eliminating things that make violence worse. Some who reach this point, despite our best efforts, will not respond to anything.

No other type of violence illustrates failure better than predatory psychopathy. With the help of intensive long-term therapy and hormone blockers that attenuate the sex drive, some sex offenders can control their unacceptable urges. If their paraphilia has taken on a lethal tone, however, even chemical castration is unlikely to reform them. The behavior of predatory psychopaths is not only extremely rewarding but irrevocably intertwined with a natural drive. Therapy and hormones are as effective as trying to knock down a brick wall by throwing a pebble at it.

The only solution to murderous sexual aggression—like all other antisocial aggression—is to prevent it, by adopting a zero-tolerance policy on childhood sexual abuse and by learning to recognize the early warning signs. Boys who are loners with a precocious interest in sex, who set fires, torture animals, express little regard for other people, or are fascinated with violent sexual images in movies or magazines should set off alarm bells. The clock is ticking; if we want to have any chance of catching them before they go over the edge, we have to act decisively. Once they're violent adults, it's too late.

Serial sexual killers are the most intractable of aggressors. Once their behavior is set, we can do little or nothing about it, for it has become their sexual orientation, an essential feature of identity. Biology does not yet have an answer for refractory, dangerous psychopaths. At this point, we can only keep predator and prey apart, fund more research, and learn from our mistakes. We have no option but prison for confirmed predators.

BEHAVIORAL VETERINARIAN Karen Overall observes that there's a fail-safe when her attempts to reform aggressive dogs fail: the death penalty. When intervention doesn't work, she says, "we can just euthanize the patients who don't comply." But she doesn't feel good about it. "Death," she emphasizes, "is not an acceptable end point."

Increasingly, we're turning to death as the solution to intractable humans. If we kill them, it's said, at least they won't have any more victims. This is true. But if death is not an acceptable end point for dogs, should we really settle for this as an acceptable last resort for human beings?

Victims' Rights: A Biological Mandate

The dance of violence has two participants. While the violent act is fresh in our minds, awash in reporters and television cameras, investigations and tabloids, we're as interested in victims as we are in perpetrators. We want to stick our hands into their lives and feel around for clues, touch their pain. But back home, after the dust has settled, it's a different story. Now victims, like funerals, are an unwelcome reminder of our vulnerability. We don't know what to say to them, and so we say nothing. At best, victims of violence are forgotten; at worst, they mutate into an annoyance, exhorted to forgive, forget, and get on with it, so we can stop feeling guilty.

But victims can't just "get on with it" by force of will. The consequences of violent actions don't stop at broken bones; they penetrate to the innermost reaches of the brain, and they do lasting damage. A solution to the problem of violence that does not include protection, respect, and assistance for victims is incomplete.

We might learn something by comparing the behavior of chimpanzees toward victims to that of another primate, the long-tailed macaque. Macaques mind their own business. The victim of an attack elicits little sympathy, even from close relatives. In the tense period immediately following a fight, victims are as likely to be ignored by their neighbors as they are to receive a friendly gesture.[76,77] Left to fend for themselves, they risk being attacked again—not only by the aggressor, but by her kin—unless they take steps to repair the breech or "redirect" their social stress into attacking another troop member. Today's victims are tomorrow's aggressors—unless they're tomorrow's victims as well.

When two chimpanzees fight, on the other hand, it's everybody's business. Within minutes of a confrontation, bystanders rush in to console the combatants, especially the victims. Perpetrators aren't neglected, but victims are hugged, kissed, stroked, or groomed four times more frequently than their attackers.[9]

An arm around the shoulders would be a start, but victims of violence need more than token gestures. Without active intervention, the trauma will annex priority storage in memory. Every detail of the situation will be red-flagged for easy retrieval. The environment will become an enemy; daily life, a gauntlet of "trauma reminders"; watchful anxiety, the beginning of a circle that can only lead downward.

NIMH's ROBERT POST makes no secret of his frustration with feeble reactions to traumatic stress. "It's criminal," he fumes. "Think about what happens in a heart attack. Our whole health care system is geared toward getting that person into the hospital and starting all sorts of treatment to keep the patient's condition from progressing to a massive second heart attack or death. Now contrast that with what happens after a posttraumatic experience. We have to tackle this as a psychiatric emergency and begin emergency treatments right from the beginning."

What would an emergency treatment plan for traumatized patients look like? Experts agree that the first priority is to provide an atmosphere of safety: a calm, low-key environment, minimal stimulation, physical comforting. Some, like Post, would maintain that we can—and should—go even further. "We need to take memory formation 'off-line,' he says. "And we may be able to do that with the right kind of pharmacotherapy, designed specifically for acute treatment." He's not talking about sedating victims to calm them down. What Post and other stress researchers propose is taking advantage of what we now know about the neurochemistry of emotional memory and using it to interrupt the prioritization of the details of the traumatic experience by the amygdala—disconnecting strong emotions from strong memories. Stripped of their painful link with memory, trauma reminders cease to be menacing shadows or men with guns and shrink back to what they really are—the approaching night, a lost commuter.

A small boy sets out with his mother for a visit to the hospital where his father works as a laboratory technician.[78] Today the father has arranged a very special privilege for his son: he will be permitted to observe a practice disaster drill taking place at the hospital. The boy and his mother check the traffic, cross the busy street in front of their apartment building, and walk to the hospital, where the boy watches a surgical team practice the drill and marvels at the sophisticated machinery. The drill runs a little later than expected, so the boy stays with his father while his mother leaves to pick up his younger sister at preschool.

Now imagine that as the boy and his mother are crossing the street, he falls and is caught in a dreadful accident. Critically injured, he's rushed to the hospital, where a brain scan shows severe internal bleeding. Doctors struggle to stabilize his condition, while a skilled surgical team reattaches his severed feet. The boy's distraught mother rushes off to collect his sister, leaving his father to keep a bedside vigil.

Try to remember both stories a week from now. If you find that the second version sticks in your head more clearly than the first, you're not alone. Larry Cahill and two colleagues at the University of California–Irvine, James Mc-Gaugh and Bruce Prins, found that people who heard the emotionally charged version of the story—and saw a series of slides illustrating the narrative—remembered significantly more detail one week later than subjects who heard the neutral story.[78] But if the team treated individuals who heard the emotional story with propranolol, a drug that blocks transmission at norepinephrine synapses, an hour before they heard the story, they remembered far less of their distressing experience—about as much as subjects who heard the first, emotionally neutral version.

Propranolol did not induce blanket amnesia, wiping the slate of memory clean. Instead, it accomplished something far more subtle: it prevented emotion from tagging the memory with survival labels, so that the story was logged in as if it were a commonplace event. Blocking norepinephrine transmission deprived it of its emotional valence; as a result, the violent details of the accident lost some of their power to commandeer the learning process.

This is the sort of emergency pharmacotherapy Robert Post has in mind. It's intervention designed to block traumatic memories at their source, an experimental procedure that could allow physicians to prevent trauma from ruining a victim's life.

Prevention is a priority because once the link between environmental cues and fearful reactions solidifies, it's not easy to break. Each encounter with a relevant cue hooks onto the emotional image, recreating the terror of the original event. And each time the former victim reexperiences that fear, the emotional valence of the cue escalates even further.

When drug treatment isn't initiated until symptoms of PTSD are apparent, pharmacology has mixed results. The SSRIs that work so well in other stress response disorders—not yet widely tested in PTSD—hold promise.[79,80] The future offers even more precise control of stress responses, in the form of drugs currently under development that block or moderate the action of corticotrophin-releasing factor, the peptide that triggers the stress hormone relay.[80]

Cognitive therapy can also break the link between trauma reminders and fearful reactions. According to Edna Foa, a psychiatrist at the Medical College of Pennsylvania, CBT, combined with pharmacotherapy; other anxiety-reducing psychotherapeutic techniques, such as relaxation training, distraction, and biofeedback; or exposure therapy, popularly characterized as "reliving the trauma," corrects "the two main erroneous cognitions that underlie PTSD, that is 'the world is extremely dangerous' and 'I (the victim) am extremely incompetent.' "[81] By engaging the analytical skills of the frontal cortex, the therapist helps the patient to create new memories in place of the old, to replace the intense anxiety provoked by fearful memories with new interpretations and more rational responses.

Free Will Versus the Will to Live

The American public worries almost as much about the spiritual implications of a biological perspective on behavior as they do about the ethical and social implications. Acknowledging our physical nature makes us less than human, many argue. Others, locked into a punishment mentality, insist that linking aggression to neural pathways, brain cells, and biochemistry will allow violent offenders to evade responsibility for their actions, that acknowledging the physical impact of brutal, chaotic environments is tantamount to an "abuse excuse." When you locate behavior in the brain, or dare to call some forms of aggression "maladaptive," you might just as well issue a universal pardon.

Recognizing the biological roots of aggressive behavior does not mean tolerating outrageous acts. It does mean demanding that people take responsibility for their bodies. We already ask for biological responsibility in other situations where willful negligence has the potential to cause harm. We require some drivers to wear eyeglasses and others to move to the passenger seat because they've had too much to drink. We insist that schoolchildren with communicable diseases stay home until they're well. Restaurant workers must wash their hands. We ask people to practice safer sex.

Biology is an explanation, not an excuse. And biology is also an opportunity to set back the clock, to undo the biological damage caused by a lifetime of accumulated insults.

DOES BIOLOGY obviate free will? If by "free will" we mean the liberty to act totally independently of the laws of nature, to chart our own destiny by sheer force of will, I do not believe that we enjoy this privilege. But if free will means a choice between recognizing, accepting, and accommodating our physical limitations or choosing to ignore them, this is a freedom we certainly have. Collectively, we have the freedom to set the boundaries for acceptable aggression anywhere we choose, as well as the freedom to decide the fate of the violent: reintegration or a lifetime in exile.

We do not have to give up on morality to accept physiology. Conscience, logic, and moral reasoning can buffer the negative effects of stress, trauma, social upheaval, isolation. But our will to survive is very strong. Our brains are constructed to respond to threat, not contemplate it. In the face of a perceived challenge to survival or well-being, rapid-response circuits circumvent executive function and proceed directly from environment to emotion to response, not because they are primitive, powerful forces welling up from the depths of some long-forgotten place, but simply because they are faster. The time to invoke conscience is before—not during—a crisis. In the quiet space between emergencies, we can contemplate more appropriate responses or better ways of avoiding known threats. Reason works better when it is not under duress.

The real moral lesson of biology is humility. Neural mechanisms that shape behavior from the earliest days of development are universal human processes common to all people. This leaves the self-righteous in an uncomfortable position, for if bad behavior is shaped by natural processes, so is good behavior. Perhaps the real fear some have about a biological perspective on violence is not that violent individuals will escape punishment, but that their own virtue will go unrewarded.

EVEN AN ANIMAL indifferent to the plight of others can appreciate that peace is preferable to war. A long-tailed macaque that has been the victim of an attack can redirect aggression to a new victim. Or the aggrieved individual can reduce social stress by actively seeking out the aggressor and working to patch up the relationship. Reconciliation has advantages both for the victim and for other members of the troop. Victims who reconcile with aggressors are less likely to be attacked again than victims who redirect their aggression or victims who do nothing. In addition, they're less likely to attack another themselves. By burying the hatchet, victims not only lower their own stress levels but also maintain peace within the community.

Reconciliation and affiliation are as much a part of our behavioral repertoire as aggression. In his book *Peacekeeping Among Primates,* Frans de Waal shifts the focus of his discussion of aggression from the conflict itself to the subsequent efforts to repair the social fabric, citing dozens of examples of reconciliation among monkeys, bonobos, and chimpanzees. De Waal concludes that "making peace is as natural as making war."[82]

We are biologically capable of much more than retaliation. We can choose affiliation, and find safety and accuracy in a collective view of the world; or we can choose reconciliation and solve problems rather than perpetuate them. When we choose to deescalate, we aren't giving up, going along, or acting irresponsibly. We're simply acting naturally.

Notes

Chapter 1. Seeds of Controversy

1. Bucks County, Pennsylvania, Health Department. The exact figure is 7,320.

2. Figure courtesy of Susan Hauser, Director, A Woman's Place.

3. W. Roush, Conflict marks crime conference, *Science 269* (1995): 1808–1809.

4. Presentation to the National Advisory Mental Health Council, February 11, 1992.

5. J. Palca, NIH wrestles with furor over conference, *Science 257* (1992): 739.

6. E. Marshall, NIH told to reconsider crime meeting, *Science 262* (1993): 23–24.

7. N. Angier, Disputed meeting to ask if crime has genetic roots, *New York Times,* September 19, 1995, C1–C6.

8. D. Oaks, Reply to David Wasserman from Support Coalition Source: [e-mail message, online], September 21, 1995, available from Internet: chrp@efn.org.

9. E. R. Kandel, Brain and behavior, in *Principles of Neural Science,* ed. E. R. Kandel, J. H. Schwartz, and T. M. Jessell (Norwalk, CT: Appleton and Lange, 1991), 5–17.

10. S. Finger, *Origins of Neuroscience: A History of Explorations into Brain Function* (New York: Oxford University Press, 1994).

11. M. A. B. Brazier, *A History of Neurophysiology in the Nineteenth Century* (New York: Raven Press, 1988).

12. M. B. Sampson, *Rationale of Crime, and Its Appropriate Treatment; Being a Treatise On Criminal Jurisprudence Considered in Relation to Cerebral Organization* (New York: D. Appleton & Co., 1846.

13. C. Lombroso, Criminal man, in *The Heritage of Modern Criminology,* ed. S. F. Sylvester (Cambridge, MA: Schenkman, 1972).

14. J. Q. Wilson and R. J. Herrnstein, *Crime and Human Nature* (New York: Simon & Schuster, 1985).

15. E. Clarke and L. S. Jacyna, *Nineteenth-Century Origins of Neuroscientific Concepts* (Berkeley: University of California Press, 1987).

16. C. Darwin, The Descent of Man, and Selection in Relation to Sex, in *The Portable Darwin,* ed. D. M. Porter and P. W. Graham (New York: Penguin Books, 1993).

17. D. B. Paul, *Controlling Human Heredity: 1865 to the Present* (Atlantic Highlands, NJ: Humanities Press, 1995).

18. D. J. Kevles, *In the Name of Eugenics: Genetics and the Uses of Human Heredity* (Berkeley: University of California Press, 1985).

19. A. Roper, *Ancient Eugenics* (Oxford: Oxford University Press, 1913).

20. M. Grant, *The Passing of the Great Race*, 1916, quoted in D. Paul, *Controlling Human Heredity* (Atlantic Highlands, NJ: Humanities Press, 1995).

21. E. F. Keller, *Refiguring Life* (New York: Columbia University Press, 1995).

22. D. Johnston, Federal agents detain man who is believed to be Unabom suspect, *New York Times*, April 4, 1996, A1–B12.

23. D. Howlett, Recluse matches FBI's profile, *USA Today*, April 4, 1996, 1A–7A.

24. T. Kaczynski, Industrial society and its future, *Washington Post*, September 19, 1995.

25. F. de Waal, *Good Natured: Origins of Right and Wrong in Humans and Other Animals* (Cambridge, MA: Harvard University Press, 1996).

26. J. Archer, *The Behavioural Biology of Aggression* (Cambridge: Cambridge University Press, 1988).

27. J. P. Scott and E. Fredricson, The causes of fighting in mice and rats, *Physiology and Zoology 24* (1951): 273–309.

28. W. Craig, Why do animals fight? *International Journal of Ethics 31* (1928): 264–278.

29. N. H. Azrin, R. R. Hutchinson, and D. F. Hake, Extinction-induced aggression, *Journal of the Experimental Analysis of Behavior 9* (1966): 191–194.

30. R. E. Ulrich and N. H. Azrin, Reflexive fighting in response to aversive stimulation, *Journal of the Experimental Analysis of Behavior 5* (1962): 511–520.

31. R. E. Ulrich and B. Symannek, Pain as a stimulus for aggression, in *Aggressive Behaviour*, ed. S. Garattini and E. B. Sigg (New York: Wiley, 1969).

32. J. Alcock, *Animal Behavior: An Evolutionary Approach* (Sunderland, MA: Sinauer, 1993).

33. K. Lorenz, *On Aggression* (New York: Harcourt, Brace, & World, 1966).

34. E. O. Wilson, *Naturalist* (Washington, D.C.: Island Press, 1994).

35. R. L. Trivers, The evolution of reciprocal altruism, *Quarterly Review of Biology 46* (1971): 35–57.

36. L. A. Stevens, *Explorers of the Brain* (New York: Knopf, 1971).

37. M. A. B. Brazier, *A History of Neurophysiology in the Seventeenth and Eighteenth Centuries: From Concept to Experiment* (New York: Raven Press, 1984).

38. E. S. Valenstein, *Brain Control: A Critical Examination of Brain Stimulation and Psychosurgery* (New York: Wiley, 1973).

39. E. S. Valenstein, *Great and Desperate Cures* (New York: Basic Books, 1986).

40. H. Narabayashi, T. Nagao, and Y. Saito, Stereotaxic amygdalotomy for behavioral disorders, *Archives of Neurology 9* (1963): 1–16.

41. V. H. Mark and F. R. Ervin, *Violence and the Brain* (New York: Harper & Row, 1970).

42. V. H. Mark, W. H. Sweet, and F. R. Ervin, Role of brain disease in riots and urban violence, *Journal of the American Medical Association 201* (1967): 895.

43. J. Longman, Du Pont, accused of a killing, holds off police at his home, *New York Times*, January 27, 1996, A1–A5.

44. J. Longman, P. Belluck, and J. Nordheimer, A life in pieces; for du Pont heir, question was control, *New York Times*, February 4, 1996, A1–A12.

45. M. Bowden, Heir had long history of odd behavior, *Philadelphia Inquirer*, September 25, 1996.

46. B. Ordine and R. Vigoda, Judge finds du Pont unfit to stand trial, *Philadelphia Inquirer*, September 25, 1996, A1–A18.

47. B. Ordine and R. Vigoda, Du Pont murder trial will plumb his state of mind, *Philadelphia Inquirer*, January 19, 1997, B1–B2.

48. R. Vigoda and B. Ordine, Du Pont is not rational, lawyers say, *Philadelphia Inquirer,* September 21, 1996, 1996, A1–A4.

49. L. M. Friedman, *Crime and Punishment in American History* (New York: Basic Books, 1993).

50. *Durham v. United States,* 214 Fed. 2d 862 (1954).

51. R. B. Schmitt, Insanity pleas fail a lot of defendants as fear of crime rises, *Wall Street Journal,* February 29, 1996, A1–A8.

52. F. Butterfield, Dispute over insanity defense is revived in murder trial, *New York Times,* March 2, 1992, A6.

53. P. McLaughlin, Our Freddies, our selves, *Philadelphia Inquirer Magazine,* October 29, 1995, 23.

54. American Psychiatric Association, *Violence and Mental Illness* (Washington, D.C.: APA, 1994), 1–4.

Chapter 2. The Vicious Circle

1. S. Kauffman, *At Home in the Universe: The Search for the Laws of Self-Organization and Complexity* (New York: Oxford University Press, 1995).

2. Society for Neuroscience, Washington, D.C.

3. J. H. Schwartz and E. R. Kandel, Synaptic transmission mediated by second messengers, in *Principles of Neural Science,* ed. E. R. Kandel, J. H. Schwartz, and T. M. Jessell (Norwalk, CT: Appleton & Lange, 1991), 173–193.

4. E. F. Keller, *Refiguring Life* (New York: Columbia University Press, 1995).

5. E. R. Kandel, J. H. Schwartz, and T. M. Jessell, *Principles of Neural Science* (Norwalk, CT: Appleton and Lange, 1991).

6. T. M. Jessell and S. Schacher, Control of cell identity, in *Principles of Neural Science,* ed. E. R. Kandel, J. H. Schwartz, and T. M. Jessell (Norwalk: Appleton & Lange, 1991), 887–907.

7. S. F. Gilbert, *Developmental Biology,* 3d ed. (Sunderland, MA: Sinauer, 1991).

8. J. H. Martin and T. M. Jessell, Development as a guide to the regional anatomy of the brain, in *Principles of Neural Science,* ed. E. R. Kandel, J. H. Schwartz, and T. M. Jessell (Norwalk: Appleton and Lange, 1991), 296–308.

9. T. M. Lamb, A. K. Knecht, and W. C. Smith, Neural induction by the secreted polypeptide noggin, *Science 262* (1993): 713–718.

10. R. Levi-Montalcini, NGF: An uncharted route, in *The Neurosciences: Paths of Discovery,* ed. A. Worden (Cambridge, MA: MIT Press, 1975), 245–265.

11. S. Cohen, R. Levi-Montalcini, and V. Hamburger, A nerve-growth-stimulating factor isolated from sarcomas 37 and 180, *Proceedings of the National Academy of Sciences, USA 40* (1954): 1014–1018.

12. R. Levi-Montalcini and B. Booker, Destruction of the sympathetic ganglia in mammals by an antiserum to a nerve-growth protein, *Proceedings of the National Academy of Sciences USA 46* (1960): 384–391.

13. E. M. Johnson, P. D. Gorin, L. D. Brandeis, et al., Dorsal root ganglion neurons are destroyed by exposure in utero to maternal antibodies to nerve growth factor, *Science 210* (1980): 916–918.

14. F. Hefti, A. Dravid, and J. Hartikka, Chronic intraventricular injections of nerve growth factor elevate hippocampal choline acetyltransferase activity in adult rats with partial septo-hippocampal lesions, *Brain Research 293* (1984): 305–311.

15. W. C. Mobley, J. L. Rutkowski, G. I. Tennekoon, et al., Choline acetyltransferase ac-

tivity in striatum of neonatal rats increased by nerve growth factor, *Science 229* (1985): 284–287.

16. M. Bothwell, Functional interactions of neurotrophins and neurotrophin receptors, *Annual Review of Neuroscience 18* (1995): 223–253.

17. G. R. Lewin and Y.-A. Barde, Physiology of the neurotrophins, *Annual Review of Neuroscience 19* (1996): 289–317.

18. P. H. Patterson and L. L. Y. Chun, The induction of acetylcholine synthesis in primary cultures of dissociated rat sympathetic neurons, *Developmental Biology 56* (1977): 263–280.

19. M. Comb, S. E. Hyman, and H. M. Goodman, Mechanisms of trans-synaptic regulation of gene expression, *Trends in Neuroscience 10* (1987): 473–478.

20. S. E. Hyman and E. J. Nestler, Initiation and adaptation: A paradigm for understanding psychotropic drug action, *American Journal of Psychiatry 153* (1996): 151–162.

21. M. Sheng and M. E. Greenberg, The regulation and function of c-*fos* and other immediate early genes in the nervous system, *Neuron 4* (1990): 477–485.

22. E. R. Kandel, Nerve cells and behavior, in *Principles of Neural Science,* ed. E. R. Kandel, J. H. Schwartz, and T. M. Jessell (Norwalk, CT: Appleton & Lange, 1991), 18–32.

23. J. H. Schwartz, Chemical messengers: Small molecules and peptides, in *Principles of Neural Science,* ed. E. R. Kandel, J. H. Schwartz, and T. M. Jessell (Norwalk, CT: Appleton & Lange, 1991), 213–224.

24. B. Lu, M. Yokoyama, C. Dreyfus, et al., Depolarizing stimuli regulate nerve growth factor gene expression in cultured hippocampal neurons, *Proceedings of the National Academy of Sciences 88* (1991): 6289–6292.

25. F. Zafra, E. Castren, and H. Thoenen, Interplay between glutamate and gamma-aminobutyric acid transmitter systems in the physiological regulation of brain-derived neurotrophic factor and nerve growth factor synthesis in hippocampal neurons, *Proceedings of the National Academy of Sciences 88* (1991): 10037–10041.

26. E. S. Levine, C. Dreyfus, I. B. Black, et al., Brain-derived neurotrophic factor rapidly enhances synaptic transmission in hippocampal neurons via postsynaptic tyrosine kinase receptors, *Proceedings of the National Academy of Sciences USA 92* (1995): 8074–8077.

27. M. Clark, R. M. Post, and S. R. B. Weiss, Regional expression of c-*fos* mRNA in rat brain during the evolution of amygdala kindled seizures, *Molecular Brain Research 11* (1991): 55–64.

28. R. M. Post, S. R. B. Weiss, and M. A. Smith, Sensitization and kindling: Implications for the evolving neural substrates of posttraumatic stress disorder, in *Neurobiological and Clinical Consequences of Stress,* ed. M. J. Friedman, D. S. Charney, and A. Y. Deutch (Philadelphia: Lippincott-Raven, 1995), 203–224.

29. G. V. Goddard, D. C. McIntyre, and C. K. Leech, A permanent change in brain function resulting from daily electrical stimulation, *Experimental Neurology 25* (1969): 295–330.

30. R. Racine, Kindling: The first decade, *Neurosurgery 3* (1978): 234–252.

31. B. Fogle, *The Encyclopedia of the Dog* (New York: DK Publishing, 1995).

32. D. L. Cheney and R. M. Seyfarth, *How Monkeys See the World: Inside the Mind of Another Species* (Chicago: University of Chicago Press, 1990).

33. M. J. Tovee, What are faces for? *Current Biology 5* (1995): 480–482.

34. T. M. Field, Social perception and responsivity in early infancy, in *Review of Human Development,* ed. T. M. Field, A. Huston, H. C. Quay, L. Troll, and G. E. Finley (New York: Wiley 1982), 20–31.

35. E. Z. Tronick, Emotions and emotional communication in infants, *American Psychologist 44* (1989): 112–119.

36. A. Fogel, Peer versus mother directed behavior in one- to three-month-old infants, *Behavioral Development 2* (1980): 215–226.

37. J. M. Da Costa, On irritable heart: A clinical study of functional cardiac disorder and its consequences, *American Journal of Medical Science 61* (1871): 17.

38. A. Kardiner, *The Traumatic Neuroses of War* (New York: Hoeber, 1941).

39. R. A. Kulka, W. E. Schlenger, J. A. Fairbank, et al., *Trauma and the Vietnam War Generation: Report of Findings from the National Vietnam Veterans Readjustment Study* (New York: Brunner/Mazel, 1990).

40. Centers for Disease Control Vietnam Experience Study, Health status of Vietnam veterans: I. Psychosocial characteristics, *Journal of the American Medical Association 259* (1988): 2701–2707.

41. J. Goldberg, W. R. True, S. A. Eisen, et al., A twin study of the effects of the Vietnam War on posttraumatic stress disorder, *Journal of the American Medical Association 263* (1990): 1227–1232.

42. R. C. Kessler, K. A. McGonagle, Z. Shanyang, et al., Lifetime and 12-month prevalence of DSM-III-R psychiatric disorders in the United States, *Archives of General Psychiatry 51* (1994): 8–18.

43. American Psychiatric Association, *Diagnostic and Statistical Manual of Mental Disorders,* 4th ed. (Washington, D.C.: American Psychiatric Association, 1994).

44. A. C. McFarlane, The prevalence and longitudinal course of PTSD: Implications for the neurobiological models of PTSD, *Annals of the New York Academy of Sciences 821* (1997): 10–23.

Chapter 3. From Wilderness to Lab Bench

1. F. de Waal, *Peacemaking Among Primates* (Cambridge, MA: Harvard University Press, 1989).

2. Yerkes Regional Primate Research Center.

3. J. Goodall, *The Chimpanzees of Gombe: Patterns of Behavior* (Cambridge, MA: Belknap Press of Harvard University Press, 1986).

4. W. K. Stevens, Logging sets off an apparent chimp war, *New York Times,* May 13, 1997, C3.

5. F. de Waal, *Good Natured: Origins of Right and Wrong in Humans and Other Animals* (Cambridge, MA: Harvard University Press, 1996).

6. F. Aureli and C. P. van Schaik, Post-conflict behavior in long-tailed macaques (*Macaca fascicularis*): I. The social events, *Ethology 89* (1991): 89–100.

7. F. Aureli and C. P. van Schaik, Post-conflict behavior in long-tailed macaques (*Macaca fascicularis*): II. Coping with the uncertainty, *Ethology 89* (1991): 101–114.

8. F. Aureli, C. P. van Schaik, and J. A. R. A. M. van Hooff, Functional aspects of reconciliation among captive long-tailed macaques (*Macaca fascicularis*), *American Journal of Primatology 19* (1989): 39–51.

9. S. A. Barnett, An analysis of social behavior in wild rats, *Proceedings of the Zoological Society of London 130* (1958): 107–152.

10. R. J. Blanchard and D. C. Blanchard, Aggressive behavior in the rat, *Behavioral Biology 21* (1977): 197–224.

11. R. J. Blanchard, D. C. Blanchard, L. K. Takahashi, et al., Attack and defensive behavior in the albino rat, *Animal Behavior 25* (1977): 622–634.

12. S. A. Barnett, Grouping and dispersive behaviour among wild rats, in *Aggressive Behavior: Proceedings of an International Symposium on the Biology of Aggressive Behaviour, Held at*

the Istituto di Ricerche Farmacologiche "Mario Negri," Milan, May 2–4, 1968, ed. S. Garattini and E. B. Sigg (New York: Wiley, 1969), 3–14.

13. S. A. Barnett, *The Rat: A Study in Behavior* (Chicago: Aldine, 1963).

14. P. J. Mitchell, Prediction of antidepressant activity from ethological analysis of agonistic behavior in rats, in *Ethology and Psychopharmacology,* ed. S. J. Cooper and C. A. Hendrie (New York: Wiley, 1994), 85–109.

15. E. C. Grant, An analysis of the social behaviour of the male laboratory rat, *Behaviour 21* (1963): 260–281.

16. E. C. Grant and M. R. A. Chance, Rank order in caged rats, *Animal Behavior 6* (1958): 183–194.

17. E. C. Grant and J. H. Mackintosh, A comparison of the social behavior of some common laboratory rodents, *Behaviour 21* (1963): 246–259.

18. A. P. Silverman, Ethological and statistical analysis of drug effects on the social behaviour of laboratory rats, *British Journal of Pharmacology 24* (1965): 579–590.

19. K. A. Miczek, E. Weerts, M. Haney, et al., Neurobiological mechanisms controlling aggression: Preclinical developments for pharmacotherapeutic interventions, *Neuroscience and Biobehavioral Reviews 18* (1994): 97–110.

20. K. A. Miczek, M. Haney, J. Tidey, et al., Temporal and sequential patterns of agonistic behavior: Effects of alcohol, anxiolytics, and psychomotor stimulants, *Psychopharmacology 97* (1989): 149–151.

21. J. Knipe-Brown, No charges filed against woman who killed spouse, *Doylestown (Pa.) Intelligencer,* 1994, 1.

22. A. Bell, Killing husband was desperate attempt to end abuse, *Bucks County Courier Times,* April 28, 1992, 1A–2A.

23. J. Gross, Abused women who kill now seek way out of cells, *New York Times,* September 15, 1992, 8.

24. J. M. Klein, The fate of women who strike back, *Philadelphia Inquirer,* May 2, 1992, 1A–6A.

25. L. M. Friedman, *Crime and Punishment in American History* (New York: Basic Books, 1993).

26. A. Browne, *When Battered Women Kill* (New York: The Free Press, 1987).

27. Pennsylvania Coalition Against Domestic Violence, health statistics for abused women, 1990.

28. R. J. Blanchard, Pain and aggression reconsidered, in *Biological Perspectives on Aggression,* ed. R. J. Blanchard, D. C. Blanchard, and K. J. Flannelly (New York: Alan R. Liss, 1984), 1–26.

29. R. J. Blanchard, D. C. Blanchard, and L. K. Takahashi, Reflexive fighting in the albino rat: Aggressive or defensive behavior? *Aggressive Behavior 3* (1977): 145–155.

30. J. E. LeDoux, Emotion, memory, and the brain, *Scientific American 270* (1994): 32–39.

31. J. E. LeDoux, Information flow from sensation to emotion: Plasticity in the neural computation of stimulus value, in *Neuroscience: Foundation of Adaptive Networks,* ed. M. Gabriel and J. Moore (Cambridge, MA: MIT Press, 1990), 3–51.

32. J. E. LeDoux, In search of an emotional system in the brain: Leaping from fear to emotion and consciousness, Paper presented at the McDonnell-Pew Conference on Cognitive Neuroscience, Lake Tahoe, CA, July 1993.

33. B. A. Campbell and J. Jaynes, Reinstatement, *Psychology Reviews 73* (1966): 478–480.

34. R. J. Blanchard, D. C. Blanchard, and K. J. Flannelly, Social stress, mortality and aggression in colonies and burrowing habitats, *Behavioral Processes 11* (1985): 209–215.

35. D. C. Blanchard, Behavioral correlates of chronic dominance-subordination relationships of male rats in a seminatural situation, *Neuroscience and Biobehavioral Reviews* (1990): 455–462.

36. D. C. Blanchard, R. R. Sakai, and B. McEwen, Subordination stress: Behavioral, brain, and neuroendocrine correlates, *Behavioral Brain Research 58* (1993): 113–121.

37. D. C. Blanchard, R. L. Spencer, and S. M. Weiss, Visible burrow system as a model of chronic social stress: Behavioral and neuroendocrine correlates, *Psychoneuroendocrinology 20* (1995): 117–134.

38. C. R. McKittrick, D. C. Blanchard, and R. J. Blanchard, Serotonin receptor binding in a colony model of chronic social stress, *Biological Psychiatry 37* (1995): 383–393.

39. R. Yehuda and S. M. Antelman, Criteria for rationally evaluating animal models of posttraumatic stress disorder, *Biological Psychiatry 33* (1993): 479–486.

40. R. M. Post, S. R. B. Weiss, and M. A. Smith, Sensitization and kindling: Implications for the evolving neural substrates of posttraumatic stress disorder, in *Neurobiological and Clinical Consequences of Stress*, ed. M. J. Friedman, D. S. Charney, and A. Y. Deutch (Philadelphia: Lippincott-Raven, 1995), 203–224.

41. R. M. Post, S. R. B. Weiss, and M. Smith, Kindling versus quenching, *Annals of the New York Academy of Sciences 821* (1997): 285–304.

42. B. A. Sorg and P. W. Kalivas, Stress and neuronal sensitization, in *Neurobiological and Clinical Consequences of Stress*, ed. M. J. Friedman, D. S. Charney, and A. Y. Deutch (Philadelphia: Lippincott-Raven, 1997), 83–123.

43. R. S. Pynoos, R. F. Ritzmann, and A. M. Steinberg, A behavioral animal model of posttraumatic stress disorder featuring repeated exposure to situational reminders, *Biological Psychiatry 39* (1996): 129–134.

44. K. E. Moyer, *The Physiology of Hostility* (Chicago: Markham, 1971).

45. K. E. Moyer, Kinds of aggression and their physiological basis, *Communications in Behavioral Biology 2* (1968): 65–87.

46. J. Archer, *The Behavioural Biology of Aggression* (Cambridge: Cambridge University Press, 1988).

47. D. Reis, Brain monoamines in aggression and sleep, *Clinical Neurosurgery 18* (1971): 471–502.

48. A. E. Pusey and C. Packer, Infanticide in lions, in *Infanticide and Parental Care*, ed. S. Parmigiani, F. vom Saal, and B. Svare (London: Harwood Academic Press, 1992).

49. J. Goodall, Life and death at Gombe, *National Geographic* (May 1979): 592–621.

50. R. I. Simon, *Bad Men Do What Good Men Dream* (Washington, D.C.: American Psychiatric Press, 1996).

Chapter 4. All the Right Connections

1. I. McHarg, *Design with Nature* (Reprint, New York: Wiley, 1992 [1969]).

2. Library Company of Philadelphia, *Philadelphia Almanac and Citizen's Manual* (Philadelphia: Library Company of Philadelphia, 1993).

3. M. Matza and D. Rubin, Sadness and fear visit Penn on the heels of homicide, *Philadelphia Inquirer*, November 3, 1996, A1–A8.

4. L. Sitton and M. Matza, Head of Penn tries to reassure parents, *Philadelphia Inquirer*, October 3, 1996, R1–R6. In 1996, the University of Pennsylvania added a campuswide electronic surveillance system to its defenses, in response to an eight-week crime spree in which a university biochemist was stabbed and killed while protecting his girlfriend, a twenty-one-year-old undergraduate was shot, and more than thirty armed robberies took place on or near

the campus. The amount spent on security measures would pay the annual tuition bill of nearly five hundred students.

5. R. F. Weigley, ed, *Philadelphia: A 300-Year History* (New York: Norton, 1982).

6. A man who did not shrink from controversy, Ferrier was a born advocate. Pressed to accede more credit for the discovery of the motor cortex to Fritsch and Hitzig (see Chapter 1), he stood his ground against no less a scientific figure than Thomas Huxley; prosecuted—unsuccessfully—under the Cruelty to Animals Act, he became one of the first scientists to confront the growing British antivivisection movement.

7. M. A. B. Brazier, *A History of Neurophysiology in the Nineteenth Century* (New York: Raven Press, 1988).

8. S. Finger, *Origins of Neuroscience: A History of Explorations into Brain Function* (New York: Oxford University Press, 1994).

9. R. A. Poldrak and J. D. Gabrieli, Functional anatomy of long-term memory, *Journal of Clinical Neurophysiology 14* (1997): 294–310.

10. R. Ivry, Cerebellar timing systems, *International Review of Neurobiology 41* (1997): 555–573.

11. J. A. Fiez, Cerebellar contributions to cognition, *Neuron 16* (1996): 13–15.

12. M. A. B. Brazier, *A History of Neurophysiology in the Seventeenth and Eighteenth Centuries: From Concept to Experiment* (New York: Raven Press, 1984).

13. P. Broca, Anatomie comparée des circonvolutions cérébrales. Le grand lobe limbique et la scissure limbique dans la série des mammifères, *Revue d'Anthropologie ser. 2, 1* (1878): 385–498.

14. J. W. Papez, A proposed mechanism of emotion, *Archives of Neurology and Psychiatry 38* (1937): 725–743.

15. P. D. MacLean, *The Triune Brain in Evolution: Role in Paleocerebral Functions* (New York: Plenum, 1990).

16. P. D. MacLean, Some psychiatric implications of physiological studies on the frontotemporal portion of the limbic system (visceral brain), *Electroencephalography and Clinical Neurophysiology 4* (1952): 407–418.

17. I. Kupfermann, Hypothalamus and limbic system: Peptidergic neurons, homeostasis, and emotional behavior, in *Principles of Neural Science,* ed. E. R. Kandel, J. H. Schwartz, and T. M. Jessell (East Norwalk, CT: Appleton and Lange, 1991), 735–749. Cannon formulated the concept of *homeostasis,* the coordination of physiological processes to maintain an internal equilibrium, as part of his work on the autonomic nervous system (see Chapter 6). The sympathetic division of this system is responsible for the well-known "fight-or-flight" reaction—the combined physiological and behavioral responses to danger.

18. A. Brodal, *Neurological Anatomy in Relation to Clinical Medicine* (New York: Oxford University Press, 1981).

19. L. W. Hamilton, *Basic Limbic System Anatomy of the Rat* (New York: Plenum Press, 1976).

20. W. R. Hess, *Biological Order and Brain Organization* (New York: Springer-Verlag, 1981).

21. L. A. Stevens, *Explorers of the Brain* (New York: Knopf, 1971).

22. M. Wasman and J. P. Flynn, Directed attack elicited from hypothalamus, *Archives of Neurology 6* (1962): 60–67.

23. C. C. Chi and J. P. Flynn, Neural pathways associated with hypothalamically elicited behavior in cats, *Science 171* (1971): 703–706.

24. A. Siegel and C. B. Pott, Neural substrates of aggression and flight in the cat, *Progress in Neurobiology 31* (1988): 261–283.

25. A. Siegel and H. M. Edinger, Role of the limbic system in hypothalamically elicited attack behavior, *Neuroscience and Biobehavioral Reviews 7* (1983): 395–407.

26. A. F. Mirsky and A. Siegel, The neurobiology of violence and aggression, in *Understanding and Preventing Violence,* ed. A. J. Reiss, Jr., K. A. Miczek, and J. A. Roth (Washington, D.C.: National Academy Press, 1994), 59–172.

27. S. A. G. Fuchs and A. Siegel, The organization of hypothalamic pathways mediating affective defense behavior in the cat, *Brain Research 330* (1985): 77–92.

28. J. de Olmos, G. F. Alheid, and C. A. Beltramino, Amygdala, in *The Rat Nervous System,* ed. G. Paxinos (Orlando, FL: Academic Press, 1985), 223–334.

29. J. E. Krettek and J. L. Price, Amygdaloid projections to subcortical structures within the basal forebrain and brainstem in the rat and cat, *Journal of Comparative Neurology 178* (1978):225–254.

30. M. Carpenter, *Human Neuroanatomy,* 7th ed. (Baltimore: Williams & Wilkins, 1976).

31. E. C. Crosby and T. Humphrey, Studies of the vertebrate telencephalon. II. The nuclear pattern of the anterior olfactory nucleus, tuberculum olfactorium and the amygdaloid complex in adult man, *Journal of Comparative Neurology 74* (1941): 309–352.

32. G. F. Alheid, G. W. Van Hoesen, and L. Heimer, Functional neuroanatomy, in *Comprehensive Textbook of Psychiatry,* ed. H. I. Kaplan and B. J. Sadock (Baltimore: Williams & Wilkins, 1989). In the anatomical literature, the basal nucleus is often subdivided into a large-celled lateral portion called, confusingly enough, the "basolateral nucleus" and a small-celled portion closer to the midline of the brain, known as the "basomedial nucleus."

33. J. E. LeDoux, Emotional memory systems in the brain, *Behavioral Brain Research 58* (1993): 69–79.

34. J. E. LeDoux, Brain mechanisms of emotion and emotional learning, *Current Opinion in Neurobiology 2* (1992): 191–197.

35. J. E. LeDoux, Cognitive-emotional interactions in the brain, *Cognition and Emotion 3* (1989): 267–289.

36. D. G. Amaral, J. L. Price, A. Pitkanen, et al., Anatomical organization of the primate amygdaloid complex, in *The Amygdala: Neurobiological Aspects of Emotion, Memory, and Mental Dysfunction,* ed. J. P. Aggleton (New York: Wiley-Liss, 1992), 1–66.

37. J. E. LeDoux, The amygdala: Contributions to fear and stress, *Seminars in the Neurosciences 6* (1994): 231–237.

38. E. Halgren, Emotional neurophysiology of the amygdala within the context of human cognition, in *The Amygdala: Neurobiological Aspects of Emotion, Memory, and Mental Dysfunction,* ed. A. P. Aggleton (New York: Wiley-Liss, 1992), 191–228.

39. F. de Waal, *Peacemaking Among Primates* (Cambridge, MA: Harvard University Press, 1989).

40. S. Brown and E. A. Schäfer, An investigation into the functions of the occipital and temporal lobes of the monkey's brain, *Philosophical Transactions of the Royal Society of London (Biology) 179* (1888): 303–327.

41. H. Klüver and P. C. Bucy, "Psychic blindness" and other symptoms following bilateral temporal lobectomy in rhesus monkeys, *American Journal of Physiology 119* (1937): 352–353.

42. H. Klüver and P. C. Bucy, Preliminary analysis of functions of the temporal lobes in monkeys, *Archives of Neurology and Psychiatry 42* (1939): 979–1000.

43. H. Klüver, Functional differences between the occipital and temporal lobes with special reference to the interrelations of behavior and extracerebral mechanisms, in *Cerebral Mechanisms in Behavior,* ed. A. Jeffress (New York: Wiley, 1951), 150–160.

44. L. Weiskrantz, Behavioral changes associated with ablation of the amygdaloid complex in monkeys, *Journal of Comparative Physiology and Psychology 4* (1956): 381–391.

45. A. S. Kling and L. A. Brothers, The amygdala and social behavior, in *The Amygdala: Neurobiological Aspects of Emotion, Memory, and Mental Dysfunction,* ed. J. P. Aggleton (New York: Wiley-Liss, 1992).

46. J. E. LeDoux, Emotion and the amygdala, in *The Amygdala: Neurobiological Aspects of Emotion, Memory, and Mental Dysfunction,* ed. J. P. Aggleton (New York: Wiley-Liss, 1992).

47. A. Kling, J. Lancaster, and J. Benitone, Amygdalectomy in the free-ranging vervet (*Cercopithecus aethiops*), *Journal of Psychiatric Research 7* (1970): 191–199.

48. H. Terzian and G. D. Ore, Syndrome of Klüver and Bucy reproduced in man by bilateral removal of the temporal lobes, *Neurology 5* (1953): 373–380.

49. E. T. Rolls, Neurophysiology and functions of the primate amygdala, in *The Amygdala: Neurobiological Aspects of Emotion, Memory, and Mental Dysfunction,* ed. J. P. Aggleton (New York: Wiley-Liss, 1992), 143–165.

50. I. Kupfermann, Localization of higher cognitive and affective functions: The association cortices, in *Principles of Neural Science,* ed. E. R. Kandel, J. H. Schwartz, and T. M. Jessell (Norwalk, CT: Appleton & Lange, 1991), 823–838.

51. A. R. Damasio, *Descartes' Error* (New York: Avon Books, 1994).

52. J. M. Harlow, Recovery from the passage of an iron bar through the head, *Bulletin of the Massachusetts Medical Society 2* (1868): 3–20.

53. H. Damasio, T. Grabowski, R. Frank, et al., The return of Phineas Gage: The skull of a famous patient yields clues about the brain, *Science 264* (1994): 1102–1105.

54. J. E. LeDoux, Information flow from sensation to emotion: Plasticity in the neural computation of stimulus value, in *Neuroscience: Foundation of Adaptive Networks,* ed. M. Gabriel and J. Moore (Cambridge, MA: MIT Press, 1990), 3–51.

55. J. E. LeDoux, Emotion, memory, and the brain, *Scientific American 270* (1994): 32–39.

56. T. W. Jarrell, C. G. Gentile, L. M. Romanski, et al., Involvement of cortical and thalamic auditory regions in retention of differential bradycardia conditioning to acoustic conditioned stimuli in rabbits, *Brain Research* (1987): 285–294.

57. University of Pennsylvania, David Mahoney Institute of Neurological Sciences.

58. E. R. Kandel, Brain and behavior, in *Principles of Neural Science,* ed. E. R. Kandel, J. H. Schwartz, and T. M. Jessell (Norwalk, CT: Appleton and Lange, 1991), 5–17.

59. A. G. Reeves and F. Plum, Hyperphagia, rage, and dementia accompanying a ventromedial hypothalamic neoplasm, *Archives of Neurology 20* (1969): 616–624.

60. D. Tranel and B. T. Hyman, Neuropsychological correlates of bilateral amygdala damage, *Archives of Neurology 47* (1990): 349–355.

61. J. Martinius, Homicide of an aggressive adolescent boy with right temporal lesion: A case report, *Neuroscience and Biobehavioral Reviews 7* (1983): 419–422.

62. J. H. Martin, J. C. M. Brust, and S. Hilal, Imaging the living brain, in *Principles of Neural Science,* ed. E. R. Kandel, J. J. Schwartz, and T. M. Jessell (Norwalk, CT: Appleton and Lange, 1991), 309–324.

63. T. A. Ketter, M. S. George, and T. A. Kimbrell, Functional brain imaging, limbic function, and affective disorders, *Neuroscientist 2* (1996): 55–65.

64. N. D. Volkow and L. Tancredi, Current and future applications of SPECT in clinical psychiatry, *Journal of Clinical Psychiatry 53* (1992): 26–28. In contrast to PET, SPECT uses a radioactive tracer labeled with a gamma-ray-emitting isotope. Used most often to chart cerebral blood flow, SPECT studies carried out with labeled psychoactive drugs as tracers can also image neurotransmitter receptors or track the regional uptake and breakdown of the drug.

65. J. W. Prichard, The nuclear magnetic resonance revolution in basic and clinical neuroscience, *Neuroscientist 1* (1995): 84–94.

66. D. Steinberg, Brain imaging assumes greater power, precision, *The Scientist 12* (1998): 1–5.

67. N. D. Volkow and L. Tancredi, Neural substrates of violent behavior: A preliminary study with position emission tomography, *British Journal of Psychiatry 151* (1987): 673–688.

68. A. Raine, *The Psychopathology of Crime* (New York: Academic Press, 1993).

69. J. D. Bremner, T. M. Scott, R. C. Delaney, et al., Deficits in short-term memory in posttraumatic stress disorder, *American Journal of Psychiatry 150* (1993): 1015–1019.

70. J. D. Bremner, P. Randall, and T. M. Scott, MRI-based measurement of hippocampal volume in patients with combat-related posttraumatic stress disorder, *American Journal of Psychiatry 152* (1995): 973–981.

71. A second MRI study of Vietnam veterans (conducted by Roger Pitman and coworkers at the Veterans Administration Research Center in Manchester, New Hampshire) reported an even greater reduction in hippocampal volume—up to 26 percent on the left side of the brain and 22 percent on the right. The magnitude of the reduction was directly correlated with the severity of PTSD symptoms and the traumatic intensity of the combat experience; those with the most hippocampal damage had the most severe symptoms. See T. G. Gurvits, M. R. Shenton, H. Hokama, et al., Magnetic resonance imaging study of hippocampal volume in chronic, combat-related posttraumatic stress disorder, *Biological Psychiatry 40* (1996): 1091–1099.

72. R. C. Delaney, A. J. Rosen, R. H. Mattson, et al., Memory function in focal epilepsy: A comparison of nonsurgical, unilateral temporal lobe, and frontal lobe samples, *Cortex 16* (1980): 103–117.

73. M. N. Starkman, S. S. Gebraski, S. Berent, et al., Hippocampal formation volume, memory dysfunction, and cortisol levels on patients with Cushing's syndrome, *Biological Psychiatry 32* (1992): 756–765.

74. J. D. Bremner, P. Randall, and E. Vermetten, Magnetic resonance imaging–based measurement of hippocampal volume in posttraumatic stress disorder related to childhood physical and sexual abuse—A preliminary report, *Biological Psychiatry 41* (1997): 23–32.

75. M. B. Stein, C. Hanna, and C. Koverola, Structural brain changes in PTSD, *Annals of the New York Academy of Sciences 821* (1997): 76–98.

76. M. B. Stein, C. Koverola, C. Hanna, et al., Hippocampal volume in women victimized by childhood sexual abuse, *Psychology and Medicine 27* (1997): 951–959. Bremner's studies found a number of differences between abuse victims and combat veterans. In the veterans, the observed reduction in hippocampal volume was greater on the right side of the brain; among the abuse victims, the reduction was greater on the left. The volume of the left temporal lobe as a whole was actually increased in abused patients relative to control subjects—an effect not noted in the veteran sample—and short-term memory deficits in this population were not correlated with the extent of hippocampal damage observed on their MRI scans. Bremner and his colleagues suggest that these differences may reflect differential effects of trauma experienced in childhood (abuse victims) or as an adult (combat veterans), as well as developmental differences in the ability of the brain to compensate for stress-related damage.

Both Bremner and Steiner acknowledge that the changes in hippocampal volume may have preceded, rather than followed, the traumatic event. However, as explained in more detail in Chapter 6, the theory that trauma leads to neuronal damage is consistent with a growing body of literature documenting the unusual vulnerability of the hippocampus to excessive levels of stress hormones.

77. Oxyhemoglobin is the fresh-from-the-lungs, oxygen-delivering form of the blood protein hemoglobin; deoxyhemoglobin is the naked variant that has been stripped of its oxygen in the capillary beds.

78. J. Glanz, Magnetic brain imaging traces a stairway to memory, *Science 280* (1998): 37.

Chapter 5. Bad Chemistry

1. U.S. Postal Service data.

2. B. Falck, N. A. Hillarp, G. Thieme, et al., Fluorescence of catechol amines and related compounds condensed with formaldehyde, *Journal of Histochemistry and Cytochemistry 10* (1962): 348–354.

3. B. Falck, Observations on the possibilities of the cellular localization of monoamines by a fluorescence method, *Acta Physiologica Scandinavia 56,* Suppl. 197 (1962): 3–25.

4. J. R. Cooper, F. E. Bloom, and R. H. Roth, *The Biochemical Basis of Neuropharmacology* (New York: Oxford University Press, 1996).

5. L. W. Role and J. P. Kelly, The brain stem: Cranial nerve nuclei and the monoaminergic systems, in *Principles of Neural Science,* ed. E. R. Kandel, J. H. Schwartz, and T. M. Jessell (Norwalk, CT: Appleton and Lange, 1991), 683–699.

6. R. Y. Moore and F. E. Bloom, Central catecholamine neuron systems: Anatomy and physiology of the norepinephrine and epinephrine systems, *Annual Review of Neuroscience 2* (1979): 113–168.

7. J. H. Schwartz and E. R. Kandel, Synaptic transmission mediated by second messengers, in *Principles of Neural Science,* ed. E. R. Kandel, J. H. Schwartz, and T. M. Jessell (Norwalk, CT: Appleton & Lange, 1991), 173–193.

8. The sympathetic nervous system is one component of the autonomic nervous system, which exercises involuntary control over bodily functions, such as heart rate, blood pressure, breathing, gastrointestinal function, and metabolism. Its counterpart, the *parasympathetic nervous system,* balances sympathetic activation with an equal and opposite reaction designed to restore the status quo. In this "rest and digest" state, coordinated by the neurotransmitter acetylcholine, cardiovascular parameters return to basal values, intestinal motility and digestion resume, and blood glucose stabilizes.

9. Bob Martin, U.S. Postal Inspection Service.

10. G. Aston-Jones, C. Chiang, and T. Alexinsky, Discharge of noradrenergic locus coeruleus neurons in behaving rats and monkeys suggests a role in vigilance, *Progress in Brain Research 88* (1991): 501–520.

11. M. J. Zigmond, J. M. Finlay, and A. F. Sved, Neurochemical studies of central noradrenergic responses to acute and chronic stress, in *Neurobiological and Clinical Consequences of Stress,* ed. M. J. Friedman, D. S. Charney, and A. Y. Deutch (Philadelphia: Lippincott-Raven, 1995), 45–60.

12. G. Aston-Jones and F. E. Bloom, Norepinephrine-containing locus coeruleus neurons in behaving rats exhibit pronounced responses to non-noxious environmental stimuli, *Journal of Neuroscience 1* (1981): 887–900.

13. E. D. Abercrombie and B. L. Jacobs, Single-unit response of noradrenergic neurons in the locus coeruleus of freely moving cats. I. Acutely presented stressful and nonstressful stimuli, *Journal of Neuroscience 7* (1987): 2837–2843.

14. M. A. Pezzone, W. Lee, G. E. Hoffman, et al., Activation of brainstem catecholaminergic neurons by conditioned and unconditioned aversive stimuli as revealed by c-*fos* immunoreactivity, *Brain Research 608* (1993): 310–318.

15. P. J. Gresch, A. F. Sved, and M. J. Zigmond, Stress-induced sensitization of dopamine and norepinephrine efflux in medial prefrontal cortex of the rat, *Journal of Neurochemistry 63* (1994): 575–583.

16. T. R. Kosten, J. W. Mason, and E. L. Giller, Sustained urinary norepinephrine and epinephrine elevation in posttraumatic stress disorder, *Psychoneuroendocrinology 12* (1987): 13–20.

17. R. Yehuda, S. M. Southwick, and E. L. Giller, Urinary catecholamine excretion and severity of PTSD symptoms in Vietnam combat veterans, *Journal of Nervous and Mental Disorders 180* (1992): 321–325.

18. S. M. Southwick, R. Yehuda, and C. A. Morgan, Clinical studies of neurotransmitter alterations in post-traumatic stress disorder, in *Neurobiological and Clinical Consequences of Stress,* ed. M. J. Friedman, D. S. Charney, and A. Y. Deutch (Philadelphia: Lippincott-Raven, 1995), 335–349.

19. J. H. Krystal, A. L. Bennett, and J. D. Bremner, Toward a cognitive neuroscience of dissociation and altered memory functions in post-traumatic stress disorder, in *Neurobiological and Clinical Consequences of Stress,* ed. M. J. Friedman, D. S. Charney, and A. Y. Deutch (Philadelphia: Lippincott-Raven, 1995), 239–269.

20. S. M. Southwick, J. H. Krystal, and C. A. Morgan, Abnormal noradrenergic function in posttraumatic stress disorder, *Archives of General Psychiatry 50* (1993): 266–274.

21. B. A. van der Kolk and R. E. Fisler, The biologic basis of posttraumatic stress, *Primary Care 20* (1993): 417–432.

22. M. Davis, D. Walker, and Y. Lee, Roles of the amygdala and bed nucleus of the stria terminalis in fear and anxiety measured with the acoustic startle reflex, *Annals of the New York Academy of Sciences 821* (1997): 305–331.

23. A. M. Rasmusson and D. S. Charney, Animal models of relevance to PTSD, *Annals of the New York Academy of Sciences 821* (1997): 332–351.

24. C. A. Morgan, C. Grillon, S. M. Southwick, et al., Fear-potentiated startle in post-traumatic stress disorder, *Biological Psychiatry 38* (1995): 78–85.

25. B. Roozendaal, G. L. Quirarte, and J. L. McGaugh, Stress-activated hormonal systems and the regulation of memory storage, *Annals of the New York Academy of Sciences 821* (1997): 247–258.

26. Mark Saunders, U.S. Postal Service Office of Corporate Relations.

27. Paul Griffo, Department of Public Affairs, U.S. Postal Inspection Service.

28. B. Eichelman and N. B. Thoa, The aggressive monoamines, *Biological Psychiatry 6* (1973): 143–164.

29. F. Lamprecht, B. Eichelman, and N. B. Thoa, Rat fighting behavior: Serum dopamine-β-hydroxylase and hypothalamic tyrosine hydroxylase, *Science 177* (1972): 1214–1215.

30. J. A. Barrett, H. Edinger, and A. Siegel, Intrahypothalamic injections of norepinephrine facilitate feline affective aggression via alpha2-adrenoreceptors, *Brain Research 525* (1990): 285–293.

31. J. A. Barrett, M. B. Shaikh, H. Edinger, et al., The effects of intrahypothalamic injections of norepinephrine upon affective defense behavior in the cat, *Brain Research 426* (1987): 381–384.

32. K. Modigh, Effects of isolation and fighting in mice on the rate of synthesis of noradrenaline, dopamine, and 5-hydroxytryptamine in the brain, *Psychopharmacologia 33* (1973): 1–17.

33. B. Eichelman, Aggressive behavior: From laboratory to clinic: Quo vadit? *Archives of General Psychiatry 49* (1992): 488–492.

34. K. A. Miczek and J. F. DeBold, Hormone-drug interactions and their influence on aggressive behavior, in *Hormones and Aggressive Behavior*, ed. B. B. Svare (New York: Plenum Press, 1983).

35. K. A. Miczek, J. F. DeBold, and M. L. Thompson, Pharmacological, hormonal, and behavioral manipulations in the analysis of aggressive behavior, in *Ethopharmacological Aggression Research*, ed. K. A. Miczek, M. R. Kruk, and B. Olivier (New York: Liss, 1984).

36. G. L. Brown, F. K. Goodwin, J. C. Ballenger, et al., Aggression in humans correlates with cerebrospinal fluid amine metabolites, *Psychiatry Research 1* (1979): 131–139.

37. D. Woodman, J. Hinton, and M. O'Neill, Relationship between violence and catecholamines, *Perceptual and Motor Skills 45* (1977): 702.

38. D. Woodman and J. Hinton, Catecholamine imbalance during stress anticipation: An abnormality in maximum security hospital patients, *Journal of Psychosomatic Research 22* (1978): 477–483.

39. D. Woodman and J. Hinton, Anomalies of cyclic AMP excretion in some abnormal offenders, *Biological Psychiatry 7* (1978): 103–108.

40. F. A. Elliott, Propranolol for the control of belligerent behavior following acute brain damage, *Annals of Neurology 1* (1977): 489–491.

41. K. A. Miczek, M. Haney, J. Tidey, et al., Neurochemistry and pharmacotherapeutic management of aggression and violence, in *Understanding and Preventing Violence*, ed. A. J. Reiss, Jr., K. A. Miczek, and J. A. Roth (Washington, D.C.: National Academy Press, 1994), 245–514.

42. K. A. Miczek, The psychopharmacology of aggression, in *Handbook of Psychopharmacology*, Vol. 19, ed. L. L. Iversen, S. D. Iversen, and S. H. Snyder (New York: Plenum Press, 1987).

43. L. P. Karper and J. H. Krystal, Pharmacotherapy of violent behavior, in *Handbook of Antisocial Behavior*, ed. D. M. Stoff, J. Breiling, and J. D. Maser (New York: Wiley, 1997).

44. M. Virkkunen, A. Nuutila, F. K. Goodwin, et al., Cerebrospinal fluid monoamine metabolites in male arsonists, *Archives of General Psychiatry 44* (1987): 241–247.

45. American Psychiatric Association, *Diagnostic and Statistical Manual of Mental Disorders* (Washington, D.C.: American Psychiatric Association, 1994).

46. R. I. Simon, *Bad Men Do What Good Men Dream* (Washington, D.C.: American Psychiatric Press, 1996).

47. A. Raine, Autonomic nervous system activity and violence, in *Aggression and Violence*, ed. D. M. Stoff and R. B. Cairns (Hillsdale, NJ: Erlbaum, 1996), 145–168.

48. A. Raine, *The Psychopathology of Crime* (New York: Academic Press, 1993).

49. A. Raine, Autonomic nervous system factors underlying disinhibited, antisocial, and violent behavior, *Annals of the New York Academy of Sciences 794* (1996): 46–59.

50. A. Raine, P. H. Venables, and M. Williams, Relationships between central and autonomic measures of arousal at age 15 years and criminality at age 24 years, *Archives of General Psychiatry 47* (1990): 1003–1007.

51. A. Raine, P. H. Venables, and S. A. Mednick, Low resting heart rate at age 3 years predisposes to aggression at age 11 years: Evidence from the Mauritius Child Health Project, *Journal of the American Academy of Child and Adolescent Psychiatry 36* (1997): 1457–1464.

52. J. A. Gray, Drug effects on fear and frustration: Possible limbic site of action of minor tranquilizers, in *Handbook of Psychopharmacology*, Vol. 8, *Drugs, Transmitters, and Behavior*, ed. L. L. Iversen, S. D. Iversen, and S. H. Snyder (New York: Plenum, 1977), 433–529.

53. D. C. Fowles, The three-arousal model: Implications of Gray's two-factor learning theory for heart rate, electrodermal activity, and psychopathy, *Psychophysiology 17* (1980): 87–104.

54. J. Olds and P. M. Milner, Positive reinforcement produced by electrical stimulation of the septal area and other regions of rat brain, *Journal of Comparative Physiology and Psychology 47* (1954): 419–427.

55. J. Olds, Pleasure centers in the brain, *Scientific American 195* (1956): 105–116.

56. M. E. Olds and J. Olds, Approach-avoidance analysis of rat diencephalon, *Journal of Comparative Neurology 120* (1963): 259–295.

57. R. A. Wise, The dopamine synapse and the notion of "pleasure centers" in the brain, *Trends in Neuroscience 3* (1980): 91–94.

58. B. Eichelman, Neurochemical correlates of aggressive behavior, *Psychopharmacology Bulletin 17* (1981): 58–62.

59. M. Haney, K. Noda, R. Kream, et al., Regional 5-HT and dopamine activity: Sensitivity to amphetamine and aggressive behavior in mice, *Aggressive Behavior 16* (1990): 259–270.

60. K. A. Miczek and W. Tornatzky, Ethopharmacology of aggression: Impact on autonomic and mesocorticolimbic activity, *Annals of the New York Academy of Science 794* (1996): 60–77.

61. J. A. Slobodzian and D. Marder, Teenagers charged in I-95 blaze, *Philadelphia Inquirer,* October 10, 1996, A1–A23.

62. J. A. Slobodzian, Teenager pleads guilty in I-95 fire, *Philadelphia Inquirer,* November 11, 1996, R1.

63. S. Sataline, For the mother, the worst got worse, *Philadelphia Inquirer,* October 10, 1996, A1–A23.

64. A. Dahlstrom and K. Fuxe, Evidence for the existence of monoamine-containing neurons in the central nervous system. I. Demonstration of monoamines in cell bodies of brain stem neurons, *Acta Physiologica Scandinavia,* Suppl. 232 (1964): 1–55.

65. A. M. Johnson, The comparative pharmacological properties of the selective serotonin re-uptake inhibitors in animals, in *Selective Serotonin Re-Uptake Inhibitors,* ed. J. P. Feighner and W. F. Boyer (New York: Wiley, 1991), 37–70.

66. D. R. Nelson, D. R. Thomas, and A. M. Johnson, Pharmacological effects of paroxetine after repeated administration to animals, *Acta Psychiatrica Scandinavia 80* (1989): 21–23.

67. J. F. Stolz, C. A. Marsden, and D. N. Middlemiss, Effect of chronic antidepressant treatment and subsequent withdrawal on [3H]-5-hydroxytryptamine and [3H]-spiperone binding in rat frontal cortex and serotonin receptor mediated behavior, *Psychopharmacology 80* (1983): 150–155. Specifically, the most consistent change in receptors noted during chronic SSRI treatment has been a downregulation of the $5HT_2$ subtype of serotonin receptor. Downregulation of $5HT_1$ receptors has also been reported, as well as a reduction in the responsiveness of presynaptic receptors of the $5HT_{1A}$ subtype; at least one study, however, has reported an increase in the responsiveness of $5HT_{1A}$ receptors. Surprisingly, chronic administration of serotonin-selective antidepressants has also been shown to downregulate and desensitize the beta subtype of norepinephrine receptor. Johnson (reference 65) and Blier (*Journal of Clinical Psychiatry 7* [1987]: 24S–35S) offer comprehensive reviews of the chronic effects of serotonin-selective antidepressants.

68. L. Valzelli and S. Bernasconi, Aggressiveness by isolation and brain serotonin turnover changes in different strains of mice, *Neuropsychobiology 5* (1979): 129–135.

69. S. Garattini, E. Giacalone, and L. Valzelli, Isolation, aggressiveness, and brain 5-hydroxytryptamine turnover, *Journal of Pharmacy and Pharmacology 19* (1967): 338–339.

70. K. M. Kantak, L. R. Hegstrand, and B. Eichelman, Facilitation of shock-induced fighting following intraventricular 5,7-dihydroxytryptamine and 6-hydroxydopa, *Psychopharmacology 74* (1981): 157–160.

71. R. G. Sewall, G. A. Gallus, F. O. Gault, et al., p-chlorophenylalanine effects on shock-induced attack and pressing responses in rats, *Pharmacology, Biochemistry, and Behavior* 17 (1982): 945–950.

72. K. A. Miczek, E. Weerts, M. Haney, et al., Neurobiological mechanisms controlling aggression: Preclinical developments for pharmacotherapeutic interventions, *Neuroscience and Biobehavioral Reviews* 18 (1994): 97–110.

73. L. Valzelli, Aggressive behavior induced by isolation, in *Aggressive Behavior*, ed. S. Garattini and E. B. Sigg (New York: Wiley, 1969), 71–76.

74. M. Hamilton, Mood disorders: Clinical features, in *Comprehensive Textbook of Psychiatry*, ed. H. I. Kaplan and B. J. Sadock (Baltimore: Williams and Wilkins, 1989), 892–913.

75. A. Roy, Suicide, in *Comprehensive Textbook of Psychiatry*, ed. H. I. Kaplan and B. J. Sadock (Baltimore: Williams and Wilkins, 1989), 1414–1427.

76. Harvard Medical School, Suicide—Part I, *Harvard Mental Health Letter* 13 (1996): 1–5.

77. M. Asberg, P. Thoren, L. Traksman, et al., "Serotonin depression"—a biochemical subgroup within the affective disorders? *Science* 191 (1976): 478–480.

78. E. F. Coccaro, The biology of aggression, *Scientific American* (January–February 1995): 38–47.

79. A. Asberg, L. Traksman, and P. Thoren, 5-HIAA in the cerebrospinal fluid: A biochemical suicide predictor? *Archives of General Psychiatry* 33 (1976): 1193–1197.

80. M. Stanley, J. Virgilio, and S. Gershon, Tritiated imipramine binding sites are decreased in the frontal cortex of suicides, *Science* 216 (1982): 1337–1339.

81. M. Stanley and J. Mann, Increased serotonin-2 binding sites in frontal cortex of suicide victims, *Lancet* 1 (1983): 214–216.

82. The altered receptors are of the $5HT_2$ subtype. See J. J. Mann, M. Stanley, and P. A. McBride, Increased serotonin-2 and beta adrenergic receptor binding in suicide victims, *Archives of General Psychiatry* 43 (1986): 954–959.

83. G. L. Brown, M. H. Ebert, P. F. Goyer, et al., Aggression, suicide, and serotonin-relationships to CSF amine metabolites, *American Journal of Psychiatry* 139 (1982): 741–746.

84. E. F. Coccaro, Central serotonin and impulsive aggression, *British Journal of Psychiatry* 155 (1989): 52–62.

85. M. Linnoila, M. Virkkunen, M. Scheinin, et al., Low cerebrospinal fluid 5-hydroxyindoleacetic acid concentration differentiates impulsive from nonimpulsive violent behavior, *Life Sciences* 33 (1983): 2609–2614.

86. M. Virkkunen, A. Nuutila, F. K. Goodwin, et al., Cerebrospinal fluid monoamine metabolite levels in male arsonists, *Archives of General Psychiatry* 44 (1987): 241–247.

87. V. M. I. Linnoila and M. Virkkunen, Aggression, suicidality, and serotonin, *Journal of Clinical Psychiatry* 53 (1992): 46–51.

88. E. F. Coccaro, M. E. Berman, and R. J. Kavoussi, Relationship of prolactin response to d-fenfluramine to behavioral and questionnaire assessments of aggression in personality-disordered men, *Biological Psychiatry* 40 (1996): 157–164.

89. E. F. Coccaro, Impulsive aggression and central serotonergic system function in humans: An example of a dimensional brain-behavior relationship, *International Clinical Psychopharmacology* 7 (1992): 3–12.

90. M. J. Raleigh, G. L. Brammer, A. Yuwiler, et al., Serotonergic influences on the social behavior of vervet monkeys [*Cercopithecus aethiops* sabaeus], *Experimental Neurology* 68 (1980): 322–334.

91. M. J. Raleigh, G. L. Brammer, and M. T. McGuire, Male dominance, serotonergic systems, and the behavioral and physiological effects of drugs in vervet monkeys [*Cercopithe-*

cus aethiops sabaeus], in *Ethopharmacology: Primate Models of Neuropsychiatric Disorders,* ed. K. A. Miczek (New York: Liss, 1983), 184–198.

92. A. M. M. van Erp and K. A. Miczek, Prefrontal cortex dopamine and serotonin: Microdialysis during aggression and alcohol self-administration in rats, *Society for Neuroscience Abstracts 22* (1996): 161.

93. K. A. Miczek, personal communication.

94. C. B. Pert and S. H. Snyder, Opiate receptor: Evidence in nervous tissue, *Science 179* (1973): 1011-1014.

95. J. Hughes, T. W. Smith, H. W. Kosterlitz, et al., Identification of two related pentapeptides from the brain with potent opiate agonist activity, *Nature 258* (1975): 577–580.

96. J. H. Schwartz, Chemical messengers: Small molecules and peptides, in *Principles of Neural Science,* ed. E. R. Kandel, J. H. Schwartz, and T. M. Jessell (Norwalk, CT: Appleton & Lange, 1991), 213–224.

97. J. W. Lewis, J. T. Cannon, and J. C. Liebeskind, Opioid and nonopioid mechanisms of stress analgesia, *Science 298* (1983): 623–625.

98. K. A. Miczek, M. L. Thompson, and L. Shuster, Opioid-like analgesia in defeated mice, *Science 215* (1982): 1520–1522.

99. R. J. Rodgers and J. I. Randall, Social conflict analgesia: Studies on naloxone antagonism and morphine cross-tolerance in male DBA/2 mice, *Pharmacology Biochemistry and Behavior 23* (1985): 883–887.

100. E. M. Nikulina, J. E. Marchand, K. A. Miczek, et al., Social defeat stress induces enhanced opioid analgesia and neurochemical changes in rat periaqueductal grey, *Society for Neuroscience Abstracts 23* (1997): 126.

101. R. K. Pitman, B. A. van der Kolk, and S. P. Orr, Naloxone-reversible analgesic response to combat-related stimuli in posttraumatic stress disorder, *Archives of General Psychiatry 47* (1990): 541–544.

102. B. A. van der Kolk, M. S. Greenberg, and S. P. Orr, Endogenous opioids, stress induced analgesia, and posttraumatic stress disorder, *Psychopharmacology Bulletin 25* (1989): 417–421.

Chapter 6. Raging Hormones

1. R. W. Settle and M. L. Settle, *Saddles and Spurs: The Pony Express Saga* (Lincoln, NE: University of Nebraska Press, 1955).

2. F. Reinfeld, *Pony Express* (New York: Macmillan, 1966).

3. J. I. Morgan and T. Curran, Stimulus-transcription coupling in neurons: Role of cellular immediate early genes, *Trends in Neuroscience 12* (1989): 459–462.

4. P. M. Plotsky, E. T. Cunningham, and E. P. Widmaier, Catecholaminergic modulation of corticotropin-releasing factor and adrenocorticotropin secretion, *Endocrine Reviews 10* (1989): 437–458.

5. I. Kupfermann, Hypothalamus and limbic system: Peptidergic neurons, homeostasis, and emotional behavior, in *Principles of Neural Science,* ed. E. R. Kandel, J. H. Schwartz, and T. M. Jessell (East Norwalk, CT: Appleton and Lange, 1991), 735–749.

6. Federal Bureau of Investigation, *Crime in the United States 1996* [Uniform Crime Reporting Program statistics] (Washington, D.C.: U.S. Department of Justice, 1996).

7. R. I. Simon, *Bad Men Do What Good Men Dream* (Washington, D.C.: American Psychiatric press, 1996).

8. National Center for Health Statistics.

9. J. Archer, *The Behavioural Biology of Aggression* (Cambridge: Cambridge University Press, 1988).

10. P. F. Brain, Hormonal aspects of aggression and violence, in *Understanding and Preventing Violence,* ed. A. J. Reiss, Jr., K. A. Miczek, and J. A. Roth (Washington, D.C.: National Academy Press, 1994), 173–244.

11. D. J. Albert, M. L. Walsh, and R. H. Jonik, Aggression in humans: What is its biological foundation? *Neuroscience and Biobehavioral Reviews 17* (1993): 405–425.

12. M.-F. Bouissou, Androgens, aggressive behavior, and social relationships in higher mammals, *Hormone Research 18* (1983): 43–61.

13. D. J. Albert, R. H. Jonik, N. V. Watson, et al., Hormone-dependent aggression in male rats is proportional to serum testosterone concentration but sexual behavior is not, *Physiology and Behavior 48* (1990): 409–416.

14. T. Schuurman, Hormonal correlates of agonistic behavior in adult male rats, *Progress in Brain Research 53* (1980): 415–420.

15. P. F. Brain and M. Haug, Hormonal and neurochemical correlates of various forms of animal "aggression," *Psychoneuroendocrinology 17* (1992): 537–551.

16. K. A. Miczek and J. F. DeBold, Hormone-drug interactions and their influence on aggressive behavior, in *Hormones and Aggressive Behavior,* ed. B. B. Svare (New York: Plenum Press, 1983).

17. A. F. Dixson, Androgens and aggressive behavior in primates: A review, *Aggressive Behavior 6* (1980): 37–67.

18. T. P. Gordon, R. M. Rose, and I. S. Bernstein, Seasonal rhythm in plasma testosterone levels in the rhesus monkey (*Macaca mulatta*): A three year study, *Hormones and Behavior 7* (1976): 229–243.

19. J. D. Loy, K. Loy, G. Keifer, et al., The behavior of gonadectomized rhesus monkeys, *Contributions in Primatology 20* (1984): 1–141.

20. R. T. Rada, D. R. Laws, and R. Kellner, Plasma testosterone levels in the rapist, *Psychosomatic Medicine 38* (1976): 257–268.

21. R. T. Rada, D. R. Laws, R. Kellner, et al., Plasma androgens in violent and nonviolent sex offenders, *Bulletin of the American Academy of Psychiatry and Law 11* (1983): 149–158.

22. J. Dabbs, R. Frady, T. S. Carr, et al., Saliva testosterone and criminal violence in young adult prison inmates, *Psychosomatic Medicine 49* (1988): 174–182.

23. M. Virkkunen, R. Rawlings, R. Tokola, et al., CSF biochemistries, glucose metabolism, and diurnal activity rhythms in alcoholic, violent offenders, fire setters, and healthy volunteers, *Archives of General Psychiatry 51* (1994): 20–27.

24. J. Bain, R. Langevin, R. Dickey, et al., Sex hormones in murderers and assaulters, *Behavioral Science and the Law 5* (1987): 95–101.

25. R. Tricker, R. Casaburi, T. W. Storer, et al., The effects of supraphysiological doses of testosterone on angry behavior in healthy eugonadal men: A clinical research center study, *New England Journal of Medicine 81* (1996): 3754–3758.

26. R. M. Rose, I. S. Bernstein, and P. A. Gordon, Consequences of social conflict on plasma testosterone levels in rhesus monkeys, *Psychosomatic Medicine 37* (1975): 50–61.

27. R. M. Rose, T. P. Gordon, and I. S. Bernstein, Plasma testosterone levels in the male rhesus: Influences of sexual and social stimuli, *Science 178* (1972): 643–645.

28. A. Mazur and T. A. Lamb, Testosterone, status, and mood in human males, *Hormones and Behavior 14* (1980): 236–246.

29. M. Elias, Serum cortisol, testosterone, and testosterone-binding globulin responses to competitive fighting in human males, *Aggressive Behavior 7* (1981): 215–224.

30. R. M. Rose, P. G. Bourne, R. O. Poe, et al., Androgen response to stress: II. Excretion of testosterone, epitestosterone, androsterone, and etiocholalanolone during basic combat training and under threat of attack, *Psychosomatic Medicine 32* (1969): 418–436.

31. J. Ehrenkrantz, E. Bliss, and M. H. Sheard, Plasma testosterone: Correlation with aggressive behavior and social dominance in man, *Psychosomatic Medicine 36* (1974): 469–475.

32. Bureau of Justice Statistics.

33. J. A. Fox, *Trends in Juvenile Violence* [report to the U.S. attorney general], quoted in F. Butterfield, Republicans challenge notion of separate jails for juveniles, *New York Times,* June 24, 1996.

34. Federal Bureau of Investigation.

35. R. Lacayo, Law and order, *Time Magazine,* January 15, 1996, 48–56.

36. F. S. Vom Saal, Models of early hormonal effects on intrasex aggression in mice, in *Hormones and Aggressive Behavior,* ed. B. B. Svare (New York: Plenum, 1983), 197–222.

37. R. E. Whalen and F. Johnson, To fight or not to fight: The question is "whom"? in *Hormones, Brain, and Behavior in Vertebrates,* ed. J. Balthazart (Basel: Karger, 1990).

38. L. K. Takahashi and R. K. Lore, Play fighting and the development of agonistic behavior in male and female rats, *Aggressive Behavior 9* (1983): 217–227.

39. R. J. Primus and C. K. Kellogg, Gonadal hormones during puberty organize environment-related social interaction in the male rat, *Hormones and Behavior 24* (1990): 311–323.

40. E. J. Susman, B. K. Worrall, and E. Murowchick, Experience and neuroendocrine parameters of development: Aggressive behavior and competencies, in *Aggression and Violence,* ed. D. M. Stoff and R. B. Cairns (Hillsdale, NJ: L Erlbaum, 1996), 267–289.

41. E. J. Susman, G. Inoff-Germain, and E. D. Nottelmann, Hormones, emotional dispositions, and aggressive attributes in young adolescents, *Child Development 58* (1987): 1114–1134.

42. E. D. Nottelman, E. J. Susman, G. Inoff-Germain, et al., Developmental processes in early adolescence: Relationships between adolescent adjustment problems and chronologic age, pubertal stage, and puberty-related serum hormone levels, *Journal of Pediatrics 110* (1987): 473–480.

43. D. Olweus, A. Mattsson, and D. Schalling, Testosterone, aggression, physical, and personality dimensions in normal adolescent males, *Psychosomatic Medicine 42* (1980): 253–269.

44. D. Olweus, A. Mattsson, and H. Low, Circulating testosterone levels and aggression in adolescent males: A causal analysis, *Psychosomatic Medicine 50* (1988): 261–272.

45. R. S. Pynoos, A. M. Steinberg, and E. M. Ornitz, Issues in the developmental neurobiology of traumatic stress, *Annals of the New York Academy of Sciences 821* (1997): 176–193.

46. B. D. Ayres, California child molesters face "chemical castration," *The New York Times,* August 26, 1996, A1–A10.

47. F. S. Berlin, G. K. Lehne, H. M. Malin, et al., The eroticized violent crime: A psychiatric perspective with six clinical examples, *Journal of Sexual Addiction and Compulsivity 1* (1997): 1–11.

48. U. Laschet and L. Laschet, Antiandrogens in the treatment of sexual deviations of men, *Journal of Steroid Biochemistry 6* (1975): 821–826.

49. J. Money, C. Wiedeking, P. C. Walker, et al., Combined antiandrogenic and counselling program for treatment of 46 XY and 47 XYY sex offenders, in *Hormones, Behavior, and Psychopathology,* ed. E. J. Sachar (New York: Raven Press, 1976), 105–120.

50. F. S. Berlin and E. Krout, Pedophilia: Diagnostic concepts, treatment, and ethical considerations, *American Journal of Forensic Psychiatry 7* (1986): 13–30.

51. F. S. Berlin, W. P. Hunt, H. M. Malin, et al., A five-year-plus follow-up survey of criminal recidivism within a treated cohort of 406 pedophiles, 111 exhibitionists, and 109 sexual aggressives: Issues and outcome, *American Journal of Forensic Psychiatry 12* (1991): 5–28.

52. U.S. Department of Justice, Women are violent crime victims at a lower rate than men, but the difference is narrowing (Washington, D.C., 1996).

53. National Center on Child Abuse and Neglect, *In fact . . . Answers to frequently asked questions on child abuse and neglect* (Washington, D.C., 1995).

54. K. E. Hood, Intractable tangles of sex and gender in women's aggressive development: An optimistic view, in *Aggression and Violence: Genetic, Neurobiological, and Biosocial Perspectives,* ed. D. Stoff and R. B. Cairns (Hillsdale, NJ: Erlbaum, 1996), 309–335.

55. S. H. Verhovek, Dead women waiting: Who's who on death row, *New York Times,* February 8, 1998, 1–3.

56. F. Butterfield, A fatal fire, a girl in prison, and a tangle of justice issues, *New York Times,* December 3, 1996, A1–A20.

57. G. Reynolds, Officials say motive in mother's slaying was to take the fetus, *New York Times,* November 21, 1995, A12.

58. D. Ivey, Woman opens fire on Penn State campus, one dead, *America Online,* September 17, 1996, 1–2.

59. D. P. Southall, M. C. Plunkett, M. W. Banks, et al., Covert video recordings of life-threatening child abuse: Lessons for child protection, *Pediatrics 100* (1997): 735–760.

60. R. B. Cairns, B. D. Cairns, H. J. Neckerman, et al., Growth and aggression: I. Childhood to early adolescence, *Developmental Psychology 25* (1989): 320–330.

61. R. B. Cairns and B. D. Cairns, *Lifelines and Risks: Pathways of Youth in Our Time* (Cambridge: Cambridge University Press, 1994).

62. K. Bjorkqvist, K. Osterman, and K. M. J. Lagerspetz, Sex differences in covert aggression among adults, *Aggressive Behavior 20* (1994): 27–33.

63. K. E. Hood, Female aggression in [albino ICR] mice: Development, social experience and the effects of selective breeding, *International Journal of Comparative Psychology 2* (1988): 27–41.

64. K. E. Hood and R. B. Cairns, A developmental genetic analysis of aggressive behavior in mice: II. Cross-sex inheritance, *Behavior Genetics 18* (1988): 605–619.

65. J. Dabbs, R. B. Ruback, R. L. Frady, et al., Saliva testosterone and criminal violence among women, *Personality and Individual Differences 9* (1988): 269–275.

66. T. Banks and J. Dabbs, Salivary testosterone and cortisol in a delinquent and violent urban subculture, *Journal of Social Psychology 136* (1996): 49–56.

67. K. Dalton, Menstruation and crime, *British Medical Journal 2* (1961): 1752–1753.

68. P. T. d'Orban and K. Dalton, Violent crime and the menstrual cycle, *Psychology and Medicine 10* (1980): 353–359.

69. K. Dalton, *The Premenstual Syndrome* (Springfield, IL: Charles C. Thomas, 1964).

70. S. H. Van Goozen, N. H. Frijda, V. M. Wiegant, et al., The premenstrual phase and reactions to aversive events: A study of hormonal influences on emotionality, *Psychoneuroendocrinology 21* (1996): 479–497.

71. G. Inoff-Germain, G. S. Arnold, and E. D. Nottelmann, Relations between hormone levels and observational measures of aggressive behavior of young adolescents in family interactions, *Developmental Psychology 24* (1988): 129–139.

72. J. W. Finkelstein, E. J. Susman, V. Chinchilli, et al., Testosterone (T) or conjugated estrogen (E) cause increases in aggressive behavior in hypogonadal boys or girls [abstract], *American Pediatric Society,* 1995.

73. In humans, the principal glucocorticoid hormone is *cortisol;* the rat equivalent is *corticosterone.*

74. S. M. McCann, An introduction to neuroendocrinology: Basic principles and historical considerations, in *Neuroendocrinology,* ed. C. B. Nemeroff (Boca Raton, FL: CRC Press, 1990), 1–17.

75. P. Petrusz and I. Merchenthaler, The corticotropin-releasing factor system, in *Neuroendocrinology,* ed. C. B. Nemeroff (Boca Raton, FL: CRC Press, 1989), 129–183.

76. W. Vale, J. Spiess, C. Rivier, et al., Characterization of a 41-residue ovine hypothalamic peptide that stimulates secretion of corticotropin and beta-endorphin, *Science 213* (1981): 1394–1396.

77. G. F. Koob, The behavioral neuroendocrinology of corticotropin-releasing factor, growth hormone-releasing factor, somatostatin, and gonadatropin-releasing hormone, in *Neuroendocrinology,* ed. C. B. Nemeroff (Boca Raton, FL: CRC Press, 1989), 353–364.

78. B. S. McEwen, Adrenal steroid actions on brain: Dissecting the fine line between protection and damage, in *Neurobiological and Clinical Consequences of Stress,* ed. M. J. Friedman, D. S. Charney, and A. Y. Deutch (Philadelphia: Lippincott-Raven, 1995), 135–147.

79. B. S. McEwen and E. Stellar, Stress and the individual, *Archives of Internal Medicine 153* (1993): 2093–2101.

80. B. McEwen, Protective and damaging effects of stress mediators, *New England Journal of Medicine 338* (1998): 171–179.

81. J. Welch, G. Farrar, A. Dunn, et al., Central 5-HT1A receptors inhibit adrenocortical secretion, *Journal of Neuroendocrinology 57* (1993): 272–281.

82. B. S. McEwen, Effects of the steroid/thyroid hormone family on neural and behavioral plasticity, in *Neuroendocrinology,* ed. C. B. Nemeroff (Boca Raton, FL: CRC Press, 1992), 333–351.

83. R. M. Sapolsky, Adrenocortical function, social rank, and personality among wild baboons, *Biological Psychiatry 28* (1990): 862–878.

84. R. M. Sapolsky, *Why Zebras Don't Get Ulcers* (New York: W. H. Freeman, 1994).

85. R. M. Sapolsky and J. Ray, Styles of dominance and their physiological correlates among wild baboons, *American Journal of Primatology 18* (1989): 1–9.

86. E. O. Johnson, T. C. Kamilaris, and C. S. Carter, The biobehavioral consequences of psychogenic stress in a small, social primate *(Callithrix jacchus jacchus), Biological Psychiatry 40* (1996): 317–337.

87. E. J. Susman and A. Ponirakis, Hormones-context interactions and antisocial behavior in youth, in *Unlocking Crime: The Biosocial Key,* ed. A. Raine, D. Farrington, and P. Brennan (New York: Plenum, 1997), 1–35.

88. Y. Delville, R. H. Melloni, and C. F. Ferris, Behavioral and neurobiological consequences of social subjugation during puberty in golden hamsters, *Journal of Neuroscience 18* (1998): 2667–2672.

89. C. D. Spielberger, Type-A behavior, anger/hostility, and heart disease, Paper presented at the Philadelphia Conference on Aggression, Philadelphia, July 19–21, 1996.

90. R. B. Williams, T. L. Haney, K. L. Lee, et al., Type A behavior, hostility, and coronary atherosclerosis, *Psychosomatic Medicine 42* (1980): 539–549.

91. D. C. Blanchard, R. L. Spencer, and S. M. Weiss, Visible burrow system as a model of chronic social stress: Behavioral and neuroendocrine correlates, *Psychoneuroendocrinology 20* (1995): 117–134.

92. D. C. Blanchard, R. R. Sakai, and B. McEwen, Subordination stress: Behavioral, brain, and neuroendocrine correlates, *Behavioral Brain Research 58* (1993): 113–121.

93. E. B. Yudko, D. C. Blanchard, M. Hebert, et al., Behavioral consequences of subordination in the visible burrow system, *Society for Neuroscience Abstracts 22* (1996): 462.

94. G. W. Arana and R. J. Baldessarini, The dexamethasone suppression test for diagnosis and prognosis in psychiatry, *Archives of General Psychiatry 42* (1985): 1193–1204.

95. J. Angst, How recurrent is depression? Paper presented at the annual meeting of the American Psychiatric Association, Washington, D. C., May 3, 1992.

96. R. Yehuda, S. M. Southwick, G. Nussbaum, et al., Low urinary cortisol excretion in PTSD, *Journal of Nervous and Mental Disorders 178* (1991): 366–369.

97. R. Yehuda, E. L. Giller, S. M. Southwick, et al., Hypothalamic-pituitary-adrenal dysfunction in posttraumatic stress disorder, *Biological Psychiatry 30* (1991): 1031–1048.

98. R. Yehuda, B. Kahana, K. Binder-Byrnes, et al., Low urinary cortisol secretion in Holocaust survivors with posttraumatic stress disorder, *American Journal of Psychiatry 152* (1995): 982–986.

99. R. Yehuda, Giller, Jr., and R. A. Levengood, Hypothalamic-pituitary-adrenal functioning in post-traumatic stress disorder: Expanding the concept of the stress response spectrum, in *Neurobiological and Clinical Consequences of Stress,* ed. M. J. Friedman, D. S. Charney, and A. Y. Deutch (Philadelphia: Lippincott-Raven, 1995), 351–365.

100. R. Yehuda, M. H. Teicher, R. Levengood, et al., Circadian rhythm of basal cortisol levels in PTSD, *Annals of the New York Academy of Sciences 746* (1994): 378–380.

101. M. Virkkunen, Urinary free cortisol in habitually violent offenders, *Acta Psychiatrica Scandinavia 72* (1985): 40–44.

102. B. Bergman and B. Brismar, Hormone levels and personality traits in abusive and suicidal male alcoholics, *Alcoholism: Clinical and Experimental Research 18* (1994): 311–316.

103. K. McBurnett, B. B. Lahey, P. J. Frick, et al., Anxiety, inhibition, and conduct disorder in children: II. Relation to salivary cortisol, *Journal of the American Academy of Child and Adolescent Psychiatry 30* (1991): 192–196.

104. M. M. Vanyukov, H. B. Moss, J. A. Plail, et al., Antisocial symptoms in preadolescent boys and in their parents: Associations with cortisol, *Psychiatric Research 46* (1993): 9–17.

105. W. Cannon, *The Wisdom of the Body* (New York: Norton, 1939).

106. P. Sterling and J. Eyer, Allostasis: A new paradigm to explain arousal pathology, *Handbook of Life Stress, Cognition and Health* (1988): 629–649.

107. R. M. Post, S. R. B. Weiss, and M. A. Smith, Sensitization and kindling: Implications for the evolving neural substrates of posttraumatic stress disorder, in *Neurobiological and Clinical Consequences of Stress,* ed. M. J. Friedman, D. S. Charney, and A. Y. Deutch (Philadelphia: Lippincott-Raven, 1995), 203–224.

108. R. M. Post, Transduction of psychosocial stress into the neurobiology of recurrent affective disorder, *American Journal of Psychiatry 149* (1992): 999–1010.

109. B. McEwen and A. M. Magarinos, Stress effects on morphology and function of the hippocampus, *Annals of the New York Academy of Sciences 821* (1997): 271–284.

110. R. M. Sapolsky, Stress, glucocorticoids, and damage to the nervous system: The current state of confusion, *Stress 1* (1996): 1–19.

111. Rats subjected to twelve hours a day of glucocorticoid abuse for as little as three months lose up to 20 percent of the cells in certain regions of the hippocampus—and the same pattern of damage occurs in the brains of rats subjected to repeated restraint stress and primates stressed by social subordination. See references 109 and 110 for more details.

112. Y. I. Sheline, P. W. Wang, and H. G. Mokhtar, Hippocampal atrophy in recurrent major depression, *Proceedings of the National Academy of Sciences USA 93* (1996): 3908–3913.

Chapter 7. Out of Their Minds

1. J. Grafman, K. J. Holyoak, and F. Boller, Preface to *Annals of the New York Academy of Sciences 769* (1995), ix.

2. E. R. Kandel, S. A. Siegelbaum, and J. H. Schwartz, Synaptic transmission, in *Principles of Neural Science,* ed. E. R. Kandel, J. H. Schwartz, and T. M. Jessell (Norwalk, CT: Appleton and Lange, 1991), 123–134.

3. M. Sheng and M. E. Greenberg, The regulation and function of c-*fos* and other immediate early genes in the nervous system, *Neuron 4* (1990): 477–485.

4. J. R. Cooper, F. E. Bloom, and R. H. Roth, *The Biochemical Basis of Neuropharmacology* (New York: Oxford University Press, 1996).

5. F. M. Benes, Is there a neuroanatomic basis for schizophrenia? An old question revisited, *Neuroscientist 1* (1995): 104–115.

6. The pyramidal cells themselves are also thought to use glutamate as their neurotransmitter.

7. J. H. Martin, The collective electrical behavior of cortical neurons: The electroencephalogram and the mechanisms of epilepsy, in *Principles of Neural Science,* ed. E. R. Kandel, J. H. Schwartz, and T. M. Jessell (Norwalk, CT: Appleton and Lange, 1991), 777–791.

8. R. Brown, A. Crane, and P. Goldman, Regional distribution of monoamines in the cerebral cortex and subcortical structures of the rhesus monkey: Concentrations and in vitro synthesis rates, *Brain Research 168* (1979): 133–150.

9. P. S. Goldman-Rakic, M. S. Lidow, J. F. Smiley, et al., The anatomy of dopamine in monkey and human prefrontal cortex, *Journal of Neural Transmission 36* (Suppl.) (1992): 163–177.

10. J. F. Smiley, S. M. Williams, K. Szigeti, et al., Light and electron microscopic characterization of dopamine-immunoreactive processes in human cerebral cortex, *Journal of Comparative Neurology 321* (1992): 325–335.

11. M. D'Esposito and M. Grossman, The physiological basis of executive function and working memory, *Neuroscientist 2* (1996): 345–352.

12. A. Baddeley, *Working Memory* (London: Oxford University Press, 1986).

13. P. S. Goldman-Rakic, Circuitry of primate prefrontal cortex and regulation of behavior by representational memory, in *Handbook of Physiology—The Nervous System V, Higher Functions of the Brain,* ed. F. Plum (Bethesda, MD: American Physiological Society, 1987), 373–417.

14. P. S. Goldman-Rakic, Architecture of the prefrontal cortex and the central executive, *Annals of the New York Academy of Sciences 769* (1995): 71–83.

15. A. R. Damasio, *Descartes' Error* (New York: Avon Books, 1994).

16. A. R. Damasio, The brain binds entities and events by multiregional activation from convergence zones, *Neural Computation 1* (1989): 123–132.

17. J. L. Cummings, Anatomic and behavioral aspects of frontal-subcortical circuits, *Annals of the New York Academy of Sciences 769* (1995): 71–83.

18. R. Vigoda and B. Ordine, Du Pont case: Saga of a mind, *Philadelphia Inquirer,* February 2, 1997, E3.

19. B. Ordine and R. Vigoda, Du Pont jury starts sifting evidence to assess his sanity, *Philadelphia Inquirer,* February 19, 1997, A1–A7.

20. DuPont's fate rests in hands of jury, *Bucks County Courier Times,* February 19, 1997, A1.

21. R. Vigoda, Du Pont is psychotic, defense doctor says, *Philadelphia Inquirer,* August 24, 1996, B1–B5.

22. R. Vigoda and B. Ordine, Du Pont is not rational, lawyers say, *Philadelphia Inquirer,* September 21, 1996, A1–A4.

23. R. Vigoda and B. Ordine, Defense doctors: Du Pont feared he was target of conspiracy, *Philadelphia Inquirer,* February 9, 1997, E1–E8.

24. B. Ordine and R. Vigoda, Du Pont is guilty, says alternate, *Philadelphia Inquirer,* February 19, 1997, A7.

25. National Institute of Mental Health, *Schizophrenia: Questions and Answers,* DHHS Publication No. (ADM) 86–1457, 1986, available from Internet: www.nimh.nih.gov/publicat/schizo.htm#schiz/.

26. J. A. Talbott, Current perspectives in the United States on the chronically mentally ill, in *Recent Advances in Schizophrenia,* ed. A. Kales, C. N. Stefanis, and J. Talbott (New York: Springer-Verlag, 1990), 279–295.

27. J. A. Grebb and R. Cancro, Schizophrenia: Clinical features, in *Comprehensive Textbook of Psychiatry,* 5th edition, ed. H. I. Kaplan and B. J. Sadock (Baltimore: Williams and Wilkins, 1989), 757–777.

28. A. Kales, J. D. Kales, and A. Vela-Bueno, Schizophrenia: Historical perspectives, in *Recent Advances in Schizophrenia,* ed. A. Kales, C. N. Stefanis, and J. Talbott (New York: Springer-Verlag, 1990), 3–23.

29. T. C. Manschreck, Delusional (paranoid) disorders, in *Comprehensive Textbook of Psychiatry,* 5th edition, ed. H. I. Kaplan and B. J. Sadock (Baltimore: Williams and Wilkins, 1989), 816–829.

30. J. C. Perry and G. E. Vaillant, Personality disorders, in *Comprehensive Textbook of Psychiatry,* 5th edition, ed. H. I. Kaplan and B. J. Sadock (Baltimore: Williams and Wilkins, 1989), 1352–1395.

31. J. W. Swanson, C. E. Holzer, V. K. Ganju, et al., Violence and psychiatric disorder in the community: Evidence from the Epidemiologic Catchment Area Surveys, *Hospital and Community Psychiatry 41* (1990): 761–770.

32. H. J. Steadman, E. P. Mulvey, J. Monahan, et al., Violence by people discharged from acute psychiatric inpatient facilities and by others in the same neighborhoods, *Archives of General Psychiatry 55* (1998): 393–401.

33. P. J. Resnick, The psychiatric prediction of violence, Workshop conducted at the annual meeting of the American Psychiatric Association, New York, May 4–9, 1996.

34. E. Mulvey, Assessing the evidence of a link between mental illness and violence, *Hospital and Community Psychiatry 45* (1994): 7.

35. B. G. Link and A. Stueve, Evidence bearing on mental illness as a possible cause of violent behavior, *Epidemiology Reviews 17* (1995): 172–181.

36. M. Krakowski, J. Volavka, and D. Rizer, Psychopathology and violence: A review of the literature, *Comprehensive Psychiatry 27* (1986): 131–148.

37. T. H. McGlashan, Schizophrenia: Psychodynamic theories, in *Comprehensive Textbook of Psychiatry,* 5th edition, ed. H. I. Kaplan and B. J. Sadock (Baltimore: Williams and Wilkins, 1989), 745–756.

38. I. Creese, D. R. Burt, and S. H. Snyder, Dopamine receptor binding predicts clinical and pharmacological potencies of antischizophrenic drugs, *Science 192* (1976): 481–483.

39. D. R. Weinberger, The biological basis of schizophrenia: New directions, *Journal of Clinical Psychiatry 58* (Suppl.) (1997): 22–27.

40. H. Jakob and H. Beckmann, Prenatal developmental disturbances in the limbic allocortex in schizophrenics, *Journal of Neurological Transmission 65* (1986): 303–326.

41. S. E. Arnold, B. T. Hyman, G. W. van Hoesen, et al., Some cytoarchitectural abnormalities of the entorhinal cortex in schizophrenia, *Archives of General Psychiatry 48* (1991): 625–632.

42. F. M. Benes, J. McSparran, E. D. Bird, et al., Deficits in small interneurons in prefrontal and cingulate cortices of schizophrenic and schizoaffective patients, *Archives of General Psychiatry 48* (1991): 990–1001.

43. S. Akbarian, W. E. Bunney Jr., S. G. Potkin, et al., Altered distribution of nicotamide-adenine dinucleotide phosphate-diaphorase cells in frontal lobe of schizophrenics implies disturbances of cortical development, *Archives of General Psychiatry 50* (1993): 169–177.

44. D. R. Weinberger, On the plausibility of "the neurodevelopmental hypothesis" of schizophrenia, *Neuropsychopharmacology 14* (1996): 1S–11S.

45. D. Y. Kimberg and M. J. Farah, A unified account of cognitive impairments following frontal lobe damage: The role of working memory in complex, organized behavior, *Journal of Experimental Psychology 122* (1993): 411–428.

46. B. B. Lahey and R. Loeber, Attention-deficit/hyperactivity disorder, oppositional defiant disorder; conduct disorder; and adult antisocial behavior: A life span perspective, in *Handbook of Antisocial Behavior,* ed. D. M. Stoff, J. Breiling, and J. D. Maser (New York: Wiley, 1997), 51–74.

47. D. Magnusson, The patterning of antisocial behavior and autonomic reactivity, in *Aggression and Violence: Genetic, Neurobiological, and Biosocial Perspectives,* ed. D. M. Stoff and R. B. Cairns (Hillsdale, NJ: Erlbaum, 1996).

48. R. A. Barkley, T. L. Shelton, C. Crosswait, et al., Preliminary findings of an early intervention program with aggressive hyperactive children, *Annals of the New York Academy of Sciences 794* (1996): 277–289.

49. S. R. Pliszka, J. T. McCracken, and J. W. Mass, Catecholamines in attention-deficit hyperactivity disorder: Current perspectives, *Journal of the American Academy of Child and Adolescent Psychiatry 35* (1996): 264–272.

50. M. Posner and S. E. Peterson, The attention system of the brain, *Annual Review of Neuroscience 13* (1990): 25–42.

51. R. A. Barkley, G. Grodzinsky, and G. J. DuPaul, Frontal lobe functions in attention deficit disorder with and without hyperactivity: A review and research report, *Journal of Abnormal Child Psychology 20* (1992): 163–188.

52. A. J. Zametkin, T. E. Nordahl, M. Gross, et al., Cerebral glucose metabolism in adults with hyperactivity of childhood onset, *New England Journal of Medicine 323* (1990): 1361–1366.

53. J. T. McCracken, A two-part model of stimulant action on attention-deficit hyperactivity disorder in children, *Journal of Neuropsychiatry 3* (1991): 201–209.

54. D. F. Musto, Opium, cocaine, and marijuana in American history, *Scientific American* (July 1991): 30–37.

55. G. G. Nahas, *Cocaine: The Great White Plague* (Middlebury, VT: Paul S. Eriksson, 1989).

56. S. L. Brady, Violence associated with acute cocaine use in patients admitted to a medical emergency department, *NIDA Research Monographs 103* (1990): 44–59.

57. *National Household Survey on Drug Abuse* (Washington, D.C.: National Institute on Drug Abuse, 1985).

58. P. Herridge and M. S. Gold, The new user of cocaine: Evidence from 800-COCAINE, *Psychiatric Annals 18* (1988): 521–522.

59. R. K. Siegal, Cocaine smoking, *Journal of Psychoactive Drugs 14* (1982): 272–341.

60. N. S. Miller, M. S. Gold, and J. C. Mahler, Violent behaviors associated with cocaine use: Pharmacological mechanisms, *International Journal of the Addictions 26* (1991): 1077–1088.

61. In San Diego, an old drug comes back, *New York Times,* February 20, 1997, A22.

62. C. Goldberg, Way out west and under the influence, *New York Times,* March 16, 1997, E16.

63. W. Kim, Crank, *Times Magazine,* June 22, 1998, 14–20.

64. National Institute on Drug Abuse, Facts about methamphetamine, *NIDA Notes* (November–December 1996): 1–2.

65. K. A. Miczek, The psychopharmacology of aggression, in *Handbook of Psychopharmacology,* vol. 19, ed. L. L. Iversen, S. D. Iversen, and S. H. Snyder (New York: Plenum Press, 1987).

66. D. Johnson, Good people go bad in Iowa, and a drug is being blamed, *New York Times,* February 22, 1996, A1–A19.

67. N. Swan, Response to escalating methamphetamine abuse builds on NIDA-funded research, *NIDA Notes* (November–December 1996): 1–4.

68. M. C. Ritz, R. J. Lamb, S. R. Goldberg, et al., Cocaine receptors on dopamine transporters are related to self-administration of cocaine, *Science 237* (1987): 1219–1223.

69. M. J. Kuhar, M. C. Ritz, D. Grigoriadis, et al., A cocaine receptor associated with dopamine transport and drug self-administration, in *Cocaine: Pharmacology, Physiology, and Clinical Strategies,* ed. J. M. Lakoski, M. P. Galloway, and F. J. White (Boca Raton, FL: CRC Press, 1992), 191–202.

70. S. E. Hyman, Why does the brain prefer opium to broccoli? *Harvard Review of Psychiatry 2* (1994): 43–46.

71. R. A. Wise, Addictive drugs and brain stimulation reward, *Annual Review of Neuroscience 19* (1996): 319–340.

72. S. L. Satel, S. Southwick, and F. H. Gawin, Clinical features of cocaine-induced paranoia, *American Journal of Psychiatry 148* (1991): 495–498.

73. S. E. Hyman and E. J. Nestler, Initiation and adaptation: A paradigm for understanding psychotropic drug action, *American Journal of Psychiatry 153* (1996): 151–162.

74. E. J. Nestler, Molecular basis of addictive states, *Neuroscientist 1* (1995): 212–220.

75. E. J. Nestler and G. K. Aghajanian, Molecular and cellular basis of addiction, *Science 278* (1997): 58–63.

76. Genes targeted by drugs of abuse include those coding for synthetic enzymes, second-messenger proteins, and the parent proteins that give rise to opiate peptides.

77. F. Weiss, L. H. Parsons, and A. Markou, Neurochemistry of cocaine withdrawal, in *The Neurobiology of Cocaine,* ed. R. P. Hammer, Jr. (Boca Raton, FL: CLC Press, 1995).

78. F. H. Gawin and M. E. Khalsa-Denison, Is craving mood-driven or self-propelled? Sensitization and "street" stimulant addiction, in *Neurotoxicity and Neuropathology Associated with Cocaine Abuse,* ed. M.D. Majewska, NIDA Research Monograph 163 (1996), 224–250.

79. Harvard Medical School, Post-traumatic stress disorder—part 1, *Harvard Mental Health Letter 12* (1996): 1–4.

80. S. M. Stine and T. R. Kosten, Complications of chemical abuse and dependency, in *Neurobiological and Clinical Consequences of Stress,* ed. M. J. Friedman, D. S. Charney, and A. Y. Deutch (Philadelphia: Lippincott-Raven, 1995), 447–464.

81. M. E. McFall, R. W. MacKay, D. M. Donovan, Combat-related PTSD and psychosocial adjustment problems among substance abusing veterans, *Journal of Nervous and Mental Disorders 179* (1991): 33–38.

82. A. Matsakis, *Post-traumatic Stress Disorder: A Complete Treatment Guide* (Oakland, CA: New Harbinger Publications, 1994).

83. B. A. Sorg and P. W. Kalivas, Stress and neuronal sensitization, in *Neurobiological and Clinical Consequences of Stress,* ed. M. J. Friedman, D. S. Charney, and A. Y. Deutch (Philadelphia: Lippincott-Raven, 1995), 83–102.

84. B. A. Horger and R. H. Roth, The role of mesoprefrontal dopamine neurons in stress, *Critical Reviews in Neurobiology 10* (1996): 395–418.

85. S. M. Antelman, A. R. Caggiula, S. Gershon, et al., Stressor-induced oscillation: A possible model of the bidirectional symptoms in PTSD, *Annals of the New York Academy of Sciences 821* (1997): 296–304.

86. B. A. Sorg and P. W. Kalivas, Effects of cocaine and footshock on extracellular dopamine levels in the ventral striatum, *Brain Research 559* (1991): 29–36.

87. M. Haney, S. Maccari, M. LeMoal, et al., Social stress increases the acquisition of cocaine self-administration in male and female rats, *Brain Research 698* (1995): 46–52.

88. B. J. Bushman and H. M. Cooper, Effects of alcohol on human aggression: An integrative research review, *Psychological Bulletin 107* (1990): 341–354.

89. P. F. Brain, R. L. Miras, and M. S. Berry, Diversity of animal models of aggression: Their impact on the putative alcohol/aggression link, *Journal of Studies of Alcohol 11* (Suppl.) (1993): 140–145.

90. J. Roizen, Issues of epidemiology of alcohol and violence, in *Alcohol and Interpersonal Violence: Fostering Multidisciplinary Perspectives,* ed. S. E. Martin, NIAAA Research Monograph 24 (1993), 3–36.

91. K. Pernanen, *Ethanol in Human Violence* (New York: Guilford, 1991).

92. D. Murdoch, R. O. Pihl, and D. Ross, Alcohol and crimes of violence: Present issues, *International Journal of Addiction 25* (1990): 1065–1081.

93. H. M. T. Barros and K. A. Miczek, Neurobiological and behavioral characteristics of alcohol-heightened aggression, in *Aggression and Violence: Genetic, Neurobiological, and Biosocial Perspectives,* ed. D. M. Stoff and R. B. Cairns (Hillsdale, NJ: Erlbaum, 1996), 237–263.

94. K. A. Miczek, E. M. Weerts, and J. F. DeBold, Alcohol benzodiazepine-GABA-A receptor complex and aggression: Ethological analysis of individual differences in rodents and primates, *Journal of Studies on Alcohol 11* (Suppl.) (1993): 170–179.

95. K. A. Miczek, personal communication.

96. M. K. Ticku, Drug modulation of GABA$_A$-mediated transmission, *Seminars in the Neurosciences 3* (1991): 211–218.

97. G. F. Koob, Drugs of abuse: Anatomy, pharmacology and function of reward pathways, *Trends in Pharmacological Sciences 13* (1992): 177–184.

98. M. Linnoila, J. DeJong, and M. Virkkunen, Family history of alcoholism in violent offenders and impulsive fire setters, *Archives of General Psychiatry 46* (1989): 613–616.

99. M. Virkkunen and M. Linnoila, Brain serotonin, type II alcoholism and impulsive violence, *Journal of Studies on Alcohol 77* (Suppl.) (1993): 163–169.

100. M. Virkkunen, E. Kallio, R. Rawlings, et al., Personality profiles and state aggressiveness in Finnish alcoholic violent offenders, fire setters, and healthy volunteers, *Archives of General Psychiatry 51* (1994): 28–33.

101. J. Evenden, The effects of ethanol on three tests of impulsive decision making in the rat, *Society for Neuroscience Abstracts 22* (1996): 699.

102. K. A. Miczek, J. F. DeBold, and A. M. M. vanErp, Neuropharmacological characteristics of individual differences in alcohol effects on aggression in rodents and primates, *Behavioral Pharmacology 5* (1994): 407–421.

103. D. R. Cherek, Effects of alcohol on human aggressive behavior, *Journal of Studies on Alcohol 46* (1985): 321–328.

104. R. J. Blanchard, L. Magee, R. Veniegas, et al., Alcohol and anxiety: Ethopharmacological approaches, *Progress in Neuro-psychopharmacology and Biological Psychiatry 17* (1993): 171–182.

105. C. A. Chiriboga and D. M. Ferriero, Neurologic complications of maternal drug use, in *Principles of Child Neurology,* ed. B. O. Berg (New York: McGraw Hill, 1996), 1363–1374.

106. J. R. West, W-J. A. Chen, and N. J. Pantazis, Fetal alcohol syndrome: The vulnerability of the developing brain and possible mechanisms of damage, *Metabolic Brain Disease 9* (1994): 291–322.

107. D. A. Frank, B. S. Zuckerman, H. Amaro, et al., Cocaine use during pregnancy: Prevalence and correlates, *Pediatrics 82* (1988): 888–895.

108. H. Amaro, B. Zuckerman, and H. Cabral, Drug use among adolescent mothers: Profile of risk, *Pediatrics 84* (1989): 144–151.

109. B. A. Blanchard, S. Steindorf, S. Wang, et al., Prenatal ethanol exposure alters ethanol-induced dopamine release in nucleus accumbens and striatum in male and female rats, *Alcoholism: Clinical and Experimental Research 17* (1993): 974–981.

110. E. L. Abel and R. J. Sokol, Incidence of fetal alcohol syndrome and economic impact of FAS-related anomalies, *Drug and Alcohol Dependence 19* (1987): 51–70.

111. A. P. Streissguth, H. M. Barr, and P. D. Sampson, Moderate prenatal alcohol exposure: Effects on child IQ and learning problems at age 7 ½ years, *Alcoholism: Clinical and Experimental Research 14* (1990): 662–669.

112. R. T. Brown, C. D. Coles, I. E. Smith, et al., Effects of prenatal alcohol exposure at school age. II. Attention and behavior, *Neurotoxicology and Teratology 13* (1991): 369–376.

113. L. C. Mayes, R. H. Granger, M. H. Bornstein, et al., The problem of prenatal cocaine exposure: A rush to judgment [commentary], *Journal of the American Medical Association 267* (1992): 406–408.

114. B. Zuckerman and D. A. Frank, Crack kids: Not broken, *Pediatrics 89* (1992): 337–339.

115. P. M. Whitaker-Azmitia, Altered brain maturation in rodents, Paper presented at Cocaine: Effects on the Developing Brain, Washington, D.C., September 16–19, 1997.

116. P. Levitt, In utero cocaine exposure causes specific cellular and molecular alterations in cortical development, Paper presented at Cocaine: Effects on the Developing Brain, Washington, D.C., September 16–19, 1997.

117. E. Friedman, D_1 dopamine receptor dysfunction in prenatal cocaine exposure, Paper presented at Cocaine: Effects on the Developing Brain, Washington, D.C., September 16–19, 1997.

118. M. S. Lidow, Effect of cocaine exposure on cerebral cortical development in nonhuman primates, Paper presented at Cocaine: Effects on the Developing Brain, Washington, D.C., September 16–19, 1997.

119. B. E. Kosofsky, Structural and functional correlates of cocaine-induced brain maldevelopment, Paper presented at Cocaine: Effects on the Developing Brain, Washington, D.C., September 16–19, 1997.

120. I. J. Chasnoff, Prenatal cocaine exposure: Implications for cognitive and behavioral development at six years of age, Paper presented at Cocaine: Effects on the Developing Brain, Washington, D.C., September 16–19, 1997.

121. The National Clearinghouse on Child Abuse and Neglect Information estimates that at least half and perhaps as many as 80 percent of child abuse cases involve substance abuse by one or both parents. Children living in alcohol-abusing families are four times more likely to be abused, five time more likely to be neglected.

122. C. G. Dowling, Women behind bars, part I, *Life Magazine,* May 1997, 77–90.

123. M. A. Hofer, On the nature and consequences of early loss, *Psychosomatic Medicine 58* (1996): 570–581.

124. M. J. Meaney, J. Diorio, D. Francis, et al., Early environmental regulation of forebrain glucocorticoid receptor gene expression: Implications for adrenocortical responses to stress, *Developmental Neuroscience 18* (1996): 49–72.

125. D. Francis, J. Diorio, P. LaPlante, et al., The role of early environmental events in regulating neuroendocrine development: Moms, pups, stress and glucocorticoid receptors, *Annals of the New York Academy of Sciences 794* (1996): 136–152.

126. G. W. Kraemer and A. S. Clarke, Social attachment, brain function, and aggression, *Annals of the New York Academy of Sciences 794* (1996): 125–135.

127. A. S. Clarke, D. R. Hedeker, M. H. Ebert, et al., Rearing experience and biogenic amine activity in infant rhesus monkeys, *Biological Psychiatry 40* (1996): 338–352.

128. A. S. Clarke, Social rearing effects on HPA axis activity over early development and in response to stress in young rhesus monkeys, *Developmental Psychobiology 26* (1993): 433–447.

Chapter 8. Leading the Charge

1. C. Vila, P. Savolainen, J. E. Maldonado, et al., Multiple and ancient origins of the domestic dog, *Science 276* (1997): 1687–1690.

2. D. McCaig, *Eminent Dogs, Dangerous Men* (New York: HarperCollins, 1991).

3. M. Taggart, *Sheepdog Training: An All-Breed Approach* (Loveland, CO: Alpine Publications, 1986).

4. D. T. Suzuki, A. J. F. Griffiths, J. H. Miller, et al., *An Introduction to Genetic Analysis* (New York: W. H. Freeman and Company, 1989).

5. Unlike white rabbits and laboratory rats, white lions are not albinos. Albinism is the total lack of body color; albinos not only have white fur, but also white skin and red (i.e., nonpigmented) eyes. White lions, on the other hand, still have pigment—just far less of it. Their white fur usually ages to cream or ivory, their skin is tan, and their eyes, like those of tawny lions, are gold.

6. In fact, white fur is so reticent that the existence of white lions wasn't verified until 1975, when graduate student Chris McBride photographed a pair of white cubs at the Timbavati Game Reserve in South Africa. See C. McBride, *The White Lions of Timbavati* (London: Paddington Press, 1977).

7. Philadelphia Zoo Public Relations Office.

8. Lassie was a classic sable and white collie. Shaula's sire, Van-M Night Train, was a tricolor—white and tan markings on a black background.

9. Collie Club of America, *The New Collie* (New York: Howell Book House, 1986).

10. R. F. Weaver and P. W. Hedrick, *Basic Genetics: A Contemporary Perspective* (Dubuque, Iowa: Wm. C. Brown Publishers, 1991).

11. K. H. Muench, *Genetic Medicine* (New York: Elsevier, 1988).

12. M. J. Welsh and A. E. Smith, Cystic fibrosis, *Scientific American* (December 1995): 52–59.

13. H. G. Brunner, MAOA deficiency and abnormal behaviour: Perspectives on an association, in *Genetics of Criminal and Antisocial Behaviour,* ed. G. R. Bock and J. A. Goode (West Sussex, England: Wiley, 1996), 155–167.

14. H. G. Brunner, M. R. Nelen, and X. O. Breakefield, Abnormal behavior associated with a point mutation in the structural gene for monoamine oxidase A, *Science 262* (1993): 578–580.

15. H. G. Brunner, M. R. Nelen, and P. van Zandvoort, X-linked borderline mental retardation with prominent behavioral disturbance: Phenotype, genetic localization, and evidence for disturbed monoamine metabolism, *American Journal of Human Genetics 52* (1993): 1032–1039.

16. J. Gelernter, Genetic association studies in psychiatry: Recent history, in *Handbook of Psychiatric Genetics,* ed. K. Blum and E. P. Noble (Boca Raton, FL: CRC Press, 1997).

17. Because MAOA processes amines in cheese, red wine, and certain processed foods, as well as over-the-counter medications, such as some cold preparations, enzyme deficiency could potentially allow blood pressure to rise high enough to precipitate a fatal hypertensive crisis.

18. E. S. Lander and N. J. Schork, Genetic dissection of complex traits, *Science 265* (1994): 2037–2045.

19. E. Hancock, The short course that's long on influence, *Johns Hopkins Magazine* (November 1996): 48–54.

20. I. I. Gottesman, Twins: En route to QTLs for cognition, *Science 276* (1997): 1522–1523.

21. K. Paigen, A miracle enough: The power of mice, *Nature Medicine 1* (1995): 215–220.

22. W. N. Frankel, Studying neurological diseases as complex genetic traits: Lessons from epilepsy, in *1996 Short Course 1 Syllabus,* ed. J. S. Takahashi (Washington, D.C.: Society for Neuroscience, 1996), 14–25.

23. D. Llewellyn-Jones, *Fundamentals of Obstetrics and Gynecology,* 6th edition (New York: Mosby, 1994).

24. R. E. Behrman and V. C. I. Vaughan, *Nelson Textbook of Pediatrics,* 12th edition (Philadelphia: Saunders, 1983).

25. G. Carey, Family and genetic epidemiology of aggressive and antisocial behavior, in *Aggression and Violence,* ed. D. M. Stoff and R. B. Cairns (Hillsdale, NJ: Erlbaum, 1996), 3–21.

26. G. Carey, Genetics and violence, in *Understanding and Preventing Violence,* ed. A. J. Reiss, Jr., K. A. Miczek, and J. A. Roth (Washington, D.C.: National Academy Press, 1994), 21–58.

27. K. O. Christiansen, Threshold of tolerance in various population groups illustrated by results from Danish criminological twin study, in *Ciba Foundation Symposium on the Mentally Ill Abnormal Offender,* ed. A. V. S. de Rueck and R. Porter (London: Churchill, 1968), 107–116.

28. B. Hutchings and S. A. Mednick, Criminality in adoptees and their adoptive and biological parents: A pilot study, in *Biosocial Bases of Criminal Behavior,* ed. S. A. Mednick and K. O. Christiansen (New York: Gardner Press, 1977), 127–141.

29. M. Bohman, Some genetic aspects of alcoholism and criminality: A population of adoptees, *Archives of General Psychiatry 35* (1978): 269–276.

30. B. Fogle, *The Encyclopedia of the Dog* (New York: DK Publishing, 1995).

31. R. Middaugh, University of Kansas, personal communication.

32. S. Chess, Temperament theory and clinical practice: The functional significance for psychiatrists and allied professions, Paper presented at the American Psychiatric Association annual meeting, May 4–9, 1996.

33. J. Kagan, S. J. Reznick, and N. Snidman, The physiology of behavioral inhibition in children, *Child Development 58* (1998): 1459–1473.

34. J. Kagan, Sanguine and melancholic temperaments in children, *Harvard Mental Health Letter 12* (1996): 4–6.

35. A. Caspi, B. Henry, R. O. McGee, et al., Temperamental origins of child and adolescent behavior problems: From age three to age fifteen, *Child Development 66* (1995): 55–68.

36. D. L. Newman, A. Caspi, T. E. Moffitt, et al., Antecedents of adult interpersonal functioning: Effects of individual differences in age 3 temperament, *Developmental Psychology 33* (1997): 206–217.

37. T. J. Bouchard, Genes, environment, and personality, *Science 264* (1994): 1700–1701.

38. R. C. Cloninger, R. Adolfsson, and N. M. Svrakic, Mapping genes for human personality, *Nature Genetics 12* (1996): 62–63.

39. R. P. Ebstein, O. Novick, and R. Umansky, Dopamine D4 receptor (D4DR) exon III

polymorphism associated with the human personality trait of novelty seeking, *Nature Genetics 12* (1996): 78–80.

40. J. Benjamin, L. Li, C. Patterson, et al., Population and familial association between the D4 dopamine receptor gene and measures of Novelty Seeking, *Nature Genetics 12* (1996): 81–84.

41. H. H. Goldsmith and I. I. Gottesman, Heritable variability and variable heritability in developmental psychopathology, in *Frontiers of Developmental Psychopathology,* ed. M. F. Lenzenweger and J. J. Haugaard (New York: Oxford University Press, 1995), 5–43.

42. R. Plomin, M. J. Owen, and P. McGuffin, The genetic basis of complex human behaviors, *Science 264* (1994): 1733–1739.

43. L. A. McInnes and N. B. Freimer, Mapping genes for psychiatric disorders and behavioral traits, *Current Opinion in Genetics and Development 5* (1995): 376–381.

44. S. C. Maxson, B. E. Ginsberg, and A. Trattner, Interaction of Y-chromosomal and autosomal gene(s) in the development of intermale aggression in mice, *Behavior Genetics 9* (1979): 219–225.

45. P. L. Roubertoux, M. Carlier, and H. Degrelle, Co-segregation of intermale aggression with the pseudoautosomal region of the Y chromosome in mice, *Genetics 135* (1994): 225–230.

46. S. C. Maxson, Issues in the search for candidate genes in mice as potential animal models of human aggression, in *Genetics of Criminal and Antisocial Behavior,* ed. G. R. Bock and J. A. Goode (New York: Wiley, 1996), 21–30.

47. C. C. Mann, Behavioral genetics in transition, *Science 264* (1994): 1686–1689.

48. M. A. Menza, L. I. Golbe, R. A. Cody, et al., Dopamine-related personality traits in Parkinson's disease, *Neurology 43* (1993): 505–508.

49. S. Pinker, *How the Mind Works* (New York: Norton, 1997).

50. J. L. Gariépy, M. H. Lewis, and R. B. Cairns, Genes, neurobiology, and aggression: Time frames and functions of social behaviors in adaptation, in *Aggression and Violence,* ed. D. M. Stoff and R. B. Cairns (Hillsdale, NJ: Erlbaum, 1996), 41–63.

51. R. B. Cairns, D. J. MacCombie, and K. E. Hood, A developmental-genetic analysis of aggressive behavior in mice: I. Behavioral outcomes, *Journal of Comparative Psychology 97* (1983): 69–89.

52. J. L. Gariépy, K. E. Hood, and B. D. Cairns, A developmental-genetic analysis of aggressive behavior in mice (*Mus musculus*): III. Behavioral mediation by heightened reactivity or increased immobility? *Journal of Comparative Psychology 102* (1988): 392–399.

53. R. B. Cairns, J. L. Gariépy, and K. E. Hood, Development, microevolution, and social behavior, *Psychological Review 97* (1990): 49–65.

54. Over successive generations, the propensity of the low-aggression line to freeze on contact steadily increased, while the reactivity of high-aggression animals stabilized. The divergence between the two lines therefore was more a function of the increase in social immobility in low-aggression animals than a substantial increase in the number of attacks by the high-aggression line.

55. R. B. Cairns, Aggression from a developmental perspective: Genes, environments and interactions, in *Genetics of Criminal and Antisocial Behavior,* ed. G. R. Bock and J. A. Goode (West Sussex, England: Wiley, 1996), 45–60.

56. M. J. Lyons, A twin study of self-reported criminal behaviour, in *Genetics of Criminal and Antisocial Behavior,* ed. G. R. Bock and J. A. Goode (West Sussex, England: Wiley, 1996), 61–75.

57. M. J. Lyons, L. J. Eaves, M. Y. Tsuang, et al., Differential heritability of adult and juvenile antisocial traits, *Psychiatric Genetics 3* (1993): 117.

58. R. B. Cairns, K. E. Hood, and J. Midlam, On fighting in mice: Is there a sensitive period for isolation effects? *Animal Behavior 33* (1985): 166–180.

59. R. B. Cairns and S. D. Scholz, Fighting in mice: Dyadic escalation and what is learned, *Journal of Comparative and Physiological Psychology 85* (1973): 540–550.

60. R. J. Nelson, G. E. Demas, P. L. Huang, et al., Behavioral abnormalities in male mice lacking neuronal nitric oxide synthase, *Nature 378* (1995): 383–386.

61. M. Hendricks, The mice that roared, *Johns Hopkins Magazine* (1996): 42–46.

62. National Center for Health Statistics.

63. T. M. Dawson, V. L. Dawson, and S. H. Snyder, A novel neuronal messenger in brain, nitric oxide, *Annals of Neurology 32* (1994): 297–311.

64. S. R. Vincent and H. Kimura, Histochemical mapping of nitric oxide in the rat brain, *Neuroscience 46* (1992): 755–784.

65. S. H. Snyder and D. S. Bredt, Biological roles of nitric oxide, *Scientific American* (May 1992): 22–29.

66. P. L. Huang and M. C. Fishman, Genetic analysis of nitric oxide synthase isoforms: Targeted mutation in mice, *Journal of Molecular Medicine* (1996): 414–421.

67. D. Brunner and R. Hen, Insights into the neurobiology of impulsive behavior from serotonin receptor knockout mice, *Annals of the New York Academy of Sciences 836* (1997): 81–105.

68. S. Ramboz, F. Saudou, and D. A. Amara, 5-HT1B receptor knock out—behavioral consequences, *Behavioral Brain Research 73* (1996): 305–312.

69. F. Saudou, D. A. Amara, A. Dierich, et al., Enhanced aggressive behavior in mice lacking 5-HT$_{1B}$ receptor, *Science 265* (1994): 1875–1878.

70. R. S. Oosting, K. L. Stark, S. Ramboz et al., Targeting aggressive behavior: Constitutive, inducible, and tissue-specific knockouts, in *1996 Short Course 1 Syllabus,* ed. J. S. Takahashi (Washington, D.C.: Society for Neuroscience, 1996), 50–58.

71. F. Balkwill, *DNA Is Here to Stay* (Minneapolis, MN: Carolrhoda Books, 1992).

Chapter 9. Picking Up the Pieces

1. Federal Bureau of Investigation, *Crime in the United States 1996* [Uniform Crime Reporting Program statistics] (Washington, D.C.: U.S. Department of Justice, 1996).

2. Doctor group says violence imperils nation, *New York Times,* November 6, 1995.

3. G. Lardner, Study says 34% in U.S. have seen women beaten by men, *Philadelphia Inquirer,* April 20, 1993.

4. National Clearinghouse on Child Abuse and Neglect, NIS-3 data show dramatic increase, in NCCAN Clearinghouse: Statistics Desk [on-line], 1994, available from Internet: nccanch@calib.com.

5. D. Carver, Workplace violence, in online coverage from the 9th Annual Psychiatric and Mental Health Congress, Medscape, 1996, available from Internet: http://www.medscape.com.

6. Aggressive driving boosting statistics, *Bucks County Courier Times,* August, 1997.

7. D. Russell, Driving ourselves to early graves, *Philadelphia Daily News,* July 7, 1997, 4–5.

8. J. Cass, Jail workers dislike get-tough laws, *Philadelphia Inquirer,* January 21, 1996, E1–E5.

9. F. deWaal, *Good Natured* (Cambridge, MA: Harvard University Press, 1996).

10. M. C. Diamond, E. R. Greer, A. York, et al., Rat cortical morphology following crowded-enriched living conditions, *Experimental Neurology 96* (1987): 241–247.

11. M. C. Diamond, R. E. Johnson, A. M. Protti, et al., Plasticity in the 904-day-old male rat cerebral cortex, *Experimental Neurology 87* (1985): 309–317.

12. M. A. Morgan, L. M. Romanski, and J. E. LeDoux, Extinction of emotional learning: Contribution of prefrontal cortex, *Neuroscience Letters 163* (1993): 109–113.

13. A. T. Beck and A. J. Rush, Cognitive therapy, in *Comprehensive Textbook of Psychiatry,* 5th edition, ed. H. I. Kaplan and B. J. Sadock (Baltimore: Williams and Wilkins, 1989), 1541–1550.

14. A. T. Beck, *Cognitive Therapy and the Emotional Disorders* (New York: International Universities Press, 1976).

15. J. M. Schwartz, P. W. Stoessel, L. R. Baxter, Jr., et al., Systematic changes in cerebral glucose metabolic rate after successful behavior modification treatment of obsessive-compulsive disorder, *Archives of General Psychiatry 53* (1996): 109–113.

16. S. Gilbert, Doctors found to fail in diagnosing addictions, *New York Times,* February 10, 1996.

17. S. Finger, *Origins of Neuroscience* (New York: Oxford University Press, 1994).

18. A. S. Lyons and R. J. Petrucelli II, *Medicine: An Illustrated History* (New York: Abrams, 1978).

19. R. Colp, History of Psychiatry, in *Comprehensive Textbook of Psychiatry,* 5th edition, ed. H. I. Kaplan and B. J. Sadock (Baltimore: Williams and Wilkins, 1989), 2132–2153.

20. K. A. Miczek, The psychopharmacology of aggression, in *Handbook of Psychopharmacology,* vol. 19, ed. L. L. Iversen, S. D. Iversen, and S. H. Snyder (New York: Plenum Press, 1987), 183–328.

21. K. A. Miczek, M. Haney, J. Tidey, et al., Neurochemistry and pharmacotherapeutic management of aggression and violence, In *Understanding and Preventing Violence,* vol. 2 (Washington, D.C.: National Academy Press, 1994), 245–514.

22. J. R. Lion, C. Azcarate, and H. Hoepke, "Paradoxical rage reactions" during psychotropic medication, *Diseases of the Nervous System 36* (1975): 557–558.

23. D. L. Gardner and R. W. Cowdry, Alprazolam-induced dyscontrol in borderline personality disorder, *American Journal of Psychiatry 142* (1985): 98–100.

24. U. S. Department of Justice, Sixteen states executed 56 offenders last year, Press release, Washington, D.C., 1996.

25. L. Cahill, The neurobiology of emotional memory: Evidence from human studies, Paper presented at Psychobiology of Posttraumatic Stress Disorder, New York, September 7–10, 1996.

26. S. H. Verhovek, "No frills" for prisoners? Wardens balk, *New York Times,* February 1, 1996.

27. Sean Kelley, Program Director, Eastern State Penitentiary.

28. N. Johnston, *Crucible of Good Intentions* (Philadelphia: Philadelphia Museum of Art, 1994).

29. L. Valzelli, Aggressive behavior induced by isolation, in *Aggressive Behavior,* ed. S. Garattini and E. B. Sigg (New York: Wiley, 1969), 70–75.

30. L. Valzelli and S. Bernasconi, Aggressiveness by isolation and brain serotonin turnover changes in different strains of mice, *Neuropsychobiology 5* (1979): 129–135.

31. K. M. J. Lagerspetz, Aggression and aggressiveness in laboratory mice, in *Aggressive Behavior,* ed. S. Garattini and E. B Sigg (New York: Wiley, 1969), 77–85.

32. A. S. Welch and B. L. Welch, Effect of stress and para-chlorophenylalanine upon brain serotonin, 5-hydroxyindolacetic acid and catecholamines in grouped and isolated mice, *Biochemical Pharmacology 17* (1968): 699–708.

33. N. Johnson, Beaver College, Glenside, PA, personal communication.

34. J. P. Scott, *Aggression* (Chicago: University of Chicago Press, 1969).

35. D. O. Lewis, on-line interview with WGBH/Frontline, May 14, 1997.

36. J. A. King, Perinatal stress and impairment of the stress response: Possible link to nonoptimal behavior, *Annals of the New York Academy of Sciences 794* (1996): 104–112.

37. R. M. Sapolsky and M. J. Meany, Maturation of the adrenocortical stress response: Neuroendocrine control mechanisms and the stress hyporesponsive period, *Brain Research Reviews 11* (1986): 65–76.

38. E. R. de Kloet, P. Rosenfeld, J. A. M. van Eckelen, et al., Stress, glucocorticoids, and development, *Progress in Brain Research 73* (1988): 101–120.

39. G. W. Sippell, H. G. Dorr, F. Bidlingmaier, et al., Plasma levels of aldosterone, corticosterone, 11-deoxycorticosterone, progesterone, 17-hydroxyprogesterone, cortisol, and cortisone during infancy and childhood, *Pediatric Research 14* (1980): 39–46.

40. W. P. Klein and N. G. Simon, Timing of neonatal testosterone exposure in the differentiation of estrogenic regulatory systems for aggression, *Physiology and Behavior 50* (1991):91–93.

41. Morbidity and Mortality Weekly Report [on-line], Centers for Disease Control, April 25, 1997, available from World Wide Web: http://www.cdc.gov.

42. R. Mathias, NIDA survey provides first national data on drug use during pregnancy, *NIDA Notes,* January–February 1995 [on-line], available from World Wide Web: http://www.nida.nih.gov.

43. C. P. O'Brien, A. T. McLellan, Myths about the treatment of addiction, *Lancet 347* (1996): 237–240.

44. J. Perrone, Red flags offer clues in spotting domestic abuse, *American Medical News,* January 6, 1992.

45. Sadly, more children die from abuse during the first two years of life than any other time.

46. B. M. Groves, How does exposure to violence affect very young children? *Harvard Mental Health Letter* (January, 1995): 8.

47. S. Turecki and S. Wernick, *The Emotional Problems of Normal Children: How Parents Can Understand and Help* (New York: Bantam, 1994).

48. J. Gray, Bill to combat juvenile crime passes House, *New York Times,* May 9, 1997, A1–A32.

49. K. L. Bierman and Conduct Problems Prevention Research Group, Integrating social-skills training interventions with parent training and family-focused support to prevent conduct disorder in high-risk populations, *Annals of the New York Academy of Sciences 794* (1996): 256–276.

50. National Institute of Justice, The cycle of violence revisited [on-line report], National Criminal Justice Reference Service, February 1996, available from World Wide Web: http://www.ncjrs.org. This study is convincing because of its size, duration, and scope. Court records were used to identify and verify cases of physical abuse, sexual abuse, and neglect; each case was then matched to a child of the same race, gender, and age who had attended the same elementary school as the victim. Researchers charted arrest rates for all 1,575 subjects over a twenty-six-year period.

51. C. S. Widom and M. G. Maxfield, A prospective examination of risk for violence among abused and neglected children, *Annals of the New York Academy of Sciences 794* (1996): 224–237.

52. National Center of Child Abuse and Neglect, National child abuse and neglect statistical fact sheet [on-line], National Clearinghouse on Child Abuse and Neglect Information, available from Internet: nccanch@calib.com.

53. U. Frith, *Autism: Explaining the Enigma* (Cambridge, MA: Blackwell, 1989).

54. L. Q. Hodgdon, Solving social-behavioral problems through the use of visually supported communication, in *Teaching Children with Autism: Strategies to Enhance Communication and Socialization,* ed. K. A. Quill (Albany, NY: Delmar, 1995), 265–286.

55. C. A. Gray, Teaching children with autism to "read" social situations, in *Teaching Children with Autism: Strategies to Enhance Communication and Socialization,* ed. K. A. Quill (Albany, NY: Delmar, 1995), 219–241.

56. M. T. Greenberg and C. A. Kosche, *Promoting Social and Emotional Development in Deaf Children: The PATHS Project* (Seattle, WA: University of Washington Press, 1993).

57. J. T. McCracken, A two-part model of stimulant action on attention-deficit hyperactivity disorder in children, *Journal of Neuropsychiatry 3* (1991): 201–209.

58. R. Barkley, University of Massachusetts, personal communication.

59. M. S. Swartz, J. W. Swanson, V. A. Hiday, et al., Violence and severe mental illness: The effects of substance abuse and nonadherence to medication, *American Journal of Psychiatry 155* (1998): 226–231.

60. J. Volavka, *Neurobiology of Violence* (Washington, D.C.: American Psychiatric Press, 1995).

61. K. M. Carroll, *A Cognitive-Behavioral Approach: Treating Cocaine Addiction,* Therapy Manuals for Drug Addiction, Manual 1 (Rockville, MD: National Institute on Drug Abuse, 1998).

62. C. P. O'Brien, Recent developments in the pharmacotherapy of substance abuse, *Journal of Consulting and Clinical Psychology 64:* 677–686.

63. C. P. O'Brien, A range of research-based pharmacotherapies for addiction, *Science 278* (1997): 66–70.

64. S. Everingham and C. Rydell, *Controlling cocaine* (Santa Monica, CA: RAND, 1994).

65. J. B. Treaster, Drug therapy: Powerful tool reaching few inside prisons, *New York Times,* July 3, 1995, 1–26.

66. M. Fava, J. F. Rosenbaum, J. A. Pava, et al., Anger attacks in unipolar depression, part 1: Clinical correlates and response to fluoxetine treatment, *American Journal of Psychiatry 150* (1993): 1158–1163.

67. The subjects in Coccaro's study did suffer from a range of personality disorders. Many, however, were not psychiatric patients, but ordinary people who referred themselves in response to a public service announcement. The group included professionals, executives, consultants, even a court official. A second study recently initiated by this group will examine the effectiveness of CBT and Prozac in men who abuse their partners.

68. E. F. Coccaro, R. J. Kavoussi, and R. L. Lauger, Serotonin function and antiaggressive response to fluoxetine: A pilot study, *Biological Psychiatry 42* (1997): 546–552.

69. E. F. Coccaro and R. J. Kavoussi, Fluoxetine and impulsive aggressive behavior in personality-disordered subjects, *Archives of General Psychiatry 54* (1997): 1081–1088.

70. J. W. Jefferson and J. H. Griest, Lithium therapy, in *Comprehensive Textbook of Psychiatry,* 5th edition, ed. H. I. Kaplan and B. J. Sadock (Baltimore: William and Wilkins, 1989), 1655–1662.

71. M. Sheard, J. Marini, C. Bridges, et al., The effect of lithium on impulsive aggressive behavior in man, *American Journal of Psychiatry 133* (1976): 1409–1413.

72. R. W. Cowdry and D. L. Gardner, Pharmacotherapy of borderline personality disorder: Alprazolam, carbamazepine, trifluoperazine, and tranylcypromine, *Archives of General Psychiatry 45* (1988): 111–119.

73. J. A. Mattes, Comparative effectiveness of carbamazepine and propranolol for rage outbursts, *Journal of Neuropsychiatry and Clinical Neurosciences 2* (1990): 159–164.

74. D. J. Stein, D. Simeon, M. Frenkel, et al., An open trial of valproate in borderline personality disorder, *Journal of Clinical Psychiatry 56* (1995): 506–510.

75. E. S. Barratt, M. S. Stanford, A. R. Felthous, et al., The effects of phenytoin on impulsive and premeditated aggression: A controlled study, *Journal of Clinical Psychopharmacology 17* (1997): 341–349.

76. F. Aureli and C. P. van Schaik, Post-conflict behavior in long-tailed macaques (*Macaca fascicularis*) I. The social events, *Ethology 89* (1991): 89–100.

77. F. Aureli and C. P. van Shaik, Post-conflict behavior in long-tailed macaques (*Macaca fascicularis*) II. Coping with the uncertainty, *Ethology 89* (1991): 101–114.

78. L. Cahill, B. Prins, M. Weber, et al., Beta-adrenergic activation and memory for emotional events, *Nature 371* (1994): 702–704.

79. B. A. van der Kolk, D. Dryfuss, M. Michaels, et al., Fluoxetine in post-traumatic stress disorder, *Journal of Clinical Psychiatry 55* (1994): 517–522.

80. M. J. Friedman, Drug treatment for PTSD: Answers and questions, *Annals of the New York Academy of Sciences 821* (1997): 359–371.

81. E. B. Foa, Psychological processes related to recovery from a trauma and an effective treatment for PTSD, *Annals of the New York Academy of Sciences 821* (1997): 410–424.

82. F. de Waal, *Peacemaking Among Primates* (Cambridge, MA: Harvard University Press, 1989).

Index

ablation, 18, 89

acetylcholine, 37, 41

ACTH (adrenocorticotrophic hormone), *152,* 172, 173, 223

addiction, 209–10, 286–87

ADHD (attention deficit–hyperactivity disorder), 200–205, 283–84

adolescence: aggression in boys, 158–61; aggression in girls, 170–71; stress of, 177–78

adoption studies, 238, 243

adrenal gland: in endocrine system, 151; in stress management, 171, 172, 173. *See also* hypothalamic-pituitary-adrenal axis

adrenaline (epinephrine), 119, 171, 172, 202, 284

adrenocorticotrophic hormone (ACTH), *152,* 172, 173, 223

affective aggression, 75

aggression: adaptive versus maladaptive, 76–77; animal models of, 54–78; in animals, 13–16, 56; brain structures associated with, 88, 111; in children, 279–85; classifications of, 73–76, *74;* cocaine as causing, 206–7; dominance hierarchies in, 13; dopamine in, 132; frustration as cause of, 13–14; genes in, 225–58; hypothalamus in, 90, 91, 92–93; internal and external inputs required in, 48; Lorenz on, 15–16; and monoamine oxidase type A, 234–35; nature and nurture collaborating in, 51; nature-based explanations of, 31; as necessary part of social interaction, 261; neurochemistry of, 115–49; norepinephrine in, 126–28; primate studies, 57–60; psychosurgery seen as cure for, 24–25; QTL analysis in genetics of, 244; resident-intruder model compared with pain-induced model, 65–67; in rhesus monkeys, 96; rodent studies, 60–64; serotonin in, 138–45; in social species, 56; stress disorders as congruent with, 50; testosterone in, 154–61; twin studies on, 242–43; in women, 167–71. *See also* impulsive aggression; violence

aggressive efficacy, 159

agonists, 116

alarm response, 117–22

Alcock, John, 14

alcohol, 213–18; binge drinking, 217, 220; dosage and aggression, 217; GABA receptor used by, 215–16; and impulsive behavior, 216; prenatal exposure to, 219–20, 276; and violence, 213–18

alcohol-heightened aggression (AHA), 215

allostasis, 182–83, *183,* 254

allostatic load, 182–83, *183, 184*

alpha receptors, 119, 173

altruism, 16

amine group, 117

amphetamines: dextroamphetamine, 205. *See also* methamphetamine

amygdala, 93–99, *95;* in brain anatomy, *87;* in cat predation, 92, 93; connections with frontal cortex, 104–6, *105;* connection with thalamus, 104, *105,* 106; CRF-containing cells in, 172; in dopaminergic pathways, *131,* 132; in emotion, 93–99, 104; in memory, 188; monoaminergic input to, 118; in noradrenergic pathways, *120,* 125; nuclei of, 94, *95;* in recognition of faces, 99; removal in monkeys, 96–98; in risk assessment, 189; in serotonin pathways, 136, *136,* 143; in startle response, 125

anabolic steroids, 155

anger: cocaine as causing, 207; commonalities with fear, 50; neurochemistry of, 115–49; pain as cause of, 14

animals: aggression in, 13–16, 56; animal models of aggression, 54–78; Lorenz on aggression in, 15–16. *See also* birds; cats; dogs; lions; primates; rodents

anterior thalamus, 86, *87*

antiaddiction medications, 286–87

anticonvulsants, 289

antidepressants, 137, 265, 266, 287–88

antipsychotic drugs, 197, 208, 286

antisocial behavior, 128–30; attention deficit–hyperactivity disorder associated with, 201; and childhood temperament, 242, 282; impulsive aggression contrasted with, 141; and maternal deprivation, 223; prevention as best solution to, 291; stress response system in, 181; and testosterone, 160; treatment of, 290–91; twin and adoption studies on, 238–39. *See also* antisocial personality disorder

antisocial personality disorder, 129, 130; in men, 153; as stress response disorder, 181, 183, 185, *185;* treatment of, 290–91

antitestosterone therapy, 163–65

aphasia, 19, 45

apomorphine, 133

appetites, 13, 16

Archer, John, *74,* 75–76

Arnold, Steve, 82, 88, 94, 107

arson, 135, 140, 216

Asberg, Marie, 139

association cortex, 189

Aston-Jones, Gary, 122

atavisms, 8

attention: attention deficit–hyperactivity disorder (ADHD), 200–205, 283–84; brain function in, 202–4, *203*

auditory cortex, 103–4

autism, 282

autonomic function, 88

autoreceptors, 116

aversions, 13, 14, 16

axons, 41, *42,* 118

baboons, olive, 174–76

Bard, Philip, 88

Barkley, Russell, 204, 205, 284

Barnett, S. A., 61–62, 65, 73, 74

Bartholow, Roberts, 19

basal ganglia, 131–32

basal nucleus of the amygdala, *95, 96*

basolateral subdivision of the amygdala, 94, 305n.32

battered women. *See* domestic violence

BDNF (brain-derived neurotrophic factor), 37, 39, 41

behavior: brain function in, 5; circular relation with chemistry, 148; as dialogue between experience and physiology, 116–17; genes and environment in, 227–28, 247–58; herding behavior in dogs, 225–27, 239–40; nature/nurture conflict in, 31–32; role of genes in, 29; social interactions in development of, 51–52, *53. See also* aggression

behavioral biology: effect of eugenics on, 11; evolutionary concerns as dominating, 14–15; fears of persecution of mentally ill in criticism of, 28–29; and free will, 295–96; ideological opposition to, 29; moral opposition to early attempts at, 9; spiritual implications of, 295

behavioral drug therapy. *See* psychopharmacology

behavioral genetics, 225–58

behavioral inhibition system, 130

behavioral sensitization, 72–73

behavioral therapy. *See* cognitive behavioral therapy

Bell, John, 7–8

Berlin, Fred S., 161–65, 166, 289

beta receptors, 119, 173

Bianchi, Leonardo, 100, 102

Bierman, Karen, 279, 283

binge drinking, 217, 220

biofeedback, 294

biology: developmental biology, 33–34; genetics as taking over, 11; interventions requiring attention to, 263–64;

sociobiology, 16–17, 247. *See also* behavioral biology; genetics; neuroscience

biomedical social control, 29

bipolar (manic-depressive) disorder, 288

birds: pecking order, 13; predation by, 73; sparrow territory defense, 12–13; testosterone and aggression in roosters, 154

birth, 274

Black, Ira, 41, 43

Blanchard, Caroline, 65–66, 70–71, 75, 76, 178–80, 186, 218

Blanchard, Robert, 65–66, 70–71, 75, 76, 178–80, 186, 218

blood pressure, 182

blue merle collies, 231–32

border collies, 226, 239, 240, 245–46

Boston Strangler, 166

brain: alarm and self-defense systems, 47; analgesics produced in, 145–48; anatomy of, 83–114; animal and human as similar, 78; behavior as reflection of function of, 5; case histories in study of, 106; the city as metaphor for, 82–83; as continuing to develop in adulthood, 262; cross-sectional (midsagittal) view of, *84;* dopaminergic pathways in, *131;* drugs as affecting, 208; elementary response patterns in, 45; hypothalamic-pituitary-adrenal axis, *152;* imaging studies of, 107–15; key nodes, 82, 83, 88; noradrenergic pathways in, 119, *120,* 125; network model of, 83, 107; neurochemical record-keeping in, 117; physical injury to, 43; schizophrenia as developmental error, 197–200; serotonin pathways in, *136;* stress as causing long-term damage to, 186–87; in vicious circle leading to violence, 52.

brain *(cont.)*
   See also functional localization; nora-
   drenergic pathways; *and structures by*
   name
Brain, Paul, 155, 214
brain-derived neurotrophic factor
   (BDNF), 37, 39, 41
brain imaging. *See* imaging studies
brain stem, 83, 84, *84*
Breakefield, Xandra, 234
Breggin, Peter, 3, 29
Broca, Paul, 19, 45, 85; and limbic lobe,
   85, *86*
Brown, Sanger, 96–97
Brunner, Hans G., 234, 235
Brunner's syndrome, 233–35
Bucy, Paul, 97
Bundy, Ted, 165

Cahill, Larry, 124, 294
Cairns, Robert, 247–50, 251–53, 269, 272
California: chemical castration law, 163;
   methamphetamine use in, 207
Cannon, Walter, 88, 90, 182, 210,
   304n.17
captive studies, 55–56
carbamazepine (Tegretol), 289
cardiovascular disease, 177–78
Carey, Gregory, 236, 239, 240, 243, 251
Carolina Longitudinal Study, 168
case histories, 106
castration, chemical, 163–65, 291
catatonic schizophrenia, 194, 196
catecholamines, 161, 171, 172, 173, 186
cats: as model of aggression, 91; Hess's
   stereotaxic surgery on, 90; hypothalamus in predation by, 92–93; norepinephrine and aggression in, 127; sham
   rage reaction in, 88
central nervous system: chemical messages in, 115. *See also* brain
central nucleus of the amygdala, 94, *95,*

96, 103, 104
Central Phrenological Society, 7
centromedial subdivision of the amygdala, 94
cerebellar peduncles, 84
cerebellum, *84,* 84–85
cerebral blood flow, 108, 109, *110,* 112
cerebral cortex: in connecting present
   and past, 188; higher mental functions associated with, 18. *See also*
   frontal cortex; prefrontal cortex
c-*fos,* 39, 43, 44, 121, 133
Chance, Michael, 62, 65
Chasnoff, Ira, 220–21, 277
chemical castration, 163–65, 291
Cheney, Dorothy, 45–46, 47
Chess, Stella, 240–41, 245, 246, 279
children: aggression in, 279–85; attention
   deficit–hyperactivity disorder in,
   200–205, 283–84; autism, 282; chemical castration for child molesters,
   163–65; crack babies, 219, 220–21;
   and maternal separation, 222–23; neglect of, 280; physical abuse of, 167,
   280; preventing developmental damage
   in, 274–79; sexual abuse of, 111, 112,
   167, 291, 307n.76; startle response in
   PTSD, 161, *162;* of substance-abusing
   mothers, 219–22; temperament in,
   240–42, *241,* 245, 278–79
Child Witness to Violence Project, 278
chimpanzees: behavior toward victims of
   violence, 292; "fight, then make up"
   sequence in, 59–60; murder among,
   58, *59,* 77; war among, 58–59
chlorpromazine (Thorazine), 22, 197
chronic mental illness: antipsychotic
   drugs, 197, 208, 286; delusional disorders, 194–95, 285–86; and violence,
   195. *See also* paranoia; schizophrenia
Chun, Linda, 37
cingulate gyrus, 85, 86, *86, 87,* 93, *120*

c-*jun,* 39
Clark, Robert Henry, 89
cocaine, 205–7; cost of treatment for, 287; crack babies, 219, 220–21; crack cocaine, 205, 207, 219, 220–21; dopamine function affected by, 208–9; long-term consequences of, 210–12; as pharmacological stressor, 213; transporter, 208–9
cocaine-induced sensitization, 72, 212–13
Coccaro, Emil, 140–41, 142–43, 266, 288, 331n.67
cognitive behavioral therapy (CBT), 264–65; for impulsive aggression, 287; for substance abuse, 286; for victims of trauma, 294–95
collies, 230–32
common marmosets, 176–77
community policing, 290
competitive aggression, 75–76
complex traits, 235; locating genes for, 243–45
computed tomography (CT), 107, 109
conscience, 181, 295, 296
control units (prisons), 272
coping, 171, 185
corpus callosum, *84, 85*
cortical nucleus of the amygdala, 94, *95*
corticomedial subdivision of the amygdala, 94
corticosterone, 179, 222
corticotropin releasing factor (CRF), *152,* 172, 173, 180, 181
cortisol: in children, 274, *275;* and depression, 180–81, 187; in endocrine system, 151; and hippocampal atrophy, 187; and maternal deprivation in monkeys, 223; in posttraumatic stress disorder, 181, 187; in subordination stress in baboons, 175–76
crack cocaine, 205, 207, 219, 220–21

Craig, Wallace, 13
crank. *See* methamphetamine
CRF (corticotropin releasing factor), *152,* 172, 173, 180, 181
Crick, Francis, 33
crime: by adolescent males, 158; alcohol consumption associated with, 214; arson, 135, 140, 216; conference on genetics and, 2–4; "crime gene," 255; insanity defense, 26–29, 285; Lombroso on, 8; phrenology on, 8; rate of, 259; temporary insanity defense, 27; twin and adoption studies on, 238; twin studies of, 238, 250–51. *See also* murder; sex offenders
CT (computed tomography), 107, 109
Cushing's disease, 112
cystic fibrosis, 233
cystic fibrosis transmembrane conductance regulator (CFTR), 233

DaCosta, Jacob, 48
Dahmer, Jeffrey, 165
Damasio, Antonio, 101–2, 105, 106, 191, 192
Damasio, Hanna, 101
Darwin, Charles, 9, 12
Davis, Michael, 125
Dawson, Ted, 254–56
Dawson, Valina, 256
death penalty, 291–92
DeBold, Joseph, 155
defeat: dominant baboons' response to, 175; giving-up response, 147, 180, *184;* and sexual offenses, 164; and stimulant use, 213; stress-induced analgesia easing reaction to, 147; as stressor, 133; in subordinate rats, 179; and testosterone levels in rhesus monkeys, 157
deinstitutionalization of the mentally ill, 286

delayed gratification, 216

Delgado, José, 23

delusional disorders, 194–95, 285–86

dendrites, 39, 41

Denmark, twin and adoption studies in, 238

dentate gyrus, 85, *86*

deoxyhemoglobin, 112, 308n.77

Depakote (sodium valproate), 289

Depo-Provera, 163, 164, 165

depression: antidepressants, 137, 265, 266, 287–88; cortisol levels in, 180–81; hippocampal atrophy in, 187; Prozac for treating, 137, 288; as stress response disorder, 183, 185, *185;* suicide as related to, 138–39

Desert Storm veterans, 125

Design with Nature (McHarg), 79

determinism: biological, 11, 25, 28, 29, 31; environmental, 44; and prenatal drug exposure, 221

developmental biology, 33–34

de Waal, Frans, 17, 57–58, 60, 96, 261, 296

dexamethasone, 175, 177

dextroamphetamine, 205

D4 dopamine receptor, 245

diaminobenzidene, 114

diencephalon: in brain anatomy, *84,* 85. *See also* hypothalamus

differentiation: defined, 33; environmental influences in, 38–44; in the nervous system, 34–44

dihydroxyphenylethylamine. *See* dopamine

Dilantin (phenytoin), 289

dispositional representations, 191

distraction, 294

dizygotic twins. *See* fraternal (dizygotic) twins

DNA: as medium of genetic transaction, 33. *See also* genes

dogs: border collies, 226, 239, 240, 245–46; collie coat color, 230–32; CRF's effect on, 172; herding behavior in, 225–27, 239–40

domestic violence: incidence of, 1–2, 259; during pregnancy, 277–78; women who kill their abusers, 64–65

dominance hierarchies: in chimpanzees, 58; early ethology on function of, 56; in motivating and restraining aggression, 13; pecking order, 13; subordination stress in baboons, 174–76; subordination stress in hamsters, 177; subordination stress in rats, 178–80, 186

dominant genes, 229, 232

dopamine: in aggression, 131–33; alcohol affecting, 216, 219; antiaddiction medications affecting, 287; in attention, 203, *203,* 204; in attention deficit–hyperactivity disorder, 205; cocaine transporter/interfering with transporter for, 208–9, 220; in cortical function, 189; dopaminergic pathways in the brain, *131;* drug use affecting, 208, 210, *211,* 212; methamphetamine blocking reuptake of, 209; in mouse aggression, 133, *134;* and novelty seeking, 244–45; in prefrontal cortical action, 189–91; reward associated with, 131–32, 208–9; Ritalin's effect on, 284; in schizophrenia, 197, 199; Thorazine's effect on, 197; in working memory, *190*

double dilutes, 232

Drosophila melanogaster (fruit flies), 236

drug abuse. *See* substance abuse

drug rehabilitation programs, 287

drug therapy. *See* psychopharmacology

du Pont, John, 26–27, 192–93

Durham v. United States (1954), 28

dyscontrol syndrome, 24, 141

Eastern State Penitentiary (Philadelphia), 268–72, *270, 271*

Eichelman, Burr, 126

"Elliot" case, 101–2, 192

embryology, and genetics, 33

emotions: amygdala in, 93–99, 104; brain structures associated with, 85–88, *87;* in dispositional representations, 191; frontal cortex in, 99–106; genetic influence on, 45; as linked to memory, 124–26; neurotransmitters underlying, 123; nitric oxide and, 255; propranolol affecting, 294; secondary emotion, 105. *See also* anger; fear

empathy, 185

endocrine system, 150–87; amygdala function in, 94; in stress management, 171; thyroid gland, 151. *See also* adrenal gland; hormones; hypothalamus; pituitary gland

endorphins, 146, 147

enhancer region, 41

enkephalin, 146

entorhinal cortex, *86,* 198, *198*

environment: and aggressive efficacy in adolescent males, 159; the brain as influenced by, 34; the brain processing information about, 153; in developmental biology, 33–34; in differentiation of the nervous system, 38–44; and genes in behavior, 227–28, 247–58; interventions requiring attention to, 263–64; as IQ predictor in drug-exposed children, 221; "nature" viewpoint on, 31; and quantitative traits, 236; social interventions for improving, 267–68; in temperament, 245; twin and adoption studies, 237–39; in vicious circle leading to violence, 52, *53. See also* nature/nurture debate

Epidemiologic Catchment Area Survey, 195

epilepsy, 82, 89

epinephrine (adrenaline), 119, 171, 172, 202, 284

Epstein, Charles, 11, 12

Ervin, Frank, 24, 141

estrogen, 151, 169, 170–71

ethograms, 62

ethology: animal models of aggression, 54–78; defined, 12; evolutionary concerns as dominating, 14–15. *See also* animals

eugenics, 9–11, 29

Evenden, John, 216

evolutionary psychology, 247

executive functions, 82, 191, 283

experience: behavior as dialogue between physiology and, 116–17; personal history, 153, 188

exposure therapy, 294

extinction, 68

faces: amygdala in recognition of, 99; facial expression in communication, 46–47

families: family interaction theories of schizophrenia, 196; hereditary traits in, 236; intergenerational violence in, 280. *See also* domestic violence

Farnham, Eliza, 8

Fast Track program, 283

fear: brain structures underlying, 88; commonalities with anger, 50; cortical function in, 106

fear conditioning, 67–70; amygdala and cortex in, 103–4; CRF facilitating, 172; glucocorticoids slowing, 173; prefrontal cortex in, 264; and startle response, 124–25

feedback mechanisms: in endocrine system, 151; in intraneuronal communi-

feedback mechanisms *(cont.)*
cation, 116, 122, 124; in stress management, 173; subordination stress affecting, 175
fenfluramine, 143
Ferrier, Sir David, 81–82, 100, 101,
304n.6
fetal alcohol effects (FAE), 219
fetal alcohol syndrome (FAS), 219
fibroblast growth factor (FGF), 37
Field, Tiffany, 46
field studies, 55–56
fight-or-flight response: depressive disorders and impulsive aggression as related to, 288; and maternal
deprivation in monkeys, 223; norepinephrine in, 118, 119; and pain response, 146; sympathetic nervous
system in, 89, 172, 304n.17
Finger, Stanley, 8
5-HT$_{1B}$ receptor, 257
5-hydroxyindoleacetic acid (5-HIAA),
139, 140, 216
flashbacks, 123, 124, 185
Flourens, Marie-Jean-Pierre, 8, 18, 30,
82
fluorescence, 118
Flynn, John, 91
Foa, Edna, 294
forgetfulness, 111–12, 186, 188
fornix, *86*
Fossey, Dian, 57
fraternal (dizygotic) twins, 237, 238;
criminal behavior in, 250–51; temperament and personality in, 242–43
freebase cocaine, 206
Freeman, Walter, 21–22
free will, 295–96
Fritsch, Gustav, 19, 81, 100
frontal cortex: in brain anatomy, *84, 85,
87,* 100; connections with amygdala,
104–6, *105;* damage to, 101–2; in

dopaminergic pathways, *131;* in emotion, 99–106; in perpetrators of violence, 110–11
frontal lobe, *86*
fruit flies (*Drosophila melanogaster*), 236
frustration: aggression caused by, 13–14;
frustration-induced aggression as competitive aggression, 75; resident-intruder model of aggression compared
with frustration-induced model,
65–67
Fulton, John, 20
functional localization, 81–83; electrical
stimulation research, 22–23; experimental demonstration of, 17–20; in
Gall's phrenology, 9; psychosurgical
applications of, 20–25
functional MRI (fMRI), 112, 113
Futterman, Andrew, 3, 4

GABA (gamma-aminobutyric acid), 189,
190, *190,* 197, 215–16
Gacy, John Wayne, 165
Gage, Phineas, 101
Gall, Franz Joseph, 6–8, 17, 18
Galton, Francis, 9, 10, 237
Galvani, Giovanni, 19
Galvani, Luigi, 19
gamma-aminobutyric acid (GABA), 189,
190, *190,* 197, 215–16
genes, 225–58; altruism as explained in
terms of, 16; in collie coat color,
230–32; for complex traits, 243–45;
"crime gene," 255; defeat-induced
analgesia affecting, 147; as determining behavior, 29; in differentiation,
33; elementary response patterns as
coded by, 45; and environment in behavior, 227–28, 247–58; genetic disorders, 232–33; in herding behavior
in dogs, 226–27; immediate early
genes, 39, 43, 121; mice for genetic

studies, 236–37; in nervous system differentiation, 34, 37, 38, 41, 43–44; perceptual framework as coded by, 44; pleitropic genes, 232; selfish gene hypothesis, 17; twin studies, 237–39, 242–43; in white lions, 228–30, *230, 231. See also* heredity

genetic disorders, 232–33

"Genetic Factors in Crime: Findings, Uses, and Implications" (conference), 3

"genetic hygiene," 10

genetics: behavioral genetics, 225–58; biology as taken over by, 11; conference on crime and, 2–4; and embryology, 33; eugenics, 9–11, 29; linkage analysis, 243–45; restriction fragment length polymorphism, 243. *See also* genes

genetic technology, 258

genotype, 235

Geoffroy-Saint-Hillaire, Isidore, 12

geographic information system (GIS) programs, 107

giving-up response, 147, 180, *184*

glucocorticoids: corticosterone, 179, 222; dexamethasone, 175, 177; hippocampus as affected by, 186–87, 318n.111; in hypothalamic-pituitary-adrenal axis, *152;* in stress management, 171–73; in stress response disorders, 183, *184, 185*; in subordinate rats, 179. *See also* cortisol

glucose metabolism, 108, 109, *110*

glutamate, 189, 197

Goldman-Rakic, Patricia, 190, 191, 256

gonads, 151

Goodall, Jane, 57, 58, 77

Goodwin, Frederick, 2–3, 128

Grant, Ewen, 62

Gray, J. Q., 130

growth (trophic) factors, 36, 37, 38, *40,* 41

gun control, 290

Halgren, Eric, 96

Hamburger, Viktor, 36

hamsters, 177

Harlow, John, 101

Harrison Act (1914), 206

Hen, Rene, 257

herding behavior in dogs, 225–27, 239–40

heredity: in nature-based explanation of behavior, 31. *See also* genes; heritability of a trait; nature/nurture debate

heritability of a trait, 236; twin and adoption studies of, 237–39

Hess, Walter Rudolph, 90–91

heterozygotes, 229

higher mental functions: cerebral cortex in, 18; frontal cortex in, 100, 191. *See also* reason

Hillside Strangler, 166

Hinckley, John, Jr., 28

hippocampus: in brain anatomy, 85, 86, *87;* in cat aggression, 92, 93; CRF-containing cells in, 172; monoaminergic input to, 118; in noradrenergic pathways, *120;* in serotonin pathways, 136, *136;* stress causing long-term damage to, 186–87, 318n.111; in victims of violence, 112, 307nn. 71, 76

histofluorescence, 118

Hitzig, Eduard, 19, 81, 100

Hofer, Myron, 222

homeostasis, 90, 182, *183,* 304n.17

homozygotes, 229

Hood, Kathryn E., 167, 168, 170

hormone receptors, 151

hormones: and aggression in men, 153–58; and aggression in puberty, 159–61; and aggression in women, 167–71; as chemical messengers, 150;

hormones *(cont.)*
environment as affecting, 153; stress hormones, 171–73. *See also* glucocorticoids, steroid hormones

Horsely, Victor Alexander Haden, 89

5-HT$_{1B}$ receptor, 257

Hughes, John, 145–46

5-hydroxyindoleacetic acid (5-HIAA), 139

Hyman, Steven, 38, 39

hypothalamic-pituitary-adrenal axis, *152;* in antisocial individuals, 181; the brain in regulation of, 173; in early childhood, 274–75; in PTSD, 180; and maternal separation, 222–23; subordination stress affecting, 175

hypothalamus, 88–93; in brain anatomy, *84, 86, 87;* in cat aggression, 91–93; in dopaminergic pathways, *131,* 132; in neuroendocrine function, 150, 151, *152,* 153; nuclei of, *84,* 89; in serotonin pathways, 136, *136;* in sham rage reaction, 88; in stress management, 172; and subordination stress in hamsters, 177. *See also* hypothalamic-pituitary-adrenal axis

identical (monozygotic) twins, 237–38; criminal behavior in, 250–51; temperament and personality in, 242–43

IED (intermittent explosive disorder), 141–42

IEGs (immediate early genes), 39, 43, 121

imaging studies, 107–15; of cognitive behavioral therapy, 265; of victims of violence, 111–12; in violence research, 108–15

immediate early genes (IEGs), 39, 43, 121

immigrants, eugenics movement on, 10

immune system, 173, 274

immunocytochemistry, 118

impulsive aggression: alcohol associated with, 216; anticonvulsants for treating, 289; antisocial behavior contrasted with, 141; cognitive behavioral therapy for, 287; criteria for, *142;* and intermittent explosive disorder, 141–42; lithium for treating, 288–89; Prozac for treating, 288; serotonin in, 139–43; SSRIs for treating, 289; stress response and, 177–78, 183, 185, *185*

indoleamines, 135

infancy: parents as the environment during, 278–79; preventing stress during, 274–75

infanticide, 9–10

insanity defense, 26–29, 285

intermittent explosive disorder (IED), 141–42

internal clock, 173

interventions, 259–96; biology and environment required in, 263–64; preventing developmental damage to children, 274–79; social interventions, 267–68; for victims of violence, 292–95. *See also* cognitive behavioral therapy; psychopharmacology

IQ: of crack babies, 221; and prenatal exposure to alcohol, 219

isolation: as breeding violence, 269, 272; in prisons, 268–69

Jackson, John Hughlings, 82, 163

Jacobsen, Carlyle, 20

Johnson-Reed Act (1924), 10

Julia case, 24

Kaczynski, Theodore, 12

Kagan, Jerome, 242

Kandel, Eric, 5, 106

Kardiner, Abram, 48

Keller, Evelyn Fox, 11, 33

key nodes, 82, 83, 88
"killer instinct," 16, 17, 56
kindling, 43–44, 72, 166, 186, 212
Kling, Arthur, 98
Klüver, Heinrich, 97
Klüver-Bucy syndrome, 97–98
Knockout mutants, 254–57
Kosterlitz, Hans, 145–46
Kraemer, Gary, 223
Krystal, John, 83, 112
Kuhar, Michael, 208–9

LAAM (1-acetyl-methadol), 286
language: aphasia, 19, 45; and primate
    communication, 45
lateral hypothalamus, *84,* 92–93
lateral nucleus of the amygdala, 94, *95,*
    96, 103, 104, 188
l-dopa rage, 132–33
LeDoux, Joseph, 68, 103–4, 106,
    124–25, 264
leucotomy, prefrontal, 21
Levi-Montalcini, Rita, 36
Levitt, Pat, 220
Lewin, Ludwig, 206, 207
Lewis, Dorothy Otnow, 273
Libertae Family House, 277
limbic lobe, 85, *86*
limbic system, 86–88, *87;* in connecting
    past and present, 188; glucocorticoid
    receptors in, 173
Link, Bruce, 196
linkage analysis, 243–45
Linnoila, Markku, 139–40, 143, 216
lions: males killing cubs, 75; white lions,
    228–30, *230, 231,* 325nn. 5, 6
lithium, 288–89, 290
lobotomy, prefrontal, 22, 99–100
locus ceruleus: in alarm response, 119,
    122; in attention, 202–3, *203,* 204; in
    attention deficit–hyperactivity disor-
    der, 205; CRF-containing cells in,

172; in endocrine system, 153; in no-
    radrenergic pathways, *120,* 120–21; in
    threat perception, 160, 172
lod score, 243
Lombroso, Cesare, 8, 9
long-tailed macaques, 292, 296
Lorenz, Konrad, 15–16
lust murder, 166

macaques, long-tailed, 292, 296
Mackintosh, John, 62
MacLean, Paul, 86, *87*
magnetic resonance imaging (MRI),
    107, 111–12, 114, 307n.71
magnetoencephalography (MEG),
    112–13
male violence: in adolescent boys,
    158–61; estrogen in, 171; female ag-
    gression contrasted with, 167–69; re-
    lationship to testosterone, 154–61; sex
    offenders, 161–67; 154–61. *See also*
    domestic violence
mammillary body, *87*
Mangold, Hilde, 34
manic-depressive (bipolar) disorder,
    288
MAOA (monoamine oxidase type A),
    234–35
Mark, Vernon, 24, 141
marmosets, common, 176–77
Marschak, Eve, 225, 226–27, 240,
    245–46
maternal aggression, 75, 169
maternal separation, 222–23
McCaig, Donald, 226
McEwen, Bruce, 182–83, 186
McGaugh, James, 125, 294
McHarg, Ian, 79–81, 107
Meaney, Michael, 222
medial geniculate body, 103
medial nucleus of the amygdala, 94, *95*
medication. *See* psychopharmacology

MEG (magnetoencephalography), 112–13

memory: flashbacks, 123, 124, 185; forgetfulness, 186, 188; as linked to emotion, 124–26; and stress-related damage to the hippocampus, 186–87; of trauma, 124; trauma and deficits in short-term, 112. *See also* working memory

men: as more violent than women, 153; suicide in, 153. *See also* male violence; sex offenders

Mendel, Gregor, 228–29, 235

menstruation, 169–70

mental illness: bipolar (manic-depressive) disorder, 288; deinstitutionalization, 286; insanity defense, 26–29, 285; schizophrenia, 192–200; temporary insanity defense, 27; violence associated with, 28. *See also* antisocial personality disorder; chronic mental illness; stress response disorders

mental retardation: in Brunner's syndrome, 233; prenatal exposure to alcohol as cause of, 219

mesoderm, 34

mesolimbic dopamine pathway, 132, *134,* 209, *211,* 216

methadone, 286

methamphetamine, 207–8; dopamine reuptake blocked by, 209; effect on dopamine release, long-term consequences of, 210–12

mice: alcohol and aggression in, 214–15; breeding for aggressive behavior, 248–50, *249,* 251–52; dopamine and aggression in, 133, *134;* female and male aggression compared, 168–69; 5-HT$_{1B}$ gene and aggression in, 257; for genetic studies, 236–37; isolation and aggression in, 269, 272; nitric oxide synthase mutants and aggression

in, 254–56; prenatal testosterone influence in, 158–59; in propranolol study, 127; QTL analysis in genetics of, 244; serotonin and aggression in, 137, 138, 143; stress-induced analgesia in, 147; testosterone and aggression in male, 155

Miczek, Klaus: on alcohol and aggression, 214–15; on causal relationship between biology and violence, 116, 127; on chronic drinkers, 217; on dopamine in mouse aggression, 133; monitoring serotonin levels in rats, 144; on resident-intruder aggression in rats, 62; on serotonin, 137; on testosterone and male violence, 154, 155

miscegenation statutes, 10

M'Naughten, Daniel, 27

M'Naughten test, 27, 28

Moffitt, Terri, 242

Moniz, Egas, 20–21

monkeys: alcohol dosage and aggression in, 217; community formation in common marmosets, 176–77; CRF's effect on, 172; electrical stimulation of brain of, 23; hormonal fluctuation in squirrel monkeys, 155; subordination stress in olive baboons, 174–76; victims of violence among long-tailed macaques, 292, 296. *See also* rhesus monkeys

monoamine oxidase type A (MAOA), 234–35

monoamine transmitters, 117–18; and aggression, 126–45; in partnership with endocrine system, 150. *See also* dopamine; norepinephrine; serotonin

monozygotic twins. *See* identical (monozygotic) twins

morphine, 145

Moyer, Kenneth, 74, *74,* 75

MRI (magnetic resonance imaging), 107, 111–12, 114, 307n.71

Münchausen syndrome by proxy, 167

murder: alcohol consumption associated with, 214; lust murder, 166; serial murder, 77, 78, 111, 153, 165–67; in the workplace, 259

naloxone, 145, 147, 148

narcotic drugs, 145, 209

National Institute of Child Health and Human Development, 176

National Institute of Justice, 280, 330n.50

National Institutes of Health (NIH): D4 dopamine receptor and novelty-seeking study, 245; "violence initiative" conference sponsorship, 3

National Institute of Mental Health (NIMH): norepinephrine study, 128; serotonin studies, 139; stress-induced fighting study, 126

natural selection, 12, 247, 252–53

nature/nurture debate: aggression as product of nature and nurture, 51; eugenics discrediting nature-based explanations, 11, 29; nature viewpoint dominating in early twentieth century, 31; neuroscientific progress as bridging the gap, 30; nurture viewpoint becoming dominant, 31–32; sociobiology outraging the nurture-oriented, 16–17. See also environment; heredity

Nazis, 10–11

negative eugenics, 10

neglect of children, 280

neighborhood watches, 290

nerve growth factor (NGF), 36–37, 38, 39, 41

nervous system: anatomy of, 79–114; chemistry of, 115–49; differentiation in, 34–38; experience as shaping, 117; maternal separation and development of, 222, 223; prenatal alcohol exposure and, 219–20; prenatal cocaine exposure and, 220–21; primate systems as socially biased, 45–47; in stress management, 171; survival bias of, 47–48, 295–96. See also brain; central nervous system; peripheral nervous system; sympathetic nervous system

neural crest, 37

neural tube, 34, 35, 37

neuroanatomy, 79–114. See also brain; nervous system

neurochemistry, 115–49; of aggression, 126–45; of alarm response, 117–22; endogenous analgesics, 145–48; the reward system, 130–45; serotonin, 135–45; social history in, 133, 144–45; of trauma, 122–26. See also psychopharmacology

neuroendocrine system. See endocrine system

neuroleptics, 197

neurons: anatomy of, 39, 41; binding of chemical messengers to, 40; communication between, 41, 42, 115–16

neuropeptides, 146

neuroscience: complexity recognized in, 32; as excluded from debate on violence, 5; as holistic in outlook, 32; progress in, 29–30. See also neuroanatomy; neurochemistry

neurotransmitters: behavior affecting, 116, 144–45; chemical binding of, 40; CRF as, 172; drug use affecting, 208–12; in emotional behavior, 115–49; histofluorescence studies of, 118; in nerve cell communication, 41, 42, 116; neuronal selection of, 37; nitric oxide as, 254–56. See also monoamine transmitters

neurotrophin receptors, 39

neurotrophins, 37

neurotrophin-3 (NT-3), 37

NGF (nerve growth factor), 36–37, 38, 39, 41

night terrors, 234

nitric oxide, 41, 254–56

nitric oxide synthase, 255–56

noggin (protein), 36

nonsuppression, 175

noradrenergic pathways, 119–22, *120;* amygdala in, 125; in antisocial behavior, 128–130; PTSD linked with systemic changes in, 122–125; safety as function of, 137

norepinephrine: in aggression, 126–28; in alarm response, 118–22; in cortical function, 189; in emotional memory, 125; in endocrine system, 153; and maternal deprivation in monkeys, 223; neuronal release of, 41; propranolol affecting, 127–28, 294; in PTSD patients, 122–26; in stress management, 171, 172, 173; in sympathetic nervous system, 37, 119; yohimbine interfering with receptors for, 208. *See also* noradrenergic pathways

norm of reaction, 227

novelty seeking, 242, 243, 244–45, 246

NT-3 (neurotrophin-3), 37

nucleus accumbens, 132, 209

nurture/nature debate. *See* nature/nurture debate

Olds, James, 132

olive baboons, 174–76

opiate receptors, 145

opiates, 145, 209

opioid peptides, 146, 147

optic chiasm, *84*

orbitofrontal cortex, 100, 105–6

Ore, Guiseppe Dalle, 98–99

Ornitz, Edward, 161

Overall, Karen, 256–57, 291–92

overreacting, 171

Owen, Richard, 100

Paigen, Ken, 237, 257–58

pain: anger as caused by, 14; endogenous analgesics, 145–48; as deterrence to violence, 273; pain-induced aggression as protective, 75; rapid response to, 146; resident-intruder model of aggression compared with pain-induced model, 65–67

Papez, James, 86, *87,* 88

Paracelsus, 266

parahippocampal gyrus, 85, *87*

paranoia, 194–95; cocaine as inducing, 206, 207; drug use resulting in, 210, 212

paranoid schizophrenia, 194, 195–96, 212, 286

paraphilias, 165, 291

parasympathetic nervous system, 308n.8

parental aggression, 75

parents: of aggressive children, 279–85; and child temperament, 240–42, 245, 278–79; as the environment for infants, 278–79; parental aggression, 75; parental care, 222

Parkinson's disease, 132, 245

parvocellular hypothalamic nuclei, 150

Patterson, Paul, 37

Paul, Diane, 9, 10

Peacekeeping Among Primates (de Waal), 296

pecking order, 13

Penfield, Wilder, 23

Pennsylvania, University of, 81, 303n.4

Pennsylvania system, 269

peptides, 41, 146, 150, 197

perception: at cellular level, 39; innate framework for, 44–47; lateral nucleus

of the amygdala in, 94, 96, 104; smell, 44, 96; of threat, 119, 149, 160–61, 290; vision, 44

periaqueductal gray area (PAG), 93

peripheral nervous system: amygdala control of, 94; growth factors in development of, 36; hypothalamic control of, 89

personal history, 153, 188

personality: consistency of, 241–42; innate response style and social interactions in development of, 52; psychosurgery for changing, 20–25; as result of developmental process, 245. *See also* temperament

Pert, Candace, 145

PET (positron emission tomography), 108, 109–10, *110,* 111, 265

pharmacotherapy. *See* psychopharmacology

phenotype, 228, 235

phenytoin (Dilantin), 289

Philadelphia: Eastern State Penitentiary, 268–72, *270, 271;* geography of violence in, 79–81, *80;* road rage occurrence in, 259

Philadelphia Society for the Alleviation of the Miseries of Public Prisons, 269

phrenology, 6–8, *7,* 17

physical abuse of children, 167, 280

Pitman, Roger, 147–48, 307n.71

pituitary gland: anterior and posterior lobes of, 151; in brain anatomy, *84;* in endocrine system, 150–51; in stress management, 172, 173. *See also* hypothalamic-pituitary-adrenal axis

pleasure, in drug addiction, 208

pleasure center, 132

pleitropic genes, 232

Pliszka, Steven, 202

PMS (premenstrual syndrome), 169–70

positive eugenics, 10

positron emission tomography (PET), 108, 109–10, *110,* 111, 265

Post, Robert, 43, 72, 186, 293, 294

posterior parietal cortex, 202, *203*

posttraumatic stress disorder (PTSD): analgesic response in, 147–48; animal modeling of, 71–73; brain imaging studies of, 111–12, 307n.71; cortisol in, 181, 187; hippocampal atrophy in, 187; norepinephrine in, 122–26; Prozac for treating, 288; SSRIs for treating, 294; startle response in children with, 161, *162;* stress response system in, 180–81, 183, 185, *185;* and substance abuse, 212–13; symptoms of, *50;* in Vietnam veterans, 49–50, 123–24, 147–48

predation: as aggression, 73, 74; brain structures in cat, 92–93

predatory aggression, 75, 76, 91

predatory psychopathy, 166, 291

prefrontal cortex: alcohol as affecting, 216–17; in attention, 202, *203, 204;* in attention deficit–hyperactivity disorder, 205; in brain anatomy, 86, 100; in cat aggression, 93; and changes in the environment, 106; CRF-containing cells in, 172; in dopaminergic pathway, 132; in fear response, 264; information processing by, 189–92; results of damage to, 192; in risk assessment, 189; and schizophrenia, 198; and sociality, 102

prefrontal leucotomy, 21

prefrontal lobotomy, 22, 99–100

prefrontal syndrome, 22

pregnancy: domestic violence during, 277–78; prenatal exposure to alcohol, 219–20, 276; substance abuse during, 275–77

premenstrual syndrome (PMS), 169–70

prevention: of antisocial behavior, 291;

prevention *(cont.)*
of developmental damage to children, 274–79

primates: aggression studies of, 57–60; aggressive reaction to disruption of social structure, 272; "fight, then make up" sequence in, 59–60; nervous systems as socially biased, 45–47; prefrontal cortex of, 189; reconciliation in, 60, 296; serotonin and aggression in, 143; testosterone and aggression in male, 155; testosterone and status in male, 156–58. *See also* chimpanzees; monkeys

prisons: as source of social upheaval, 272–73; failure of, 268–74; not all violent people belonging in, 285–87; solitary confinement, 272; supermax prisons, 272

prolactin, 142–43

propranolol, 127–28, 294

protective (self-defensive) aggression, 75, 127, 218

Prozac, 137, 266, 287, 288

psychic blindness, 97–98

psychopaths, 129, 166, 181, 291

psychopharmacology, 265–67; antiaddiction medications, 286–87; anticonvulsants, 289; antidepressants, 137, 265, 266, 287–88; antipsychotic drugs, 197, 208, 286; carbamazepine (Tegretol), 289; chlorpromazine (Thorazine), 22, 197; in correctional programs, 267; lithium, 288–89, 290; neuroleptics, 197; phenytoin (Dilantin), 289; Prozac, 137, 266, 287, 288; reasons for using, 289–90; as replacing psychosurgery, 22; Ritalin, 205, 284; for schizophrenia treatment, 197; selective serotonin reuptake inhibitors (SSRIs), 137, 287–88, 289, 294, 311n.67; sodium valproate (De-

pakote), 289; tranquilizers, 266; Valium, 266; for victims of violence, 293, 294; Xanax, 266

psychosurgery, 20–25; ablation, 18, 89; as not recognizing interdependence of cortex and amygdala, 104; prefrontal leucotomy, 21; prefrontal lobotomy, 22, 99–100; stereotaxic psychosurgery, 24–25, 89

PTSD. *See* posttraumatic stress disorder

puberty: and ADHD, 284; estrogen in aggression during, 171; testosterone and aggression in, 159–61. *See also* adolescence

pulvinar, 202, *203*

Pynoos, Robert, 160–61

pyramidal cells, 189, 190, *190,* 197

QTL analysis, 244

quantitative traits, 235–36; locating genes for, 243–45

race: biological approach to violence and, 3–4; psychosurgery associated with racism, 25

"race suicide," 10

Raine, Adrian, 111, 129, 130

RAND group, 287

raphe nuclei, 136, *136,* 153, 172

Rationale of Crime (Farnham), 8

rats: aggressive behavior in, 61–62, *63;* alcohol and impulsive behavior in, 216; alcohol and response to threat in, 218; alcohol dosage and aggression in, 217; cocaine and stress in, 213; CRF's effect on, 172; fear conditioning of, 68, *69,* 103–4, 124–25; frustration as cause of aggression in, 14; kindling experiments, 43–44; *l*-dopa rage in, 132–33; maternal separation in, 222; pain as inducing aggression in, 14, *15;* prenatal exposure to alcohol in, 219;

repetitive stress affecting cortisol levels in, 175; resident-intruder model of aggression in, 61, 65–67; serotonin monitoring study, 144; sexual maturity and environmental response in, 159; stress-induced fighting in, 126; subordination stress in, 178–80, 186; testosterone and aggression in male, 154, 155; visible burrow system for studying, 70–71, *71*

Rauwolfia serpentina, 265

reason: reflective delay, 266, 287; in representational knowledge, 191–92; in risk assessment, 189; working memory in, 191–92, 199

receptor autoradiography, 118

receptor binding, 39, 118, 145, 151

recessive genes, 229, 232

reciprocity in intraneuronal communication, 116, 122, 127, 210

reconciliation, in primates, 60, 296

reconstruction: as establishing a new equilibrium, 263; rehabilitation contrasted with, 262–68

reflective delay, 266, 287

rehabilitation: as one-sided, 263; reconstruction contrasted with, 262–68

reinforcement value, 132

Reis, Don, 75

relaxation training, 294

releasing factors, 150, 151

representational knowledge, 191–92

reserpine, 265

resident-intruder model, 61, 65–67

Resnick, Phillip, 192, 193

restriction fragment length polymorphism (RFLP), 243

reticular formation, 83–84

reverse genetics, 256

reverse tolerance, 212

"revolving-door justice," 260

reward system, 130–45

RFLP (restriction fragment length polymorphism), 243

rhesus monkeys: amygdala and aggression in, 96–98; maternal separation in, 223; social history and testosterone levels in male, *156,* 156–57

risk assessment: addictive drugs as affecting, 208; cognitive behavioral therapy for improving, 265; cortical influence on, 264; in dangerous environments, 267; prefrontal cortex in, 189; representational thinking in, 191–92; stress as affecting, 185

Ritalin, 205, 284

RNA polymerase, 39

road rage, 259

Robbins, Jillian, 167

rodents: aggression studies in, 60–64; burst-gap pattern in aggression in, 62; pecking orders in, 13; subordination stress in hamsters, 177. *See also* mice; rats

Rolls, Edmund, 99, 105

roosters, 154

Roper, Allan, 9

Rose, Robert, 156

Rostain, Tony, 201–2, 279, 283

Sapolsky, Robert, 174–76, 181, 186, 187

Scandinavia, twin and adoption studies in, 238

Schäfer, Edward Albert, 96–97

schizophrenia, 193–96; attention deficit–hyperactivity disorder contrasted with, 205; brain development in, 197–200; catatonic schizophrenia, 194, 196; cocaine-induced psychosis compared with, 206; disorganized schizophrenia, 194; family interaction theories of, 196; faulty working memory in, 199–200; neurochemical theories of, 197; paranoid schizophrenia,

schizophrenia *(cont.)*
  194, 195–96, 212, 286;
  treatment of, 285–86; and violence,
  195–96
Schjelderup-Ebbe, 13
Schultz, David, 26, 192–93
Scott, J. P., 273
secondary emotion, 105
second messenger proteins, 39, *40*, 118,
  119, 189
selective serotonin reuptake inhibitors
  (SSRIs), 137, 287–88, 289, 294,
  311n.67
self-defensive (protective) aggression, 75,
  127, 218
self-defensive overreaction, 123–24,
  132–33
selfish genes, 17
Selye, Hans, 174
sensitization, behavioral, 72–73
sensitization, drug-induced, 212
septum, 92
serial murder, 77, 78, 111, 153,
  165–67
serotonin, 135–45; and aggression,
  138–45; alcohol affecting, 216; in cortical function, 189; in endocrine system, 153; 5-HT$_{1B}$ gene and aggression
  in mice, 257; glucocorticoids moderating influence of, 173; in impulsive
  aggression, 139–43; isolation affecting, 272; pathways in the brain, *136;*
  in schizophrenia, 197; selective serotonin reuptake inhibitors, 137,
  287–88, 289, 294, 311n.67; in stress
  response management, 173
sex offenders, 161–67; chemical castration for, 163–65, 291; multiple motivations of, 163–65; paraphilias, 165,
  291; serial killers, 165–67, 291; testosterone and aggression in, 155; treatment for, 291

sexual abuse, 111, 112, 167, 291, 307n.76
Seyfarth, Richard, 45–46, 47
sham rage, 88
Sheard, Michael, 288–89
shell shock, 48
Siegel, Alan, 88, 90, 91–93, 126–27
Simon, Robert, 78, 129, 166
single photon emission computed tomography (SPECT), 108, 111,
  306n.64
smell, sense of, 44, 96
smoking, during pregnancy, 276
Snyder, Solomon, 145, 254–55
social interactions: aggression as necessary part of, 261; aggression as shaped
  by, 252; amygdala affecting, 98, 99; in
  behavioral development, 51–52, *53;*
  and maternal deprivation in monkeys,
  223; plasticity of social behavior,
  253–54; prefrontal cortex in, 102; primate nervous systems as socially biased, 45–47; stress in community
  formation, 176–77
social interventions, 267–68
social learning deficit, 282–83
social-skills training, 283
Society for Neuroscience, 30
sociobiology, 16–17, 247
sociopaths, 78, 129
sodium valproate (Depakote), 289
solitary confinement, 272
Sopko, Tom, 127
SPECT (single photon emission computed tomography), 108, 111,
  306n.64
speech, mapped to the brain, 19
Spemann, Hans, 34
Spurzheim, Johann, 6–7, 8
squirrel monkeys, 155
SSRIs (selective serotonin reuptake inhibitors), 137, 287–88, 289, 294,
  311n.67

Stanford, Matthew, 141, 289, 290

startle response: CRF as intensifying, 172; development of, 160–61; glucocorticoids blunting, 183; in children with PTSD, 161, *162;* rat experiments, 124–25

status: and stress response in baboons, 174–76; subordination stress in rats, 178–80; and testosterone in male primates, 156–58. *See also* dominance hierarchies

Stein, Murray, 112, 307n.76

stereotaxic psychosurgery, 24–25, 89

steroid hormones: estrogen, 151, 169, 170–71; in intraneuronal communication, 41; testosterone and male violence, 154–58. *See also* estrogen, testosterone

stimulus hunger, 129, 202

stimulus-transmission coupling, 39

stress: of adolescence, 177–78; and allostatic load, 182–83, *183, 184;* balance required in response to, 181–87; brain as affected by, 186–87; defeat as stressor, 133; hyporesponsive period in childhood, 274, *275;* in prisons, 272–73. *See also* stress response disorders; stress response system; subordination stress

stress hormones, 171–73, 223

stress hyporesponsive period, 274

stress-induced analgesia, 145–48

stress-induced fighting, 126

stress response disorders: as dysregulation of stress response system, 183, *184;* key features of, *185;* renewed interest in, 50; and substance abuse, 212–13; treatment of, 287–88. *See also* antisocial personality disorder; depression; impulsive aggression; posttraumatic stress disorder

stress response system, 174–81; in aggressive children, 283; in infancy and early childhood, 274; and maternal separation in rats, 222; SSRIs for normalizing, 289; in women, 218–19

stress test, 123

Stueve, Ann, 196

subcallosal gyrus, 85, *86*

subordination stress: in baboons, 175–76; in hamsters, 177; in rats, 178–80, 186; and stimulant use, 213

substance abuse: addiction, 209–10, 286–87; antiaddiction medications, 286–87; drug rehabilitation programs, 287; among mothers, 218–22; narcotic drugs, 209; opiates, 145, 209; during pregnancy, 275–77; sensitization, 212; and stress disorders, 212–13; tolerance, 210, *211,* 212; treatment of, 286–87; twelve-step recovery programs, 286; and violence, 195, 208. *See also* alcohol; cocaine; methamphetamine

substantia nigra, 131, *131*

suicide: men as more suicidal than women, 153; as related to depression and violence, 138–39; role of serotonin in, 139–40

superior colliculus, 202, *203*

supermax prisons, 272

superpredators, 158, 223

survival: altruism in, 16; human nervous system as biased toward, 47–48, 295–96; reinforcing behaviors that promote, 132

Susman, Elizabeth, 160, 171

Sweden, twin and adoption studies in, 238

Sweet, William, 25

sympathetic ganglia, 36

sympathetic nervous system, 308n.8; in alarm response, 122, 125; in antisocial behavior, 129, 181; in attention, *203;* in attention deficit–hyperactivity dis-

sympathetic nervous system *(cont.)*
order, 202; in emotional arousal, 119;
in response to threat, 171, 172–73;
yohimbine in dysregulation of, 124
sympathetic neurons, 36–38
synapses, 41, *42*

Taft, William Howard, 206
"Tan" case, 19
target genes, 41
Tegretol (carbamazepine), 289
temperament: in children, 240–42, *241,*
245, 278–79; environment and, 245,
246–47
temporal lobe: in brain anatomy, *86;* in
perpetrators of violence, 109–10; removal of, 96
temporary insanity defense, 27
territoriality, dominance as linked to, 13
Terzian, Hrayr, 98–99
testosterone: in adolescent aggression,
158–61; and aggression, 154–61; antitestosterone therapy, 163–65; in endocrine system, 151; estrogen
compared with, 171; in female aggression, 169; and status in male primates,
156, 156–58; stress affecting levels of,
177; in subordinate rats, 179
thalamus: in brain anatomy, *84,* 85; connection with amygdala, 104, *105,*
106; in noradrenergic pathways, *120;*
in serotonin pathways, *136*
therapy. *See* interventions
Thomas, Alexander, 240
Thorazine (chlorpromazine), 22, 197
threat: perception of, 119, 149, 160–61,
290; response to, 171–74. *See also* risk
assessment
"threat/control-over-ride" symptoms,
196
thyroid gland, 151
thyroxine, 151

time-space problem, 108, 113
tolerance (drug), 210, *211,* 212
Tovee, Martin, 46
tranquilizers, 266
transcription factors: in alarm response,
121; drug use as affecting, 210; hormone receptors as, 151; in intraneuronal communication, 116; in
receptor binding, 39, *40,* 118; steroid
hormones as, 41; stress as affecting,
186
trauma: neurochemistry of, 122–26; reminders of, 124, 293, 294; therapy for
victims of, 294–95. *See also* posttraumatic stress disorder
trauma reminders, 124, 293, 294
Tronick, Edward, 51, 52
trophic (growth) factors, 36, 37, 38, *40,*
41
tryptophan, 135, 138
Turecki, Stanley, 279
twelve-step recovery programs, 286
twin studies, 237–39; on criminal behavior, 250–51; of temperament and personality, 242–43
Type A personalities, 177–78
Type H personalities, 178, 183
tyrosine, 131

Uhl, George, 243
Unabomber, 11–12
urban riots, psychosurgery suggested as
solution to, 25

Valenstein, Elliot, 22
Valium, 266
Valzelli, Luigi, 137–38, 269, 272
vengeance, 268
ventral tegmental area, 131, *131*
vervet monkeys, 45–46, 97, 143
victims of violence: brain imaging studies
of, 111–12; interventions for, 292–95

Vietnam veterans, 48–49, 123–24, 147–48, 250–51, 307n.71

violence: and alcohol use, 213–18; amphetamine use as cause of, 207; attention deficit–hyperactivity disorder associated with, 201; as developmental process, 32–33, 50–52, *53*, 260, 274; drug use as causing, 208; as failure to respect boundary of acceptable and unacceptable aggression, 261; frustration with, 4; geography of urban, 79–81, *80*; heart rate and, 129–30; heterogeneity of, 73–76, 285; ideological opposition to biological explanations of, 29; imaging studies of, 108–15, *110*; interventions for, 259–96; medicalization of, 27–29; men as more violent than women, 153–58; mental illness associated with, 28, 195; neuroscience as excluded from debate on, 5; phrenological account of, 6–8; and schizophrenia, 195; and substance abuse, 195, 208; twin and adoption studies on, 238–39; vicious circle of physical and environmental factors leading to, 52, *53*; "violence initiative" conference, 2–4. *See also* aggression; male violence; victims of violence

Violence and the Brain (Mark and Ervin), 24, 25

Virkkunen, Matti, 139–40, 216

visible burrow system: as animal model of aggression, 70–71; and subordination stress, 178–79

vision, 44

Volkow, Nora, 109, 110, 111

Wasserman, David, 2, 3–4

Watson, James, 33

Watts, James, 21–22

white lions, 228–30, *230, 231,* 325nn. 5, 6

Wilson, E. O., 16–17

women: aggression in, 167–71; maternal aggression, 75; maternal separation, 222–23; men as more violent than, 153–58; methamphetamine use by, 207; premenstrual syndrome and aggression in, 169–70; substance abuse among mothers, 218–22. *See also* domestic violence; pregnancy

working memory, 191–92; in aggressive children, 283; in attention, 202, 203, *203;* dopamine function in, *190;* in schizophrenia, 199–200

workplace violence, 259

Wournos, Aileen Carol, 167

Wren, Christopher, 85

Xanax, 266

X chromosome, 234

Y chromosome, 234

Yehuda, Rachel, 180

Yerkes Regional Primate Center Field Station, 57–58

yohimbine, 123–24, 132, 208

Zigmond, Michael, 121, 122